Eddy Currents

Eddy Currents

Theory, Modeling, and Applications

Sheppard J. Salon
Rensselaer Polytechnic Institute
Schenectady, NY, USA

M. V. K. Chari
Rensselaer Polytechnic Institute
Burnt Hills, NY, USA

Lale T. Ergene
Istanbul Technical University
Isanbul, Turkey

David Burow
Kingsport, TN, USA

Mark DeBortoli
Clifton Park, NY, USA

IEEE PRESS
WILEY

Published by John Wiley & Sons, Inc., Hoboken, New Jersey.
Published simultaneously in Canada.

For general information on our other products and services or for technical support, please contact our Customer Care Department within the United States at (800) 762-2974, outside the United States at (317) 572-3993 or fax (317) 572-4002.

Wiley also publishes its books in a variety of electronic formats. Some content that appears in print may not be available in electronic formats. For more information about Wiley products, visit our website at www.wiley.com.

Library of Congress Cataloging-in-Publication Data:

Names: Salon, S. J. (Sheppard Joel), 1948- author.
Title: Eddy currents : theory, modeling and applications / Sheppard J. Salon [and four others].
Description: Hoboken, NJ : Wiley-IEEE Press, [2024] | Includes bibliographical references and index.
Identifiers: LCCN 2023040651 (print) | LCCN 2023040652 (ebook) | ISBN 9781119866695 (cloth) | ISBN 9781119866701 (adobe pdf) | ISBN 9781119866718 (epub)
Subjects: LCSH: Eddy currents (Electric)
Classification: LCC TK2211 .S25 2024 (print) | LCC TK2211 (ebook) | DDC 621.31/042–dc23/eng/20230906
LC record available at https://lccn.loc.gov/2023040651
LC ebook record available at https://lccn.loc.gov/2023040652

Cover Design: Wiley
Cover Image: © Sheppard J. Salon (with edits by David Burow)

Set in 9.5/12.5pt STIXTwoText by Straive, Chennai, India

Contents

About the Authors

Sheppard J. Salon, PhD, is Professor Emeritus in the Department of Electrical, Computer and Systems Engineering at Rensselaer Polytechnic Institute in Troy, New York, USA, and a founder of the Magsoft Corporation. He has published on many electrical engineering subjects and his awards and honors include an IEEE Life Fellowship and the IEEE 2004 Nicola Tesla Award.

M. V. K. Chari, PhD, now retired, was a Research Professor at Rensselaer Polytechnic Institute in Troy, New York, USA. He is a former Technical Leader at General Electric, an IEEE Life Fellow, and recipient of the 1993 Nicola Tesla Award. He has published extensively on electrical engineering subjects.

Lale T. Ergene, PhD, is a Full Professor in the Electrical Engineering Department at Istanbul Technical University, Turkey. She was an adjunct professor at Rensselaer Polytechnic Institute in Troy, New York, USA and worked at Magsoft Corporation as a consulting engineer. She is an IEEE Senior Member and advisory board member of the Scientific and Technological Research Council of Turkey. She has published widely on electrical engineering subjects.

David Burow, PhD, is Owner and Head Programmer at Genfo, Inc., a company that provides custom programming for Macintosh, Windows, and Linux operating systems. He was a Postdoctoral Researcher at Rensselaer Polytechnic Institute and has published several papers on electrical engineering subjects.

Mark DeBortoli, PhD, received a doctoral degree in Electric Power Engineering at Rensselaer Polytechnic Institute. He has over 30 years of industrial experience in the design, analysis, and testing of electrical equipment. He has taught graduate engineering courses and authored a number of technical papers and publications. He is currently an engineering consultant.

Preface

It has been over 50 years since the classic monographs of *Eddy Currents* by Jiří Lammeraner and Miloš Štafl and *The Analysis of Eddy Currents* by R. L. Stoll. Since that time there have been great advances both in eddy current computation and eddy current applications. Modern numerical methods are now commonplace in the analysis of eddy currents. We can now solve three-dimensional eddy current problems with complicated boundaries, non-linear, and anisotropic materials. Further, the continued importance of high efficiency electrical devices has made the study of eddy currents and the need for accuracy in loss evaluation more important.

While these former works are excellent in their treatment of many important eddy current problems and are well referenced in this book, they are written for an audience that is already well-versed with electromagnetic phenomena and low-frequency applications.

We hope this book will be accessible to people looking for a place to start in the study and applications of eddy currents. We begin with a very basic introduction to the principles on which eddy current analysis is based (Faraday's law, Ampere's law, and Kirchhoff's laws), and refer back to these ideas throughout the book. We have included a lot of tutorial information as well as dozens of worked out numerical examples. Each problem is followed by a discussion of the results and how the basic principles of eddy currents can be seen in the solution.

We also hope that this work will be useful as a reference for experienced engineers working in the field. We include many examples of closed form and analytical solutions as well as numerical methods and approximation methods for many practical applications. The basic ideas of the numerical modeling are presented along with examples of their use and methods of interpreting and checking the results. Numerical methods are used as well in the applications section, in which we attempt to analyze problems that have both analytical and numerical solutions.

We would like to acknowledge Philippe Wendling for his invaluable help. All of the authors also want to express their gratitude to family and friends for their patience, understanding, encouragement, and support during the writing process.

Schenectady, NY
April 2023

Sheppard J. Salon

Part I

Theory

1

Basic Principles of Eddy Currents

1.1 Introduction

The discovery of eddy currents is usually attributed to French physicist Leon Foucault. In 1855, Foucault measured the force on a rotating conducting disk. He found that it took a greater force to rotate the disk when the disk was in a magnetic field produced by an electromagnet. He also noticed that the disk was heated when spun through a magnetic field. These observations, that eddy currents can produce force and torque and also heating, are still major applications of eddy currents. The production of force or torque by eddy currents is key to the operation of induction motors and induction generators, eddy current brakes, eddy current magnetic bearings, liquid metal stirring, and electromagnetic metal forming, to name a few applications. The forces produced by eddy currents may be a hindrance. In such cases, methods of eddy current mitigation will be necessary. Eddy current heating also has many applications such as induction heating for metal treatment and induction cooktops. There are also many applications in which eddy currents result in unwanted losses, limiting the efficiency of devices and requiring more thermal management.

We will use a very broad definition of eddy currents in this treatment. We include all electric currents induced by time-varying magnetic fields and/or relative motion between conductors and magnetic fields. This also includes the redistribution of currents due to the self-field of conductors excited with external sources.

All currents induced by a time-varying magnetic field can be thought of as eddy currents, but in this book, we are mainly dealing with applications in which the current density is nonuniform in the conductor. Whether we speak of eddy currents, skin effect, or proximity effect, we are speaking of the same physical phenomenon and this is described by the same set of equations.

In this chapter, we will develop the ideas necessary to understand the eddy current phenomenon. We will begin with a very basic introduction to Faraday's law

Eddy Currents: Theory, Modeling, and Applications, First Edition.
Sheppard J. Salon, M. V. K. Chari, Lale T. Ergene, David Burow, and Mark DeBortoli.

and Lenz's law. This will give us a qualitative understanding of some of the important concepts of eddy current analysis such as skin effect and proximity effect. We then discuss the concept of resistance and reactance limited eddy currents in Section 1.4. We then give a more formal introduction to Faraday's law, emf, and potential difference and discuss the different ways voltage can be produced in a conductor. In eddy current analysis, we will be solving the diffusion equation. From electromagnetic theory, this means that we are making a quasi-static approximation. The full set of Maxwell's equations will result in the wave equation for time-varying phenomena. In Section 1.6, we will justify this approximation and study its implications.

Eddy currents can sometimes be rather difficult to visualize. There are other physical phenomena such as particle diffusion and heat transfer, that are described by the same mathematics, the diffusion equation. We will derive the expressions for eddy currents formally from Maxwell's equations in Section 2.2, but making the analogy to these other areas can help clarify some of the physics. We will first look at particle diffusion by means of random walks in Section 1.9. This will introduce the diffusion equation and some of its classic solutions. In Section 1.10 we discuss the electromagnetic diffusion problem in the time domain and the analogy is made to the particle diffusion results. We then look at the concept of skin depth or depth of penetration in electromagnetic applications for steady-state sinusoidal excitation. This concept is one of the key results of eddy current analysis and will be derived formally from Maxwell's equations in Section 2.2. As an example of skin depth and sinusoidal excitation, we turn to another branch of physics, heat transfer. Heat conduction and heat storage are described by the same set of equations as electromagnetic diffusion. We use an example of heat conduction to illustrate the concepts introduced about skin depth. Another method of understanding and computing eddy currents that we will find useful is the use of magnetically coupled circuits to model the eddy currents. This idea is introduced in Section 1.11.

With this introduction, we then present a number of applications and techniques for eddy current analysis as well as several eddy current applications of interest today. Chapters 2 and 3 present the analysis of several eddy current applications of practical interest, mainly in closed form, for rectangular conductors and for conductors with circular cross-sections, respectively. In Part II of the book, we focus on modern numerical techniques to study eddy currents. First, we introduce the most common mathematical formulations for eddy currents and the different variables used for these models. We then present the formulations for finite difference, finite element, and integral equations, with several examples that refer back to the results found in Chapters 2 and 3. In Part III of the book, we consider a number of practical applications such as electric machines, transformers, induction heating, and liquid metal stirring.

1.2 Faraday's Law and Lenz's Law

We will present a more formal introduction of Faraday's law below in Section 1.5. In this section, we will introduce Faraday's law and Lenz's law in non-mathematics terms to introduce some of basic ideas of eddy currents. Faraday's law is one of Maxwell's equations. It states that there is a voltage induced in a circuit equal to the negative time rate-of-change of the flux linking the circuit. We refer to this induced voltage as the emf (electromotive force). The "negative" sign is referred to as Lenz's law but is really part of Faraday's law. Lenz's law tells us the voltage induced in the circuit will be in a direction to circulate current that will produce flux which will *oppose* the change in flux linking the circuit. For example, if the flux linkage is increasing, the voltage will circulate current in a direction to decrease the flux linkage.

Consider the case illustrated in Figure 1.1. The figure shows a loop with an external source of magnetic flux that is varying in time in such a way as to increase the flux linking the loop. According to Lenz's law, we expect induced current in the sense shown in the figure to oppose the increase of flux linkage. Note that it is not the direction of the flux that is important but the time rate of change of the flux linking the circuit. With flux in the same direction as shown in the figure, if the flux were decreasing, the current direction would reverse.

In another example, consider the two coplanar loops in Figure 1.2. One of the loops contains a dc source. When the switch is closed, a current will flow counter-clockwise in the powered loop. This will cause flux from the first circuit to enter the plane of circuit 2 in the direction shown. We expect current in circuit 2 to circulate in the counter-clockwise direction to counter this flux; that is, this current will oppose the change of flux linkage from the powered circuit.

After some time, the system has reached steady state with dc current in loop 1 and no current in loop 2. If we now open the switch, interrupting the current in loop 1, current will be induced in loop 2 to maintain the steady-state flux linkage. Current will be induced in the clockwise direction in this case.

Figure 1.1 Loop with flux linkage increasing and induced current direction.

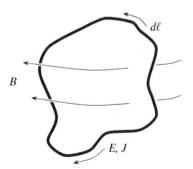

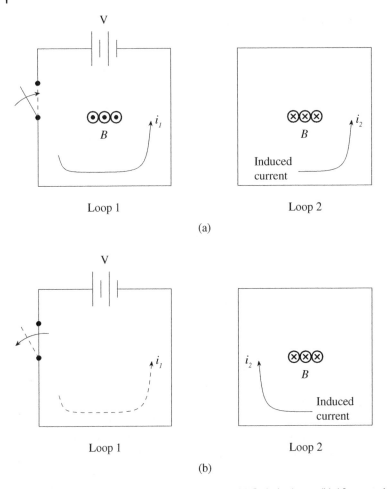

Figure 1.2 Two circuits with mutual coupling. (a) Switch closes. (b) After steady state is reached, open switch in loop 1.

We can now apply these ideas to the case of induced current in a circular wire. Referring to Figure 1.3, we have a conductor of circular cross section with time varying magnetic flux produced by an external source, in the axial direction.

Let us assume that the source flux density is uniform over the cross section and is varying sinusoidally with time. By symmetry, in this cylindrical geometry, we expect the eddy currents to follow circular trajectories and the eddy currents circulating to oppose the change in source flux. If we divide the conductor into concentric layers as shown in the figure, we can make the following argument.

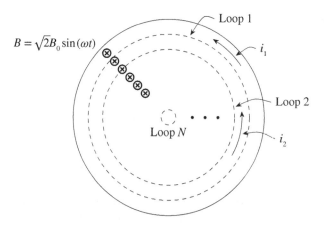

Figure 1.3 Conducting cylinder with axial magnetic flux field applied.

The induced voltage from the source in each circuit will be

$$\text{emf}_i = -j\omega B_0 S_i \tag{1.1}$$

where B_0 is the root-mean-square (RMS) value of the flux density, S_i is the cross-sectional area enclosed by each of the loops, and ω is the angular frequency of the source. This tells us that the emf in the outer circuits is greater than the emf of the inner circuits due to the increased surface area and therefore the increased flux linking the current path. The current circulating in the outer circuit, loop 1, now produces a *reaction* flux which, by Lenz's law, opposes the change in source flux. This reduces the total flux in the interior of the cylinder, which is the sum of the source and reaction flux. Circuit 2, with a smaller emf, will also have current circulating in a direction to oppose the change in source flux. This will cause a further reduction in the net flux inside circuit 2. Note that the reaction field from circuit 2 does not pass through the conductor area of loop one, but only inside the path of loop 2. This continues for loops 3 to N. As we move inward, therefore, the source emf is smaller and the effects of the reaction field are greater. The problem is complicated by the fact that each loop has a resistance and inductance, and therefore there will be a phase shift in the currents. We will deal with this issue in Section 1.7. For our purposes here, it is enough to note that we will have higher current and flux in the outer rings and lower current and flux in the inner loops. There is a characteristic length called the *depth of penetration* or *skin depth*, which we will treat more formally in Chapter 2, that describes a distance from the surface in which "most" of the current and flux is contained. We will see that this characteristic depth depends on the material properties (conductivity and permeability) and the frequency of excitation.

1.3 Proximity Effect

In the previous example, we considered the nonuniform current distribution in a circular conductor due to the self-field of the conductor. The term *proximity effect* is often used to describe the situation in which the emf produced by near-by sources influences the current distribution in the conductor. Referring to Figure 1.4, we can apply Faraday's law and Lenz's law to illustrate this effect. The figure depicts two parallel conductors with opposing currents, a go-and-return circuit.

Considering the current on the left, in the $+z$ direction, we see that the field produced by conductor 1 in conductor 2, is increasing in the $+y$ direction assuming that the current in conductor 1 is increasing. By the application of Lenz's law, the emf will tend to circulate current in conductor 2 such that the current on the left side of conductor 2 is in the $-z$ direction, while the current on the right side is in the $+z$ direction. This is a circulating current, going down one side of the conductor and returning on the other side. Integrating this circulating current density over the surface of the conductor will result in zero total current. This circulating current will subtract from the load current $(-z)$ on the right side and add on the left side. By applying this argument to conductor 1 and considering the field produced by conductor 2, we find the currents add on the right side of conductor 1 and subtract on the left side. The conclusion is that the currents crowd to the inside surfaces when the load currents are in the opposite direction.

If the load currents were in the same direction, we will find the currents crowd to the outside surfaces. In practice, if we are dealing with ac currents, the addition and subtraction of the circulating current is not a direct arithmetic addition or subtraction. Depending on the impedance of the circulating current path, there will be a phase shift that must be included. Figures 1.5 and 1.6 show the results of a finite element analysis of the two conductor problem for the case of opposing

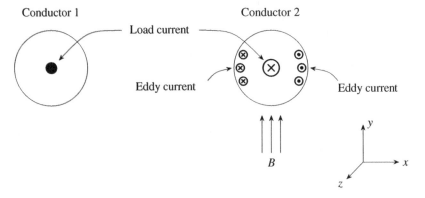

Figure 1.4 Parallel conductors with opposing currents.

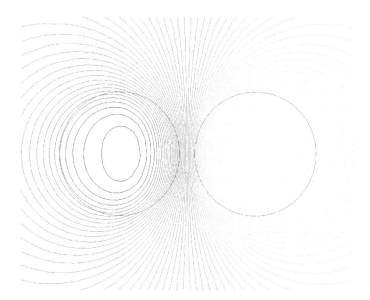

Figure 1.5   Real part of flux density with opposing currents.

Figure 1.6   Current density with
opposing currents.

current direction. Figures 1.7 and 1.8 show the results of a finite element analysis
of the two conductor problem for the case of load currents in the same direction.
We can see that the results conform to our expectations on the nonuniformity of
the current produced by the external fields.

We will discuss the losses in conductors in detail in Chapters 2 and 3, but it is
not too early to start developing an intuition about the interaction of the fields
and current in this simple circular conductor example. Let us ask the question,
can we compute the losses in one conductor ignoring the effects of the other conductor, then in a separate calculation, find the losses in that same conductor only
due to the fields produced by neighboring conductors and simply add the two
results to find the loss in the conductor under consideration. The answer is, generally speaking, we can not. If the conductors have linear material properties, for
example copper or aluminum, the electromagnetic equations are linear, and we
can indeed solve for the fields separately and add the results to find the net fields.

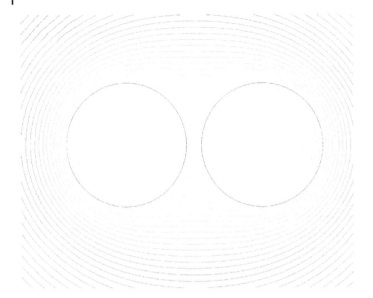

Figure 1.7 Real part of flux density with currents in the same direction.

Figure 1.8 Current density with currents in the same direction.

The losses, however, are found from the *square* of the current density and therefore we must take care. Let us again consider Figure 1.4. There are losses in conductor 2 produced by current density J_{22}, which we will define as the current density that would appear in conductor 2 without the influence of any external sources. There is also current density in conductor 2 produced by the fields due to conductor 1, which we will call J_{12}. Recall from the discussion above, the current J_{12} is a circulating current that sums to zero over the conductor. We can compute the total loss in the conductor, per unit depth, by integrating the loss density over the surface of the conductor.

$$P_2 = \frac{1}{\sigma} \int_S \left(J_{22} + J_{12} \right)^2 dS \tag{1.2}$$

where σ is the electrical conductivity of the conductor. Expanding the terms in parenthesis, we get three terms.

$$P_2 = \frac{1}{\sigma} \int_S \left(J_{22}^2 + J_{12}^2 + 2J_{22}J_{12} \right) \, dS \tag{1.3}$$

The first term is clearly the loss produced by the current that would have been generated in the conductor without the influence of external sources. The second term is the loss produced only by the external fields, without the influence of the current in conductor 2. Since both of these terms involve the square of the current density, the values being integrated are never negative. We now consider the third term

$$\frac{2}{\sigma} \int_S J_{22}J_{12} \, dS \tag{1.4}$$

The question we must answer is whether this integral vanishes. If it does, then we can simply add the two components. We know that J_{12} alone integrates to zero, but this is not sufficient. Let us consider the symmetry of the current. This is relatively easy with the regular circular geometry. The self term, J_{22}, will be symmetric around the vertical line bisecting the circular conductor. In other words, the current density on the right should be the same as the current density on the left. We call this *even symmetry*. So the result depends on whether the current density distribution J_{12} has *odd symmetry*. If conductor 1 is relatively far away compared to the radius of conductor 2, then this requirement is approximately satisfied, since the field produced by conductor 1 is relatively uniform over conductor 2. If conductor 1 is quite close to conductor 2, the term in Equation (1.4) will not vanish and there will be some error if we neglect it.

1.4   Resistance and Reactance Limited Eddy Currents

The concept of *resistance limited* and *reactance limited* eddy currents can be explained by considering the circuit of Figure 1.9.

Let us assume, we are concerned with the losses in the load resistor R_l. To increase or decrease the losses in the load, should we increase or decrease the load resistance? The answer to this question is that it depends on whether the losses are resistance limited or reactance limited. For a fixed system voltage and system impedance, we can plot the losses in the load as a function of the load resistance. This is illustrated in Figure 1.10.

If the load resistance were zero, then of course there would be no losses. Also if the resistance were infinite, again we have no losses. This means that there must be a maximum at some value of the load resistance. We will find this maximum below, but first notice that on the left side of the maximum on Figure 1.10 losses, the losses

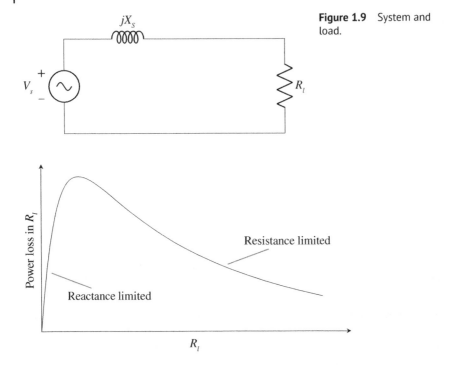

Figure 1.9 System and load.

Figure 1.10 Losses in load vs. load resistance.

rise with increasing load resistance. For small values of the load resistance (small compared to the system impedance), the load sees an ideal voltage source in series with a large impedance. This is approximately a constant current source and the losses are given by $I^2 R_l$. The slope in Figure 1.10 for small values of the resistive load has a relatively constant slope indicating the losses are proportional to R_l. The value of the load resistance has little effect on the system current. On the right side of the peak, the load resistance is greater than the system impedance ($|R_l| \gg |X_s|$). In this case, the voltage across the load is approximately constant and independent of the load resistance. The power to the load is then approximately $P_l \approx V^2/R_l$. The trajectory in the region $R_l \gg X_s$ is $1/R_l$.

To find the maximum power, consider the problem of an ideal sinusoidal voltage source, V_s, a transmission line that is a pure reactance, jX_s, and a variable resistance load, R_l. The objective is to find the optimum resistance of the load for maximum power transfer. The current magnitude is given by

$$I = \frac{V_s}{\sqrt{R_l^2 + X_s^2}} \tag{1.5}$$

The power to the load is then

$$P_l = I^2 R = \frac{V_s^2}{R_l^2 + X_s^2} R \tag{1.6}$$

Multiplying through and rearranging we have

$$P_l R_l^2 - V_s^2 R_l = -P_l X_s^2 \tag{1.7}$$

We now "complete the square" by adding $V_s^4/4P_l^2$ to each side.

$$R_l^2 - V_s^2 \frac{R_l}{P_l} + \frac{V_s^4}{4P_l^2} = \frac{V_s^4}{4P_l^2} - X_s^2 \tag{1.8}$$

$$\left(R_l - \frac{V_s^2}{2P_l} \right)^2 = \frac{V_s^4}{4P_l^2} - X_s^2 \tag{1.9}$$

This gives

$$R_l = \frac{V_s^2}{2P_l} + \sqrt{\frac{V_s^4}{4P_l^2} - X_s^2} \tag{1.10}$$

We see from Equation (1.10) that as P_l increases, the square root term decreases. The maximum value that P_l can attain is that which makes the square root equal to zero. This term can never be negative since this would give an imaginary part to the resistance. We conclude that, at maximum power

$$\frac{V_s^4}{4P_l^2} = X_s^2 \tag{1.11}$$

and

$$R_l = \frac{V_s^2}{2P_l} \tag{1.12}$$

This means that

$$P_l = \frac{V_s^2}{2R_l} \tag{1.13}$$

but we also have

$$P_l = \frac{V_s^2 R_l}{R_l^2 + X_s^2} \tag{1.14}$$

This means that for maximum power, $R_l = X_s$. This is actually a demonstration of the Maximum Power Transfer Theorem, which states that the maximum power to a load is achieved when the load and system impedances are matched.

This example illustrates a very important point regarding losses in a conductor. It is not just the load that is important but the source as well. Another qualitative example will illustrate this point. Consider the two magnetic circuits in Figure 1.11

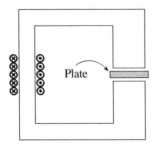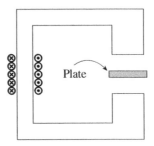

Figure 1.11 Two magnetic circuits with identical conducting plates.

with identical conducting plates. We can adjust the ac winding currents in the two magnetic circuits so the air-gap flux is the same. In this case, an observer on each plate would experience the same field at the same frequency.

In this example, the plate in the large air gap will have higher losses than the plate in the smaller air gap. The induced voltage in the two plates is the same since the impinging field is the same. The plate in the small air gap has a higher inductance than the one in the large air gap. This means the current will be limited by its higher impedance. Another way to see this is to consider the magnetic circuit as a transformer with a shorted secondary (the plate). To get the same air-gap flux density in the two cases, the ampere-turns on the primary of the large air-gap device must be greater. In a transformer, the primary and secondary ampere-turns are more or less balanced. So the large air-gap plate will have more current than the small air-gap plate to balance the primary currents. We also note that the L/R ratio of the conductor is important. In cases where the resistance dominates, the induced current is relatively in-phase with the emf. In highly inductive circuits, the current will lag the emf with an angle approaching 90°.

1.5   Electromotive Force (emf) and Potential Difference

Currents in conductors are produced by an electric field which is a measure of the force on electric charges. There are two sources of the electric field. One is due to a distribution of electric charges. This field can be described by a scalar potential. The other component of the electric field is produced by a change in magnetic flux linkage [1]. It is this second component that produces eddy currents. The induced voltage from this component is often referred to as *emf* or electromotive force. This component can not be described by a scalar potential. It is necessary to keep these two components separated in our analysis as they play different roles in the production of current. This is often a confusing point since a measurement, such as a voltmeter reading, cannot tell the difference and any measurement may include both of these effects. This point has been well-discussed in the literature and given rise to a number of papers which explore this issue [2–10]. We define the potential difference in terms of energy. A potential difference of one volt will impart one joule of energy per coulomb of charge that passes through it.

This component of electric field, which we will call E_c and is produced by a charge distribution, is a *conservative* field and the integral of this electric field component around a closed path is zero.

$$\oint_c E_c \cdot d\ell = 0 \tag{1.15}$$

The physical interpretation of Equation (1.15) is that it takes no work to move an electric charge around any closed path in the presence of a conservative electric field. The differential representation of Equation (1.15) is

$$\nabla \times E_c = 0 \tag{1.16}$$

Since the curl of the gradient of any function is zero, we deduce that this component of the electric field can be represented as the gradient of a scalar.

The second component of the electric field is described by Faraday's law. Faraday's law states that the line integral of the electric field around a closed contour is equal to the negative of the time rate of change of the magnetic flux linkage. We write this as

$$\oint_c E_\psi \cdot d\ell = -\frac{d}{dt} \oint_S B \cdot dS = -\frac{d\psi}{dt} \tag{1.17}$$

In Equation (1.17), the surface integral through which the flux crosses is the surface defined by the boundary of the line integral in the first term. To see this, consider the coil shown in Figure 1.12.

We imagine a helical surface defined by the coil as its boundary. In the figure, we see that some of the magnetic flux passes through or *links* all of the turns, or in other words, it passes through the surface more times than other tubes of flux. This flux "counts" more in the integral as it penetrates the surface multiple times. Some of the flux shown only links part of the coil and its role in emf production is therefore less. In differential form Equation (1.17) is written as

$$\nabla \times E_\psi = -\frac{dB}{dt} \tag{1.18}$$

While Faraday's law describes the electric field integral around a closed boundary, we are interested in open circuits as well. Consider the open circuit of Figure 1.13. If the flux linking the coil is increasing in the direction shown in the figure, then current, if allowed to flow in the circuit, would circulate in the clockwise direction. Since there is a small gap in the circuit, positive charges will build up at point *a* and negative charges at point *b*. Since this is a field produced

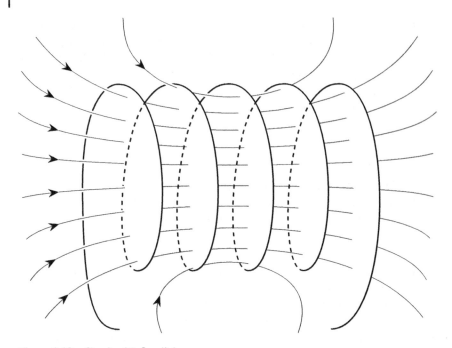

Figure 1.12 Circuit with flux linkage.

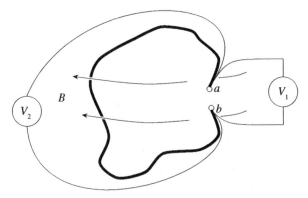

Figure 1.13 Loop with open circuit.

by static charges, it can be described by a scalar potential. The integral from a to b of the conservative component of the electric field is

$$V_a - V_b = \int_a^b E_c \cdot d\ell \tag{1.19}$$

This integral is independent of the path taken from a to b. Since it is not possible for current to flow due to the gap, when equilibrium occurs the net electric field

Figure 1.14 Circuit closed through load resistor.

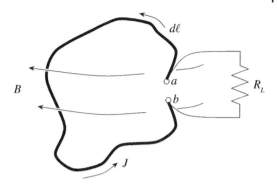

in the conductor body must be zero. If not, charges will continue to flow and we are not yet at equilibrium. Therefore

$$E_c + E_\psi = 0 \tag{1.20}$$

$$V_{ab} = \int_a^b E_c \cdot d\ell = -\int_a^b E_\psi \cdot d\ell \tag{1.21}$$

If we consider the voltmeter readings in the figure and assume all of the flux is passing through the circuit, we might expect that the two readings would be the same as the meters are both measuring between points a and b. However, voltmeter V_1 will read the negative time rate of change of the flux linking the loop while voltmeter V_2 will read 0 since there is no flux linking the closed contour made by the loop and the wires connected to the voltmeter.

Let us now allow current to flow through a load resistance R_L as shown in Figure 1.14.

Integrating around the closed loop and using points a and b as start and end points for the line integral, we have

$$\int_a^b E_\psi \cdot d\ell + \int_a^b E_c \cdot d\ell = \int_a^b \frac{J}{\sigma} \cdot d\ell \tag{1.22}$$

Using our previous definitions

$$-\frac{d\psi}{dt} + V_{ab} = -IR \tag{1.23}$$

This definition is consistent with the equivalent circuit of Figure 1.15.

We should note here that the *source* in the circuit is not localized as shown on the diagram. Also, in much IEEE literature, there are so-called *source* and *load* sign conventions. A source convention, used for generators, would result in the sign of the emf being reversed.

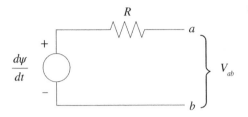

Figure 1.15 Equivalent circuit of Equation (1.23).

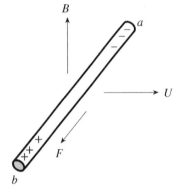

Figure 1.16 Conductor moving in a magnetic field.

An electric field can also be induced by relative motion. The force on an electric charge is described by the Lorentz equation

$$F = qU \times B \tag{1.24}$$

where U is the velocity.[1]

Consider the conductor in Figure 1.16 in which the conductor is moving to the right with velocity U. In the figure the conductor, magnetic field and velocity are all orthogonal.

$$E = \frac{F}{q} = U \times B \tag{1.25}$$

Since the electric field is the force on a unit charge, the electric field due to the charge separation is

$$E = -U \times B \tag{1.26}$$

The induced voltage is then

$$V = \int_a^b E \cdot d\ell = U\ell B \tag{1.27}$$

---

1 The reader may note the similarity to Ampere's force law, $F = i\ell \times B$ which can be related to Equation (1.24) by recognizing that the current is defined as the number of charges per second crossing a surface.

If we now consider a circuit that is stationary in a coordinate system, then by Faraday's law

$$\oint E \cdot d\ell = -\oint \frac{\partial B}{\partial t} \, dS \tag{1.28}$$

If the circuit is moving in the coordinate system, we will be interested in the force being experienced by a free charge. The force observed by a stationary observer is

$$F = q(E + U \times B) \tag{1.29}$$

This force must be the same when observed on the moving charge. We deduce then that the force on a unit charge in the moving system, E' is

$$E' = E + U \times B \tag{1.30}$$

So,

$$\oint E' \cdot d\ell = -\int \frac{\partial B}{\partial t} \, dS + \oint (U \times B) \cdot d\ell \tag{1.31}$$

The first term on the right is sometimes referred to as the *transformer* voltage, and the second term as the *motional* or *generator* voltage. Equation (1.31) is equivalent to

$$\oint E \cdot d\ell = -\frac{d\psi}{dt} \tag{1.32}$$

as will be demonstrated in the example below. It is sometimes convenient to separate these two components due to ease of computation and the different roles these may play in electro-mechanical energy conversion.

Consider the rectangular coil shown in Figure 1.17. The coil is moving to the right at velocity U. The magnetic field is in the z direction and has both space and time variation.

$$B_z(x, t) = B_0 \sin(\omega t) \sin\left(\frac{\pi x}{\tau}\right) \tag{1.33}$$

The coil has width (in the x direction) of τ and height (in the y direction) of ℓ.

The flux linkage is

$$\psi = \int B \cdot dS = \int_0^\tau \int_0^{x_1+\tau} B_0 \sin(\omega t) \sin\left(\frac{\pi x}{\tau}\right) dx \, dy \tag{1.34}$$

$$\psi = \frac{2\ell\tau B_0}{\pi} \sin(\omega t) \cos\left(\frac{\pi x_1}{\tau}\right) \tag{1.35}$$

The emf is the negative time rate of change of the flux linkage, so

$$e = -\frac{d\psi}{dt} = -\frac{\partial \psi}{\partial x} \frac{dx_1}{dt} - \frac{\partial \psi}{\partial t} \tag{1.36}$$

Evaluating the derivatives we find

$$e = 2\ell B_0 \sin(\omega t) \sin\left(\frac{\pi x_1}{\tau}\right) \frac{dx_1}{dt} - \frac{2\omega B_0 \tau \ell}{\pi} \cos\left(\frac{\pi x_1}{\tau}\right) \cos(\omega t) \tag{1.37}$$

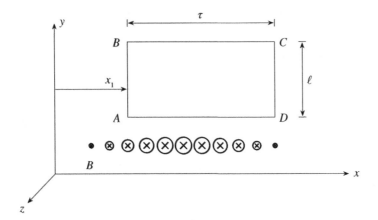

Figure 1.17 Rectangular loop moving through time and space varying field.

Now let us apply Equation (1.31). The transformer voltage is

$$-\int \frac{\partial B}{\partial t} \cdot dS = -\frac{d\psi}{dt} = -\frac{2\tau \ell B_0 \omega}{\pi} \cos\left(\frac{\pi x_1}{\tau}\right) \cos(\omega t) \qquad (1.38)$$

To obtain the motional component, we integrate around the path (*ABCDA*).

$$\oint (U \times B) \cdot d\ell = \int_A^B (U \times B) \cdot d\ell + \int_B^C (U \times B) \cdot d\ell + \int_C^D (U \times B) \cdot d\ell$$
$$+ \int_D^A (U \times B) \cdot d\ell \qquad (1.39)$$

There is no contribution from the second and fourth terms of this equation since $U \times B$ is in the y direction and the dot product with $d\ell$ is zero. The first term on the right becomes

$$\int_A^B (U \times B) \cdot d\ell = B(x_1, t) \frac{dx_1}{dt} \int_y^{y+\ell} dy = B(x_1, t)\ell \frac{dx_1}{dt} \qquad (1.40)$$

In the same way, we obtain for the third term

$$\int_C^D (U \times B) \cdot d\ell = -B(x_1 + \tau, t) \frac{dx_1}{dt} \int_{y+\ell}^y dy = B(x_1 + \tau, t)\ell \frac{dx_1}{dt} \qquad (1.41)$$

The result for the motional term is

$$\oint (U \times B) \cdot d\ell = 2\ell B_0 \sin(\omega t) \sin\left(\frac{\pi x_1}{\tau}\right) \frac{dx_1}{dt} \qquad (1.42)$$

The *emf* is then the sum of the two components and this gives the same result as Equation (1.37).

Figure 1.18 Moving surface in a time-varying field.

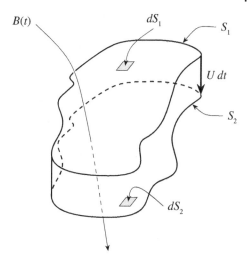

A more general approach is given by Panofsky and Philips [11]. Referring to Figure 1.18, we have a magnetic flux crossing surface *S*.

$$\frac{d\psi}{dt} = \frac{d}{dt}\int B \cdot dS \tag{1.43}$$

The surface is in location 1 at time t and position 2 at time $t + \Delta t$. The surface may also change shape during the motion. We then have

$$\frac{d\psi}{dt} = \frac{1}{\Delta t}\int \left(B_{t+\Delta t} \cdot dS_2 - B_t \cdot dS_1\right) \tag{1.44}$$

We now apply Gauss' law for the magnetic field and find the magnetic flux leaving the closed surface formed by S_1 and S_2 and the sides connecting them.

$$\int \nabla \cdot B \, dV = \int \left(B_t \cdot dS_2 - B_t \cdot dS_1\right) - \int B_t \cdot (U \, dt \times d\ell) \tag{1.45}$$

The last term describes the magnetic flux through the sides.

The value of B at time $t + \Delta t$ on surface S_2 can be expanded in terms of its value at time t.

$$B_{t+\Delta t} = B_t + \frac{\partial B}{\partial t}dt + \cdots \tag{1.46}$$

Substituting we have

$$\frac{d}{dt}\int B \cdot dS = \int \frac{\partial B}{\partial t} \cdot dS + \int B \times U \cdot d\ell + \int \frac{\nabla \cdot B}{dt}dV \tag{1.47}$$

The second term on the right represents the change of flux through the sides of the figure and the third term describes any "sources" of magnetic flux. We now apply Stoke's theorem and recognize that $dV = U \cdot dS\, dt$

$$\frac{d}{dt} \int B \cdot dS = \int \frac{dB}{dt} \cdot dS + \int \nabla \times (B \times U) \cdot dS \tag{1.48}$$

Defining E' as the electric field measured in the moving reference frame, we have

$$\int E' \cdot d\ell = -\frac{d\phi}{dt} = -\int \left(\frac{dB}{dt} + \nabla \times (B \times U) \right) \cdot dS \tag{1.49}$$

Again using Stoke's theorem

$$\nabla \times E' = -\frac{\partial B}{\partial t} + \int \nabla \times (B \times U) \tag{1.50}$$

or

$$\nabla \times \left(E' - U \times B \right) = -\frac{\partial B}{\partial t} \tag{1.51}$$

However we have seen that $E' - U \times B$ is the electric field measured by a stationary observer. We conclude then that the stationary observer finds

$$\nabla \times E = -\frac{\partial B}{\partial t} \tag{1.52}$$

1.6 Waves, Diffusion, and the Magneto-Quasi-static Approximation

If we take the complete set of Maxwell's equations in their time-harmonic form, we have Ampere's law

$$\nabla \times H = J + j\omega\epsilon E = \sigma E + j\omega\epsilon E \tag{1.53}$$

and for Faraday's law

$$\nabla \times E = -j\omega\mu H \tag{1.54}$$

In Equation (1.53), the first term on the right, σE, is called the *conduction current*. This current physically involves the movement of charge, the ampere being defined as one coulomb of charge crossing a surface per second. The second term, $j\omega\epsilon E$, is called the *displacement current*. There is no charge flow involved in displacement current. It is the time variation of the electric flux density. This current produces a magnetic field just as conduction current does.

We substitute (1.54) into (1.53), take the curl of both sides, and use the vector identity

$$\nabla \times \nabla \times H = \nabla\nabla \cdot H - \nabla^2 H \tag{1.55}$$

and the fact that $\nabla \cdot H = 0$ to obtain the homogeneous *wave equation*

$$\nabla^2 H + \mu\epsilon \left(1 - \frac{j\sigma}{\omega\epsilon} \right) \omega^2 H = 0 \tag{1.56}$$

Often, the term in parenthesis is combined with the permittivity to create an effective complex permittivity

$$\epsilon_{\text{eff}} = \epsilon \left(1 - \frac{j\sigma}{\omega\epsilon} \right) \tag{1.57}$$

Defining the propagation constant, k, as

$$k^2 = \omega^2 \mu\epsilon_{\text{eff}} \tag{1.58}$$

The wave equation becomes

$$\nabla^2 H + k^2 H = 0 \tag{1.59}$$

We will break the wave number, k, into real and imaginary parts so that

$$k = k' - jk'' \tag{1.60}$$

The real part of the wave number is

$$k' = \omega^2 \mu\epsilon \tag{1.61}$$

The imaginary part is

$$k'' = \omega\mu\sigma \tag{1.62}$$

If we had ignored the displacement current term and applied the same procedure, we would get the diffusion equation.

$$\nabla^2 H - j\mu\sigma\omega H = 0 \tag{1.63}$$

We see here that it is the relative importance of k' and k'' that will determine if the wave equation can be approximated by the diffusion equation. In fact, the magnitude of the term

$$\frac{\sigma}{\omega\epsilon},$$

the so called *loss tangent*, is the determining factor.[2]

The wave equation is of course the more general description of the phenomena. The diffusion equation, which we will use throughout this work to describe the eddy current behavior, is therefore an approximation, and we must study what conditions are necessary for this approximation to be valid.

---

2 This quantity is called the loss tangent since the arc-tangent of this quantity gives the angle between the displacement current density and the total current density. The angle is zero for a perfect insulator and 90° for a perfect conductor.

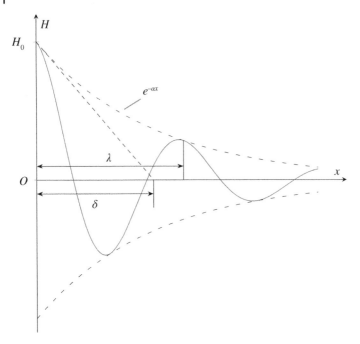

Figure 1.19 Attenuation of wave in conducting material.

If we consider a plane wave propagating in the $+x$ direction, the solution to the wave equation has the form

$$e^{-jkx} = e^{-jk'x - k''x} \tag{1.64}$$

which can be verified by substitution into (1.59). Recall that the time variation is implied in the phasor representation of the field quantities. Figure 1.19 shows the progress of the wave into the material.

Let us now explore the loss tangent. This term is the ratio of the magnitude of the conduction current, σE, and the displacement current, $j\omega\epsilon E$. If we take copper for example, the conductivity is $\sigma = 5.8 \times 10^7$ Sm^{-1}. The permittivity is approximately $\epsilon_0 = \frac{1}{36\pi} \times 10^{-9}$ Fm^{-1}. For frequencies in the same order of magnitude as power frequency, say 100 Hz, the ratio of conduction to displacement current is on the order of 10^{18}. For a higher frequency, say radio frequency, in the range of 100 MHz, we get a ratio on the order of 10^{10}. In fact, for the displacement current to equal the conduction current, we must have a frequency in the order of 10×10^{18} Hz. It seems that for good-conducting metals, the approximation is an excellent one over a very large range of frequencies, and we can safely ignore the displacement currents. Let us look a bit deeper into the wave behavior of lossy materials. For poor conductors ($\sigma \ll \omega\epsilon$), we have[3]

$$k = \omega\sqrt{\mu\epsilon}\sqrt{1 - \frac{j\sigma}{\omega\epsilon}} \approx \omega\sqrt{\mu\sigma}\left(1 - \frac{j\sigma}{2\omega\epsilon}\right) \tag{1.65}$$

In this case, $k' \approx \omega\sqrt{\mu\epsilon}$, which is the free-space (no conductivity) result. The imaginary part is $k'' \approx \frac{\sigma}{2}\sqrt{\frac{\mu}{\epsilon}}$. For low conductivity, this can be a large number. If the conductivity is large and $\sigma \gg \omega\epsilon$, then

$$k = \omega\sqrt{\mu\epsilon}\sqrt{1 - \frac{j\sigma}{\omega\epsilon}} \approx \omega\sqrt{\mu\epsilon}\sqrt{\frac{-j\sigma}{\omega\epsilon}} = \sqrt{-j\omega\sigma\mu} \tag{1.66}$$

We use $\sqrt{-j} = \frac{1-j}{\sqrt{2}}$ and find that

$$k \approx (1 - j)\sqrt{\frac{\omega\mu\sigma}{2}} \tag{1.67}$$

Note here that the real and imaginary parts of the wave number are numerically equal. The skin depth for a good conductor is (see Section 1.7)

$$\delta = \frac{1}{k''} = \sqrt{\frac{2}{\omega\mu\sigma}} \tag{1.68}$$

In a good conductor, the wavelength is approximately $\lambda \approx 2\pi\delta \approx 2\pi/k''$.

Another quantity coming from wave theory is the complex *wave impedance*. The wave impedance is defined as the ratio of the electric field and magnetic field. In this case

$$\eta = \sqrt{\frac{\mu}{\epsilon_{\text{eff}}}} = \sqrt{\frac{\mu}{\epsilon - j\sigma/(\omega\epsilon)}} \tag{1.69}$$

For a good conductor

$$\eta \approx (1 + j)\sqrt{\frac{\omega\mu}{2\sigma}} \tag{1.70}$$

We will see in Section 2.2 that, using the diffusion equation for good conductors, the real and imaginary parts of the impedance will be equal for linear materials.

The wave impedance is also useful in the evaluation of the Poynting vector, which we will use to evaluate real and reactive power. We define the Poynting vector for steady-state sinusoidal fields as

$$S = E \times H^* \tag{1.71}$$

which can be written in terms of the wave impedance as

$$S = \frac{|E|^2}{\eta^*} \tag{1.72}$$

---

3 We have used the Taylor expansion for the square root, $\sqrt{1 + \xi} \approx 1 + \xi/2$ for small values of ξ.

Let us look at the problem of an electromagnetic wave normally incident on a copper plate. The electric and magnetic field components are then parallel to the surface of the copper plate. For this example, the frequency is 100 kHz. The propagation constant is

$$k = \sqrt{j\omega\mu(\sigma + j\omega\epsilon)} = \alpha + j\beta = 4785.1 + j4785.1 \tag{1.73}$$

The wavelength is

$$\lambda = \frac{2\pi}{\beta} = 4.15 \times 10^{-4} \text{ m} \tag{1.74}$$

Compare this to the wavelength in free space which is $\lambda = c/f = 3000$ m. There is a significant compression of the wavelength as the wave travels through the conductor. The phase velocity is

$$u = \frac{\omega}{\beta} = 131.31 \text{ m s}^{-1} \tag{1.75}$$

The wave impedance is

$$\eta = \sqrt{\frac{j\omega\mu}{\sigma + j\omega\epsilon}} = 8.2502 \times 10^{-5} + j8.2502 \times 10^{-5} \ \Omega \tag{1.76}$$

The real part of the wave impedance can be considered the effective surface resistance. Using the good conductor approximation as will be introduced in Chapter 2, the effective resistance is

$$R = \frac{1}{\sigma\delta} = 8.2502 \times 10^{-5} \ \Omega \tag{1.77}$$

This is equivalent to the dc resistance if the current is confined to a skin depth, δ.

We will also discuss shielding by eddy currents in Section 3.5. We can use the attenuation constant to understand the decay of the wave as it travels through the conductor. For example, how far must a wave travel in our present example, to attenuate to 1% (40 dB) of its value at the surface? Evaluating the magnitude of the wave we find

$$e^{-\alpha d} = 0.01 \tag{1.78}$$

which gives

$$d = 9.624 \times 10^{-4} \text{ m} \tag{1.79}$$

So we see that for a wide range of problems involving waves and good conductors, we can ignore the displacement currents and in this case, we will be describing the phenomena with the diffusion equation rather than the more general wave equation. There are certain nonphysical implications of this approximation, but they will rarely play an important role in our analysis. For example, if we were to apply a step-function of excitation to a conductor, there will be a finite instantaneous response even at infinity. This is not physically possible since it would involve the wave traveling faster than the speed of light. The wave equation would have resulted in a delay which the diffusion equation does not. With these criterion, the reader can make the determination if the approximation is valid for their particular application.

1.7 Skin Depth or Depth of Penetration

In Chapter 2, we will derive the concept of *skin depth* directly from Maxwell's equations. This short introduction to this important parameter begins with the diffusion equation written for steady-state sinusoidal conditions. In this case, the variables are phasors. Phasors are complex quantities that give us information on the magnitude and phase angle of the variable. It is assumed that all variables in the problem are sinusoidally time-varying at the same frequency. In principle, we can only use phasor analysis for linear problems. If the material properties are a function of the variables in the problem, phasors are no longer valid. We will consider problems in which we have nonlinear material behavior in Section 2.6. For now, consider the one-dimensional diffusion equation written for the current density.

$$\frac{\partial^2 J}{\partial x^2} = -j\omega\mu\sigma J \tag{1.80}$$

The solution to (1.80) is

$$J = C_1 e^{\alpha x} + C_2 e^{-\alpha x} \tag{1.81}$$

where

$$\alpha = \sqrt{-j\omega\mu\sigma} \tag{1.82}$$

For the case of a conductor with infinite extent (the finite conductor case will be treated in Section 2.2), the constant C_1 must be zero to prevent the solution from blowing up at infinity. We now replace $\sqrt{-j} = e^{-j\pi/4}$ with $\frac{1-j}{\sqrt{2}}$. The solution is then

$$J(x) = J_0 e^{-\alpha x} = J_0 \left(e^{-x/\delta} e^{jx/\delta} \right) \tag{1.83}$$

where

$$\delta = \sqrt{\frac{2}{\omega\mu\sigma}} \tag{1.84}$$

The parameter δ has units of meters. For copper ($\mu = \mu_0$ H m^{-1}, $\sigma = 5.8 \times 10^7$ S m^{-1}), we can write

$$\delta = \frac{0.066085}{\sqrt{f}} \tag{1.85}$$

The frequency, in hertz, has units of inverse seconds, and writing the skin depth in this way reveals the $\sqrt{t}$ factor that will be introduced in Section 1.9 as part of the solution to the diffusion equation. The first exponential term in (1.83) describes the magnitude of the current density decaying with x. At one skin depth, the magnitude decreases to $1/e \approx 0.37$ of its value at the surface, and after 3–5 skin depths, the current density is extremely small compared to the surface value. While this first term describes the magnitude of the current density phasor, the second exponential term describes the phase shift. The term $e^{j\theta}$ has magnitude 1.0 for any value of the angle θ. We have a linear phase shift with respect to distance of one radian per skin depth. This is shown in Figure 1.20 where Equation (1.83) is plotted for copper at 60 Hz.

Knoepfel [12] has introduced a skin depth calculator which we have recreated and included as Figure 1.21. By using this diagram, one can select the frequency and conductivity, and for magnetic materials, one can select different permeabilities and find the skin depth. Note that the scale is logarithmic. As an illustration,

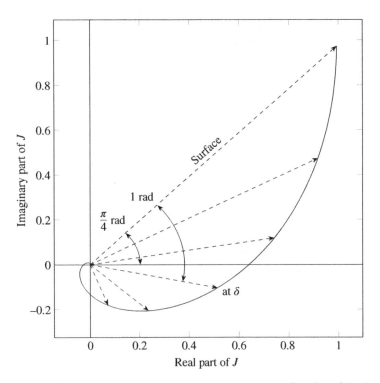

Figure 1.20 Magnitude and phase of current density as a function of depth.

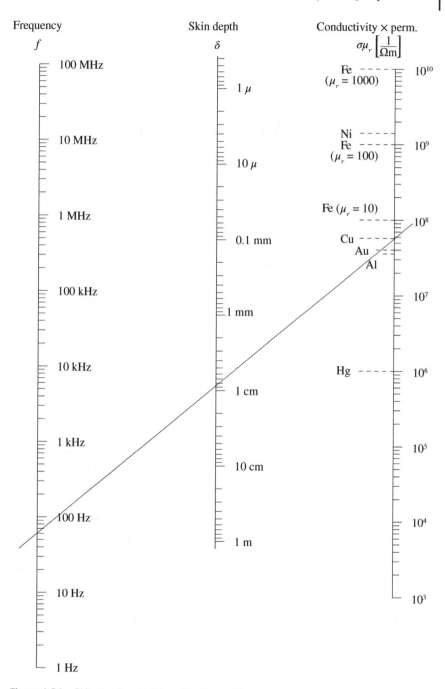

Figure 1.21 Skin depth calculator after Knoepfel.

consider copper at 60 Hz. We construct a straight line from the bottom left on the figure at 60 Hz and connect it to the scale on the right which has copper listed at $\sigma\mu_r = 5.8 \times 10^7$ S m^{-1}. The line intersects the center scale at the skin depth. The skin depth in this case is $\delta = 0.00853$ m.

1.8 Diffusion, Heat Transfer, and Eddy Currents

Heat transfer is also described by the diffusion equation. Heat transfer problems are often easier to visualize than electromagnetic diffusion and serve as a good introduction to some of the principles. There are differences of course. Temperature is a scalar while the magnetic field and the eddy current density are vectors. However, for a large class of problems, a direct analogy can be made and the basic ideas can be applied to both domains.

Fourier's law states that the flow of heat is proportional to the negative of the gradient of temperature. The direction of the transfer of thermal energy is from higher to lower temperature. This is, of course, the same as Fick's law of diffusion.

Consider a long column of material at a uniform temperature. When a constant temperature boundary condition is applied at the surface, some heat conducts through the material, and some heat is stored in the material. The heat conduction depends on the thermal conductivity (or thermal resistivity), while the heat storage depends on the specific heat and density of the material. The diffusion equation becomes [13]

$$DV^2 T = \frac{dT}{dt} \tag{1.86}$$

where D is the thermal diffusivity. This is the ratio of thermal conductivity to the volumetric specific heat. In solving complicated heat transfer problems involving conduction and storage, one of the most popular methods is the use of an equivalent circuit in which the resistors are proportional to the thermal resistivity and capacitors are proportional to the volumetric specific heat. In Section 5.1, we will use the same equivalent circuit to represent eddy currents in solid conductors.

To see the connection to eddy currents, let us consider the temperature distribution near the surface of the earth over time (see Figure 1.22). These data are simulated using information in [13] which have been verified by measurements. If we take one cycle per year as the fundamental frequency, we can model the seasonal temperature change using Equation (1.86). The solution to the diffusion equation is

$$T(y, t) = T_{\text{amb}} + A_0 e^{-\frac{y}{\delta}} \sin\left(\frac{2\pi(t - t_0)}{365} - \frac{y}{\delta} - \frac{\pi}{2}\right) \tag{1.87}$$

Here the parameter δ is given by [4]

$$\delta = \sqrt{\frac{2D}{\omega}} \tag{1.88}$$

We will see equations of the same form in eddy current applications. In this simulated example, a randomness is artificially added to the amplitude to make it more realistic. We will also consider the daily variation of the temperature, with highs and lows occurring daily. This is included by again using Equation (1.87) with the some modifications. We can then superimpose the two solutions. The changes are that the ambient temperature term, T_a is omitted since it has already been included in the yearly variation calculation. We also change the frequency to 365 cycles per year, which changes δ. A new amplitude is also used to account for the daily temperature cycles. The result at different depths is shown in Figure 1.22. There are a number of extremely important observations to be made here. The first is concerned with the exponential term in Equation (1.87). This term gives a solution that decreases exponentially with depth. If, for the fundamental frequency, we compare the temperature amplitude at the surface to the temperature amplitude at a depth $y = \delta$ the amplitudes will have a ratio of $1/e \approx 0.37$. In fact, for any two points, separated by vertical distance δ, the amplitudes will have the same ratio. The parameter δ, as we have seen, is the skin depth, or depth of penetration. After 3–5 skin depths, the fluctuation is usually considered negligible. In the current example, the thermal diffusivity, D, is 0.3×10^{-6} m² s⁻¹ which gives a skin depth

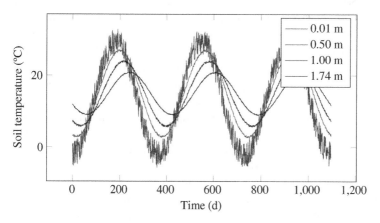

Figure 1.22 Temperature at various depths over a few years.

---

4 If we replace the parameter D with $\frac{1}{\mu\sigma}$, we obtain from Equation (1.88) the result $\delta = \sqrt{\frac{2}{\omega\mu\sigma}}$ which is the electromagnetic skin depth.

for the fundamental (one cycle per year) of $\delta_{year} = 1.735$ m. The skin depth for the 365th harmonic (one cycle per day) is $\delta_{day} = 0.091$ m. The ratio of the two exponential constants is equal to the square root of the frequency ratio ≈ 19.1. This explains why the fundamental is still very noticeable at 1 m depth but the daily fluctuation has mostly disappeared after 0.4 m. We will note the same behavior in the study of eddy currents.

The second essential point that is illustrated by this heat transfer example is the phase shift of the temperature fluctuation. Considering the argument of the sine function in Equation (1.87), we see that the phase of the function changes linearly with depth. When we move one skin depth in the y direction, the phase shifts one radian. This phase shift is evident in Figure 1.22. In observing the peaks in the temperature profile, we see that the peaks are delayed as we move down into the soil. This explains why underground pipes can freeze several days or even weeks after a cold front. This phase shift will also be an important result in the eddy current analysis and can result in a reversal in the direction of the eddy currents. These two results, the exponential decay of the amplitude and the linear variation of the phase angle with depth will exactly coincide with our findings on eddy currents.

1.9 The Diffusion Equation and Random Walks

We will see in Section 1.10 that eddy current solutions often result in Gaussian or Normal distributions. Also, we will find factors of $\sqrt{t}$ and *diffusion time constants* resulting from transient eddy current problems.

To see the connection to particle diffusion, we can take a probabilistic approach to the diffusion process. There is a vast literature on how particles diffuse through a medium, going from regions of high concentration to regions of lower concentration [14]. If we release particles into a medium, initially there will be a high concentration near the location of origin. Although the motion of individual particles is random, over time and with large numbers of particles, we eventually reach a steady state in which, from a macroscopic point of view, the concentration is uniform. The process can be described by the diffusion equation.

Consider a one-dimensional *random walk*, as illustrated in Figure 1.23. Let us say that an individual particle is located at $x = 0$ on the one-dimensional axis of Figure 1.23. In an *unbiased* random walk, the individual can move Δx in the positive direction or Δx in the negative direction. Each of these possibilities has a probability of $p = 0.5$. We will assume that each move takes place in a time interval Δt. The probability of finding our individual at a particular location at a particular time is $p(x, t)$. What is the probability of finding the individual at location x at

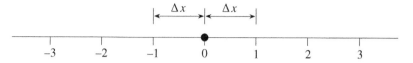

Figure 1.23 One-dimensional random walk.

time $t + \Delta t$? There are only two possibilities. Either they were at location $x - \Delta x$ or $x + \Delta x$ at time t. The total probability is then

$$p(x, t + \Delta t) = \frac{1}{2}(p(x - \Delta x, t) + p(x + \Delta x, t)) \qquad (1.89)$$

We now subtract $p(x, t)$ from each side and divide each side by Δt. Then we multiply and divide the right-hand side by $(\Delta x)^2$. This gives

$$\frac{p(x, t + \Delta t) - p(x, t)}{\Delta t} = \frac{1}{2}(\Delta x)^2 \frac{(p(x - \Delta x, t) - 2p(x, t) + p(x + \Delta x, t))}{(\Delta x)^2 \Delta t} \qquad (1.90)$$

In the limit, as Δt approaches zero, the left-hand side becomes $\frac{\partial p}{\partial t}$. For the right-hand side, we define the diffusion constant as $D = (\Delta x)^2/(2\Delta t)$. The remaining term, as Δx approaches zero, is the difference equation for the second derivative of p with respect to x. This result will be derived in Section 5.1. This gives the one-dimensional diffusion equation

$$\frac{\partial p}{\partial t} = D \frac{\partial^2 p}{\partial x^2} \qquad (1.91)$$

With a large number of trials, we will obtain the probability distribution with a maximum value at $x = 0$. The solution of Equation (1.91) is

$$p(x, t) = \frac{1}{\sqrt{4\pi t D}} e^{-\frac{x^2}{4Dt}} \qquad (1.92)$$

This is the well-known normal distribution. Note that the solution is symmetric around $x = 0$ and has a decay rate including a $1/\sqrt{t}$ term. This solution is relatively intuitive for particle distribution. We expect the initial concentration to be highest at $x = 0$ and eventually decay to a uniform distribution over a long time period. When we are dealing with thermal or electromagnetic diffusion, where it is heat or flux or current that is being redistributed, it is perhaps more difficult to picture although we will find exactly the same result. In Section 1.10, we will cover magnetic problems that are described by the diffusion time constants and exhibit the $\sqrt{t}$ behavior.

Figure 1.24 shows the normal distribution at various times. In this plot, $D = 1.0$. We can see that the distribution is very concentrated around zero at the beginning of the process and tends to a uniform distribution as time progresses.

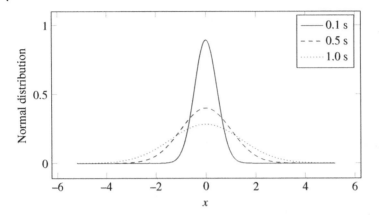

Figure 1.24 Normal distribution at various times.

1.10 Transient Magnetic Diffusion

We will now look at the problem of transient magnetic diffusion into a conducting material. This process allows us to analyze problems with arbitrary inputs, such as a step-function of magnetic field. We will see later in the book that the diffusion equation applied to eddy current analysis can be written in terms of the magnetic field, H, the current density, J, and the magnetic vector potential, A. The choice can depend on the application and the method of solution. Consider the diffusion equation in one-dimension for the magnetic vector potential.

$$\frac{\partial^2 A}{\partial x^2} = \mu\sigma \frac{\partial A}{\partial t} \tag{1.93}$$

Letting $k = \mu\sigma$, we have

$$\frac{\partial^2 A}{\partial x^2} = k \frac{\partial A}{\partial t} \tag{1.94}$$

Insight into the problem can be found by using dimensional analysis. In this problem, we have five parameters:

1. B_0: $\mathrm{Wb\,m^{-2}}$
2. A: $\mathrm{Wb\,m^{-1}}$
3. k: $\mathrm{s\,m^{-2}}$
4. x: m
5. t: s

With five parameters and three different units, we can form two dimensionless groups. Let

$$D_1 = \frac{A}{B_0\sqrt{t/k}} \tag{1.95}$$

$$D_2 = \frac{x}{\sqrt{t/k}} \tag{1.96}$$

In dimensional analysis, we take one non-dimensional group to be a function of the other, so that

$$D_1 = f(D_2) \tag{1.97}$$

and in this case

$$A(x,t) = \frac{B_0}{\sqrt{t/k}}f\left(\frac{x}{\sqrt{t/k}}\right) \tag{1.98}$$

Now we make a change of variables that will transform the equation from a partial differential equation (PDE) to an ordinary differential equation (ODE). Let

$$\eta = \frac{x}{\sqrt{4t/k}} \tag{1.99}$$

This gives

$$\frac{\partial A}{\partial x} = \frac{dA}{d\eta}\frac{\partial \eta}{\partial x} = \frac{1}{\sqrt{4t/k}}\frac{dA}{d\eta}$$

$$\frac{\partial^2 A}{\partial x^2} = \frac{d}{d\eta}\left[\frac{\partial A}{\partial x}\right]\frac{\partial \eta}{\partial x} = \frac{k}{4t}\frac{d^2 A}{d\eta^2} \tag{1.100}$$

For the derivative with respect to time, we have

$$\frac{\partial A}{\partial t} = \frac{dA}{d\eta}\frac{\partial \eta}{\partial t} = \frac{-x}{2t\sqrt{4t/k}}\frac{dA}{d\eta} \tag{1.101}$$

Substituting, we find

$$\frac{d^2 A}{d\eta^2} = -2\eta\frac{dA}{d\eta} \tag{1.102}$$

We have transformed the equation into an ODE. No matter what the values of x or t, we can express the result exclusively as a function of the variable η.

Also note, from our definition of η in Equation (1.99), that for $x = 0$ we have $\eta = 0$ and for $x = \infty$ we have $\eta = \infty$.

Now we can separate the terms.

$$\frac{d\left(\frac{dA}{d\eta}\right)}{\frac{dA}{d\eta}} = -2\eta d\eta \tag{1.103}$$

We integrate once to find

$$\ln\left(\frac{dA}{d\eta}\right) = -\eta^2 + C_1 \tag{1.104}$$

Integrating again, we get

$$A = C_1 \int_0^\eta e^{-u^2} du + C_2 \tag{1.105}$$

We now apply boundary conditions to evaluate the constants. Assume the initial value of the potential before the source is applied is A_i. We call the potential at the surface A_0. Then, from the condition $A = A_0$ at $\eta = 0$, we conclude that $C_2 = A_0$. Using the condition that at infinity we have $A = A_i$, we evaluate

$$A_i = C_1 \int_0^\infty e^{-u^2} du + A_0 \tag{1.106}$$

This integral has been evaluated [14] and we find

$$C_1 = 2\frac{A_i - A_0}{\sqrt{\pi}} \tag{1.107}$$

The general solution is

$$\frac{A - A_0}{A_i - A_0} = \frac{2}{\sqrt{\pi}} \int_0^\eta e^{-u^2} du = \text{erf}(\eta) \tag{1.108}$$

Note that the error function approaches 1 as η approaches infinity. (For information on the error function, see Appendix C.)

In this case, the initial value of potential is $A_i = 0$ and we also know the derivative at $x = 0$,

$$B_0 = -\frac{\partial A}{\partial x}\Big|_{x=0} = A_0 \frac{d\,(\text{erf}(\eta))}{d\eta}$$

$$= A_0 \frac{\frac{2}{\sqrt{\pi}} e^{-\eta^2}}{\sqrt{4t/k}}\Big|_{\eta=0} = \frac{A_0}{\sqrt{\pi t/k}} \tag{1.109}$$

Then the final solution is

$$A(x, t) = 2B_0 \sqrt{\frac{t}{\pi k}} e^{\frac{-x^2 k}{4t}} - B_0 x\, \text{erfc}\left(\frac{x}{2\sqrt{t/k}}\right) \tag{1.110}$$

where erfc is the complementary error function defined as $\text{erfc} = 1 - \text{erf}$.

To introduce an alternate approach, not using dimensional analysis, consider Figure 1.25. We will solve the PDE

$$\frac{\partial J}{\partial t} = \frac{1}{\mu\sigma} \frac{\partial^2 J}{\partial y^2} \tag{1.111}$$

Figure 1.25 Infinite conductor with applied current sheet.

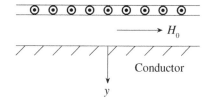

by means of the Fourier transform.[5] We transform the function of y to a function of k by means of

$$J(y, t) = \int_{-\infty}^{\infty} \frac{dk}{2\pi} e^{jky} \times F(k, t) \tag{1.112}$$

where

$$F(k, t) = \int_{-\infty}^{\infty} e^{jky} \times J(y, t)\, dy = \int_{0}^{\infty} \left(e^{jky} + e^{-jky}\right) \times J(y, t)\, dy \tag{1.113}$$

The derivative of $F(k, t)$ is then simply multiplication by jk, so

$$\frac{\partial F}{\partial t} = -\frac{k^2}{\mu\sigma} F(k, t) \tag{1.114}$$

For every value of k, we have a first-order differential equation, the solution of which is

$$F(k, t) = F(k, 0) e^{-k^2 t/(\mu\sigma)} \tag{1.115}$$

If we consider the case of a current sheet that is turned on at $t = 0^+$, in other words a step function, we can evaluate the surface condition. Just near the surface of the conductor, say at a distance $y = -\epsilon$, the current sheet can be represented by the application of H_0. This is illustrated in Figure 1.25. The current density is the derivative of H with respect to y or

$$J(y, 0) = -\frac{\partial H}{\partial y} = H_0 \delta(y - \epsilon) \tag{1.116}$$

The Fourier transform of the current sheet is then

$$F(k, 0) = \int_{0}^{\infty} \left(e^{jky} + e^{-jky}\right) H_0 \delta(y - \epsilon)\, dy = 2H_0 \cos(k\epsilon) \approx 2H_0 \tag{1.117}$$

For $t > 0$ the solution is then

$$F(k, t) = 2H_0 e^{-\frac{k^2}{\mu\sigma} t} \tag{1.118}$$

We now take the inverse transform to transform a function of k to a function of y.

$$J(y_0, t) = \int \frac{1}{2\pi} e^{jky_0} \times 2H_0 e^{-\frac{k^2}{\mu\sigma} t}\, dk \tag{1.119}$$

---

5 Crank [14] uses a similar approach to solve the analogous problem of heat conduction using Laplace transforms.

Equation (1.119) is the well-known Gaussian, or normal, distribution which was introduced in Section 1.9. Since we are only interested in the positive values of y we have just half of the distribution as the solution.

$$J(y_0, t) = \frac{2H_0}{\sqrt{4\pi Dt}} e^{-\frac{y_0^2}{4Dt}} \tag{1.120}$$

If we define a new function that has the time variation in it

$$g(t) = \sqrt{4Dt} \tag{1.121}$$

The width of the distribution increases with time following the function $g(t)$.

We can use Ampere's law to find the magnetic field by integrating $J(x, t)$.

$$
\begin{aligned}
H(y, t) &= \int_y^t J \, dy \\
&= \frac{2H_0}{\sqrt{\pi}} \int_y^\infty e^{-\frac{\xi^2}{g(t)^2}} \, d\xi \\
&= H_0 \left(1 - \text{erf}\left(\frac{y}{g(t)}\right) \right) \\
&= H_0 \, \text{erfc}\left(\frac{y}{g(t)}\right)
\end{aligned}
\tag{1.122}
$$

As an example, consider a copper block, $\sigma = 5.8 \times 10^7$ S m^{-1} and $\mu_0 = 4\pi \times 10^{-7}$ H m^{-1}. We apply a step function of $H_0 = 1.0$ A m^{-1} at $t = 0$ s. Figure 1.26 shows the current density distribution, from Equation (1.120), as a function of depth for three different times, $t = 0.1, 0.5, 1.0$ s. Note the similarity to Figure 1.24, which we found while discussing particle diffusion. For short times, the current is crowded near the surface. As time progresses, the maximum value decreases

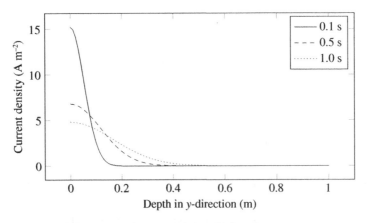

Figure 1.26 Current density vs. depth for different times.

and the distribution becomes more uniform. Although this is also a normal distribution solution, in our particular electromagnetic example, we have the solution only in the positive y region, which is only the positive half of the distribution.

1.11 Coupled Circuit Models for Eddy Currents

Using the magneto-quasi-static assumptions, it is legitimate to model eddy current phenomena by a set of coupled circuits with resistance and self and mutual inductances. Unfortunately, it is often very difficult to compute the self and mutual inductances for general geometries and doing so often involves a field solution by numerical techniques. There is however a fairly large class of problems in which the inductances are relatively easy to find and standard electrical network techniques can be used. These solutions are often easier to interpret than direct field solutions. We will use these coupled circuit methods in the study of slots in electric machines in Section 11.1. We have already introduced the concepts of skin depth and proximity effect and will now consider a coupled circuit model which illustrates these points. Consider the long go-and-return flat bus bars illustrated in Figure 1.27. All current is in the z direction and the magnetic field is in the $x - y$ plane.

As indicated in the figure, we divide the conductor into small segments in the x direction. These form a set of mutually coupled conductors that are parallel to each other. We will assume that the thickness of the conductor is small compared to the skin depth. This however is not a necessary assumption, and we could divide the plate into segments in the vertical direction as well. The closed form solution for a number of long parallel conductors is well known and a summary is given in Appendix E. If the number of segments is N, then we will have an $N \times N$ system of coupled circuit equations. The resistance per unit depth of each segment i is found as

$$R_i = \frac{1}{\sigma \Delta x \Delta y} \qquad (1.123)$$

We form a resistance matrix, which is a diagonal matrix. Each element of the diagonal has the value found from Equation (1.123). The inductance matrix will

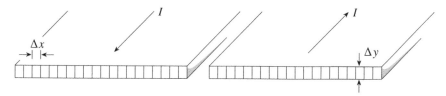

Figure 1.27 Parallel flat copper bus bars.

be a full matrix. In finding the flux linkage of a conductor in the presence of a number of long parallel conductors, we find for conductor 1

$$\lambda_1 = \frac{\mu_0}{2\pi} \left(I_1 \ln \left(\frac{1}{r_1'} \right) + I_2 \ln \left(\frac{1}{d_{12}} \right) \dots I_n \ln \left(\frac{1}{d_{1N}} \right) \right) \tag{1.124}$$

where r' is the geometric mean radius of the conductor, and d_{ij} is the distance between conductors i and j. This process is repeated for all N conductors. We now have an $N \times N$ system of simultaneous equations of the form

$$(V) = (R + j\omega L)(I) \tag{1.125}$$

In this example, we consider two parallel copper plates. The plates are 0.00125 m thick and 0.1 m wide. There is a 0.01 m space between the conductors. The conductors are divided into 160 small segments. The distances are taken as the distance from center to center of the segments. A voltage of $(1.0 + j0.0)$ V at 60 Hz is applied to the conductor on the left and $(-1.0 + j0.0)$ V to the conductor on the right. The system of 160 simultaneous equations is solved for the current in the segments. Figures 1.28 and 1.29 show the real and imaginary components of the current

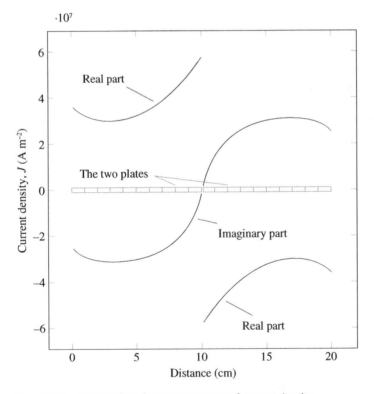

Figure 1.28 Real and imaginary components of current density.

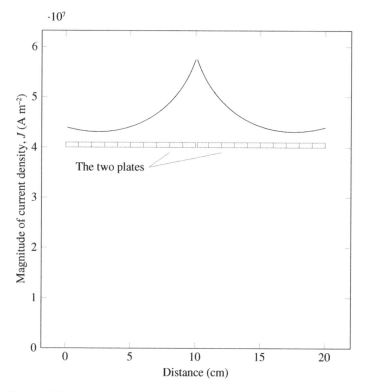

Figure 1.29 Magnitude of the current density.

density and the magnitude of the current density, respectively. For reference, a representation of the two plates is superimposed on each figure.

We note that the current crowds to the center and is anti-symmetric around the center line. This is the behavior that we expect from our discussion of the proximity effect. If the currents are in opposite directions, the currents crowd to the inner surfaces. Figure 1.30 shows the finite element solution for this example. We can see from the flux plot that the currents are crowding toward the center.

There are also other important parameters of this configuration that are now available to us. Adding up the currents in one of the bars, we find $I = (4616.7 - j3235.0)$ A. We can divide this by the applied voltage to find the impedance, which is $Z = (2.905 + j2.036 \times 10^{-4})$ Ω. To compare, we can easily find the dc resistance as $R_{dc} = 2/(\sigma \times \text{area}) = 2.76 \times 10^{-4}$ Ω. As expected, this is smaller than the ac resistance due to the nonuniformity of the current density.

If we had applied voltages in the same direction, we would have found currents crowding to the outside. This example is illustrated in Figure 1.31 for the finite element solution.

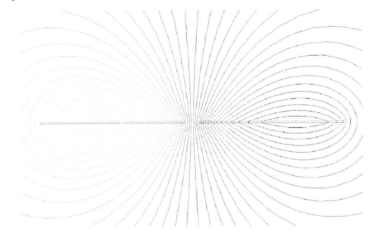

Figure 1.30 Finite element solution to the go and return example.

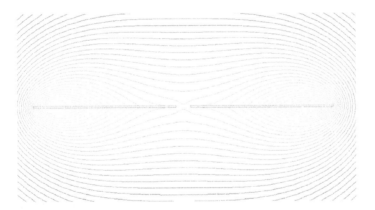

Figure 1.31 Finite element solution for currents in the same direction.

The equivalent circuit method uses relatively few equations compared to the finite element method. However, there is a lot of information in the terms of the equivalent circuit equations. They include the closed form solutions for the self and mutual flux linkages. As we will see in Section 6.1, the individual finite element equations include very little information about the field and its variations, but only about how the elements are connected and an assumed variation of potential. It is only when all of the element equations are assembled and a minimization technique is applied, that we get a valid solution.

Figure 1.32 shows a comparison of the magnitude of the current density in one plate from the equivalent circuit method and the finite element method.

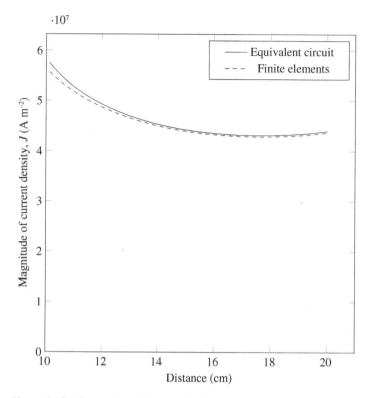

Figure 1.32 Comparison of current density magnitude in one plate.

1.12 Summary

In this brief introduction, we have introduced the basic principles that will be used in eddy current analysis. The fundamental equations are Faraday's law, Ampere's law, and Lenz's law. We saw how current is induced in conductors and circuits. We discussed the assumptions made in magneto-quasi-static analysis, showed the connection to wave phenomena, and discussed the difference between the wave equation and diffusion equation. We also introduced the idea of skin depth or depth of penetration in steady-state sinusoidal analysis. Diffusion is also prevalent in other fields, and we compared the results of our electromagnetic analysis to a heat transfer example and saw that the fundamental results were the same. We have considered the solutions to the diffusion equation in the time domain for an important example, a step function of current. This introduced the normal distribution solutions and transient skin depth. This was then contrasted with particle diffusion beginning with an entirely random process. We found that the results were analogous to our electromagnetic diffusion example. This introduction ended with an example using coupled magnetic circuits to model eddy current phenomena. The use of equivalent circuits is quite popular in eddy current analysis and often offers a more intuitive approach than the direct solution of differential equation.

2

Conductors with Rectangular Cross Sections

In Chapter 1, we have laid the foundation for the study of eddy currents. We have discussed the diffusion equation and its solution for transient and steady-state sinusoidal conditions and introduced the concepts of resistance and reactance-limited eddy current phenomena as well as the proximity effect and the skin effect. In this chapter, we discuss some practical examples that are described in Cartesian or rectangular coordinates. These examples include laminations, bus bars, plates, and solid conductors with rectangular cross sections. We also include solutions for eddy currents in materials that have nonlinear material properties (magnetic saturation) including approximations for including the effects of magnetic hysteresis. These solutions are found in closed form and are based on the theory presented in Chapter 1. There are of course many other examples of conductors with rectangular cross sections, such as those used in electric machines and transformers. We present some of these examples in Part 3 of this book as applications, where we discuss conductors in slots of electric machines and transformer windings.

2.1 Finite Plate: Resistance Limited

The first problem we will consider is the determination of resistance-limited eddy current loss in conductors of rectangular cross section. While this problem has a relatively simple solution, the result is of extreme importance in practical applications. Many electrical devices such a motors, generators, transformers, electromagnets, inductors, and actuators make use of magnetic cores, which are made of thin laminations, which are insulated from each other and then stacked in various shapes to form a magnetic circuit. As seen in Figure 2.1, the currents are circulating so that the sum of the current crossing the area perpendicular to the current is zero.

Eddy Currents: Theory, Modeling, and Applications, First Edition.
Sheppard J. Salon, M. V. K. Chari, Lale T. Ergene, David Burow, and Mark DeBortoli.
© 2024 The Institute of Electrical and Electronics Engineers, Inc. Published 2024 by John Wiley & Sons, Inc.

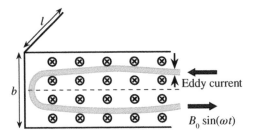

Figure 2.1   Side view of lamination with path of eddy current.

The formulas we will develop in this section are still used in the design of these magnetic components. We will also see that these devices often have windings, usually made of copper or aluminum, which use conductors with rectangular cross sections. These include motors and generators with formed coils, many power transformers, and transformers and inductors with sheet or foil windings.

To develop expressions for resistance-limited eddy currents and losses in plates, we will follow the approach of Carter [15]. Consider a non-ferromagnetic conducting plate of length ℓ, width b, and depth D as shown in Figure 2.2. An alternating magnetic flux density $B(t) = B_0 \cos(\omega t)$ impinges perpendicularly on the surface

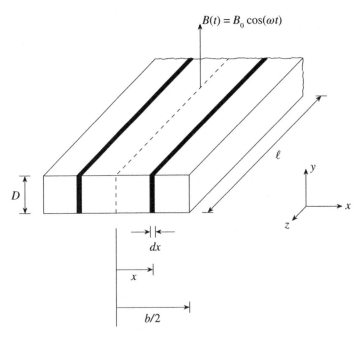

Figure 2.2   Sides of eddy current loop in non-ferromagnetic conducting plate with perpendicularly impinging magnetic flux density.

of the plate. Eddy currents arise in the plate so as to produce flux that opposes the change in $B(t)$. A section of the eddy current path, of track width dx, is indicated in the figure.

It will be assumed that the flux density produced by the eddy current is small relative to the externally-imposed flux density $B(t)$. As we discussed, since b is small compared to the skin depth, the reaction field produced by the eddy currents may be neglected. It is further assumed that the width b of the plate is small relative to the length ℓ so that the ends of the current path can be neglected.

The flux enclosed by the differential current loop in Figure 2.2 is

$$\psi(t) = B(t) \cdot \text{Area} = 2x\ell B_0 \cos(\omega t) \tag{2.1}$$

The induced voltage (emf) that drives current around the loop is the negative of the derivative of ψ with respect to time

$$\mathcal{E}(t) = -\frac{d\psi}{dt} = 2x\ell \omega B_0 \sin(\omega t) \tag{2.2}$$

and the RMS induced voltage is

$$\mathcal{E}_{\text{rms}} = \sqrt{2}x\ell \omega B_0 \tag{2.3}$$

The resistance of the current track, neglecting the ends, is

$$dR = \frac{\rho \times \text{path length}}{\text{path cross-sectional area}} = \frac{2\rho\ell}{D\,dx} \tag{2.4}$$

where ρ is the resistivity.

Now the eddy current loss in the track can be found by combining (2.3) and (2.4)

$$dW = \frac{\mathcal{E}_{\text{rms}}^2}{dR} = \frac{(x\omega B_0)^2 D\ell}{\rho}\,dx \tag{2.5}$$

and integrating to obtain the total loss in the plate yields

$$W = \int_{x=0}^{x=b/2} dW = \int_0^{b/2} \frac{(x\omega B_0)^2 D\ell}{\rho}\,dx = \frac{b^3(\omega B_0)^2 D\ell}{24\rho} \tag{2.6}$$

or when frequency is expressed in terms of $f = \omega/2\pi$

$$W = \frac{b^3(\pi f B_0)^2 D\ell}{6\rho} \tag{2.7}$$

Finally, the loss per unit volume is

$$\frac{\text{loss}}{\text{volume}} = \frac{W}{D\ell b} = \frac{(\omega B_0 b)^2}{24\rho} = \frac{(\pi f B_0 b)^2}{6\rho} \tag{2.8}$$

The current density is found as the electric field divided by the resistivity. The electric field is found by dividing the induced voltage, from (2.3), by the path length, 2ℓ:

$$J_{\text{rms}}(x) = \frac{1}{\rho}\frac{\mathcal{E}_{\text{rms}}}{2\ell} = \frac{\sqrt{2}x\omega B_0}{2\rho} \tag{2.9}$$

The strong dependence of the loss on width, b, as shown in Equation (2.6), suggests that an effective way to reduce the loss is to subdivide or laminate the plate into N-insulated sections of width b/N. The total loss is found from (2.6), replacing b with b/N and multiplying by N:

$$W = N\frac{(b/N)^3(\omega B_0)^2 D\ell}{24\rho} = \frac{b^3(\omega B_0)^2 D\ell}{24N^2\rho} \tag{2.10}$$

Comparison of (2.6) and (2.10) shows that laminating the plate into N sections reduces the total loss by a factor of N^2. Next, we will turn our attention to the case in which the reaction field plays a role in eddy current distribution.

2.2 Infinite Plate: Reactance Limited

In Section 2.1, we analyzed the problem of a rectangular plate in a sinusoidally time-varying field using the resistance-limited assumption. The more general problem is the reactance-limited case that we will now look into. There are two variations of this case, one being the semi-infinite conducting region and the other being the rectangular plate of finite width. Both can be represented by considering the large solenoid with sinusoidal current schematically shown in Figure 2.3. The coils extend infinitely in the y and z directions. Between the coils, we have a linear conducting region. This region also extends infinitely in the y and z directions. The magnetic field has only a y, or vertical, component. There is no variation of the field in the y and z directions. The ac field will induce currents

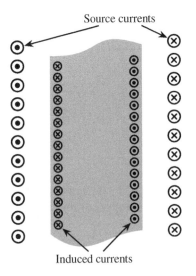

Source currents

Induced currents

Figure 2.3 Infinite conducting slab with source and induced currents.

in the conducting slab and, due to Lenz's law, the currents will be in a direction to oppose the change of magnetic field produced by the coils.

Carter [15] provides a first-principles approach to this problem. The induced currents in the conducting region have only a z component. The induced electric field is, therefore, in the z direction as well. We will first let the width of the slab grow to infinity to treat the semi-infinite domain geometry and then analyze the finite plate. In the first case, the distance between the source currents is infinite, and we can consider only the infinite half-plane (left side). The solid conductor is shown in Figure 2.4.

The material occupies the half-space $x > 0$. There is a magnetic field, H_0, applied in the y direction. The problem is infinitely deep in the z direction, i.e. into the page, which implies there is no variation in the z direction. Since the problem is infinite in the y direction as well and an infinitely long solenoid is applying the field, there is no variation of any field quantity in the y direction. The field varies sinusoidally with time at angular frequency ω. The material has constant homogeneous and isotropic properties of permeability μ and conductivity σ. In this one-dimensional case, **B** and **H** have only a y component, **E** and **J** have only a z component and these four quantities vary in the x direction.

Ampere's law states that the line integral of the magnetic field around a closed path is equal to the current (Ampere-turns) enclosed. In evaluating Ampere's law,

Figure 2.4 Half-space with applied sinusoidal magnetic fields.

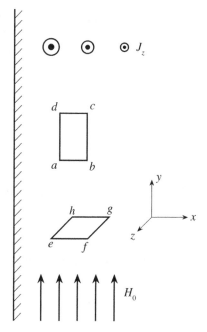

it is only the component of **H** in the direction of the integration path which is included and thus the dot product.

$$\oint \mathbf{H} \cdot d\ell = I_{\text{encl}} \tag{2.11}$$

We apply this integral around the rectangular contour *abcda* which lies in the (x, y) plane. Note that for this one-dimensional example, there is only a y component of the magnetic field and a z component of the electric field and current. Since there is no magnetic field in the x direction, there is no contribution to the line integral along *ab* and *cd* due to the dot product. In this case, the line integral is

$$\oint_{abcda} \mathbf{H} \cdot d\ell = \left(H_y + \frac{\partial H_y}{\partial x} \frac{\Delta x}{2} \right) \Delta y - \left(H_y - \frac{\partial H_y}{\partial x} \frac{\Delta x}{2} \right) \Delta y$$

$$= \frac{\partial H_y}{\partial x} \Delta x \Delta y \tag{2.12}$$

Setting this result equal to the total current enclosed, $J_z \Delta x \Delta y$, yields

$$\frac{\partial H_y}{\partial x} = J_z \tag{2.13}$$

Faraday's law states that the line integral of the electric field around a closed path is equal to the negative of the time rate of change of the magnetic flux linking the path:

$$\oint \mathbf{E} \cdot d\ell = -\frac{\partial \psi}{\partial t} \tag{2.14}$$

We apply this integral to path *efghe*. Referring to the figure, *ef* and *gh* are in the x direction while *fg* and *he* are in the z direction. There is no component of the electric field in the x direction so the line integrals along *ef* and *gh* have no contribution due to the dot product.

$$\oint_{efghe} \mathbf{E} \cdot d\ell = -\left(E_z + \frac{\partial E_z}{\partial x} \frac{\Delta x}{2} \right) \Delta z + \left(E_z - \frac{\partial E_z}{\partial x} \frac{\Delta x}{2} \right) \Delta z$$

$$= -\frac{\partial E_z}{\partial x} \Delta x \Delta z \tag{2.15}$$

In this case, with sinusoidal time variation, we can replace $\partial/\partial t$ by $j\omega$ in (2.14). We also use the relationship between flux and flux density, $\psi_m = B_y \Delta x \Delta z$. With these substitutions in Equations (2.14) and (2.15),

$$\frac{\partial E_z}{\partial x} = j\omega B_y \tag{2.16}$$

Note that the two equations we found, (2.13) and (2.16), can be obtained directly from the point, or differential, forms of Ampere's law and Faraday's law:

$$\nabla \times \mathbf{H} = \mathbf{J}$$

$$\nabla \times \mathbf{E} = -\frac{\partial \mathbf{B}}{\partial t}$$

Using the fact that **H** has only a y component and **E** only a z component and there is no variation of any quantity in the y or z directions, the same result is easily found. However, the integral form-based derivations above give more physical information and are consistent with future derivations.

The magnetic field quantities **B** and **H** are related by the constitutive equation

$$\mathbf{B} = \mu \mathbf{H} \tag{2.17}$$

and **J** and **E** are related by the constitutive equation

$$\mathbf{J} = \sigma \mathbf{E} \tag{2.18}$$

Then (2.13) can be written as

$$\frac{\partial B_y}{\partial x} = \mu \sigma E_z$$

Differentiating again with respect to x, and using the result of (2.16), we obtain

$$\frac{\partial^2 B_y}{\partial x^2} = \mu \sigma \frac{\partial E_z}{\partial x} = j \omega \mu \sigma B_y \tag{2.19}$$

This second-order homogeneous differential equation with constant coefficients is the diffusion equation written for sinusoidal time variation. Defining

$$k = \sqrt{j \omega \mu \sigma} \tag{2.20}$$

we find the general solution to (2.19) is

$$B_y(x) = a_1 e^{kx} + a_2 e^{-kx} \tag{2.21}$$

This may be checked by substitution back into (2.19).

The $\sqrt{j}$ term is a bit inconvenient. We can replace it using the Euler relationship

$$e^{j\theta} = \cos\theta + j\sin\theta$$

From this we see that

$$j = e^{j\pi/2}$$

and therefore

$$\sqrt{j} = e^{j\pi/4} = \frac{1+j}{\sqrt{2}}$$

Substituting this result into (2.20) yields

$$k = (1+j)\sqrt{\frac{\omega\mu\sigma}{2}} \tag{2.22}$$

or if we define a new parameter δ such that

$$\delta = \sqrt{\frac{2}{\omega\mu\sigma}} \tag{2.23}$$

then

$$k = \frac{1+j}{\delta} \tag{2.24}$$

and (2.21) can be written as

$$B_y(x) = a_1 e^{(1+j)x/\delta} + a_2 e^{-(1+j)x/\delta} \tag{2.25}$$

or

$$B_y(x) = a_1 e^{x/\delta} e^{jx/\delta} + a_2 e^{-x/\delta} e^{-jx/\delta} \tag{2.26}$$

In our present case of the semi-infinite plate, we may eliminate the first term because as x increases without bounds, this term will go to infinity. We therefore set $a_1 = 0$. When $x = 0$, i.e. at the surface of the plate, the flux density should be the source flux density B_0; therefore $a_2 = B_0$, and the expression for B_y becomes

$$B_y(x) = B_0 e^{-x/\delta} e^{-jx/\delta} \tag{2.27}$$

The first exponential of (2.27) indicates that the magnitude of the flux density is decreasing exponentially with x. The second exponential has a magnitude of unity and describes the phase shift of the flux density.

The quantity δ given in (2.23) is the depth of penetration or skin depth and is one of the most important parameters in the study of eddy currents. The skin depth has the units of length. At $x = \delta$, the value of the flux density (and as we will see, the current density as well) is $1/e \approx 0.37$ of its value at the surface. In $3 - 4$ skin depths, the flux density is practically zero. From the second exponential of (2.27), we see that at a depth $x = \delta$ the flux density lags the surface flux density B_0 by one radian.

The current density can be found from the flux density. Combining (2.13) and (2.17), we obtain

$$J_z = \frac{1}{\mu} \frac{\partial B_y}{\partial x} \tag{2.28}$$

Using the solution for B_y from (2.27), this gives

$$J = -\frac{B_0}{\mu} \frac{1+j}{\delta} e^{-(1+j)x/\delta}$$
$$= -J_0 e^{-(1+j)x/\delta} \tag{2.29}$$

The total loss to infinity is found by integrating the loss density into the conductor where J is the RMS current density.

$$W = \int_0^\infty \frac{1}{\sigma} |J|^2 \, dx$$
$$= \frac{1}{\sigma} |J_0|^2 \int_0^\infty e^{-2x/\delta} \, dx$$
$$= \frac{\delta}{2\sigma} |J_0|^2 \tag{2.30}$$

This is the loss per square meter of surface. The loss is the same as if the current density at the surface is constant to a depth $\delta/2$. The skin depth is, therefore, very useful in obtaining the equivalent resistance or power in a solid conductor.

2.3 Finite Plate: Reactance Limited

Now let us generalize this to the problem of a finite width plate as shown in Figure 2.5. We start by recalling the general solution that we obtained for B_y. In this case, we retain both terms:

$$B_y(x) = a_1 e^{(1+j)x/\delta} + a_2 e^{-(1+j)x/\delta}$$

Since we have a symmetric problem, the constraints are that the resultant flux density is equal to the source flux density B_0 at $x = \pm d$. These conditions are satisfied by setting

$$a_2 = a_2 = \frac{B_0}{e^{(1+j)d/\delta} + e^{-(1+j)d/\delta}}$$

and then

$$
\begin{aligned}
B_y(x) &= B_0 \frac{e^{(1+j)x/\delta} + e^{-(1+j)x/\delta}}{e^{(1+j)d/\delta} + e^{-(1+j)d/\delta}} \\
&= B_0 \frac{\cosh\left[(1+j)x/\delta\right]}{\cosh\left[(1+j)d/\delta\right]}
\end{aligned}
\tag{2.31}
$$

Due to the eddy currents, the effective flux-carrying cross-sectional area of the plate is reduced. The total flux passing through the plate per meter depth is found

Figure 2.5 Resultant fields and currents in a finite-width plate.

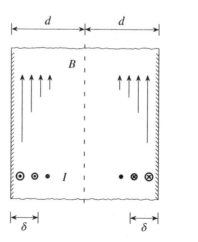
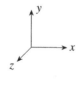

by integrating the flux density from $x = -d$ to $x = +d$:

$$\psi_m = \frac{B_0}{\cosh\left[\frac{(1+j)d}{\delta}\right]} \int_{-d}^{d} \cosh\left[\frac{(1+j)x}{\delta}\right] dx$$

$$= \frac{2B_0\delta}{1+j} \frac{\sinh\left[\frac{(1+j)d}{\delta}\right]}{\cosh\left[\frac{(1+j)d}{\delta}\right]} \tag{2.32}$$

We will now define dimensionless parameters α and β such that

$$\alpha = \frac{d}{\delta} = \beta$$

and employ the following identities

$$\sinh(\alpha + j\beta) = \sinh\alpha\cos\beta + j\cosh\alpha\sin\beta$$
$$\cosh(\alpha + j\beta) = \cosh\alpha\cos\beta + j\sinh\alpha\sin\beta$$
$$2\cos^2\alpha = 1 + \cos 2\alpha$$
$$2\sin^2\alpha = 1 - \cos 2\alpha$$
$$2\cosh^2\alpha = \cosh 2\alpha + 1$$
$$2\sinh^2\alpha = \cosh 2\alpha - 1$$

with (2.32) to write the amplitude of the flux per unit depth as

$$|\psi_m| = \sqrt{2}B_0\delta\sqrt{\frac{\cosh 2\alpha - \cos 2\alpha}{\cosh 2\alpha + \cos 2\alpha}} \tag{2.33}$$

If there were no eddy currents, the flux would be

$$\psi_0 = 2B_0 d \tag{2.34}$$

Therefore, we can say that the eddy currents have resulted in an effective reduction in the area of the laminations by a factor of

$$\left|\frac{\psi_m}{\psi_0}\right| = \frac{1}{\alpha\sqrt{2}}\sqrt{\frac{\cosh 2\alpha - \cos 2\alpha}{\cosh 2\alpha + \cos 2\alpha}} \tag{2.35}$$

This factor is plotted in Figure 2.6.

The current density is found from the flux density using

$$J = \frac{\partial H_y}{\partial x} = \frac{1}{\mu}\frac{\partial B_y}{\partial x} \tag{2.36}$$

Using the expression for $B_y(x)$ from (2.31), this becomes

$$J_z = \frac{B_0}{\mu}\left(\frac{1+j}{\delta}\right)\frac{\sinh\left[(1+j)x/\delta\right]}{\cosh\left[(1+j)d/\delta\right]} \tag{2.37}$$

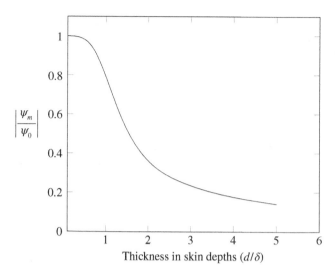

Figure 2.6 Reduction of effective area with thickness.

which can be written as

$$J_z = \frac{B_0}{\mu}\left(\frac{1+j}{\delta}\right)\left\{\frac{\sinh(x/\delta)\cos(x/\delta)+j\cosh(x/\delta)\sin(x/\delta)}{\cosh\alpha\cos\alpha+j\sinh\alpha\sin\alpha}\right\} \tag{2.38}$$

where $\alpha = d/\delta$. Using the same identities above, the magnitude is found as

$$|J_z| = \frac{\sqrt{2}B_0}{\mu\delta}\sqrt{\frac{\cosh(2x/\delta)-\cos(2x/\delta)}{\cosh 2\alpha + \cos 2\alpha}} \tag{2.39}$$

To find the losses in the plate, we integrate the loss density

$$W = \int_{-d}^{d}\frac{|J^2|}{\sigma}\,dx \tag{2.40}$$

$$W = \frac{2B_0^2}{\sigma\mu^2\delta}\left(\frac{\sinh 2\alpha - \sin 2\alpha}{\cosh 2\alpha + \cos 2\alpha}\right) \tag{2.41}$$

We have seen in (2.8) that in the case of a thin lamination with uniform flux density applied normal to the "small" dimension, we can use the resistance-limited formula for the eddy current losses. Substituting $b = 2d$ into (2.8) yields

$$W = \frac{(\omega B_0 d)^2\sigma}{6}\ \ \text{resistance-limited case} \tag{2.42}$$

Making the substitution

$$\delta^4\omega^2\sigma = \frac{4}{\mu^2\sigma} \tag{2.43}$$

into (2.42), it can be shown that the effect of eddy currents on the losses in the reactance-limited case may be determined by applying the following correction factor to the losses for the resistance-limited case.

$$K = \frac{6\delta^3}{d^3}\left[\frac{\sinh(d/\delta) - \sin(d/\delta)}{\cosh(d/\delta) + \cos(d/\delta)}\right] \tag{2.44}$$

The correction factor, K, plotted against $d/2\delta$ is shown in Figure 2.7.

For cases in which the skin depth is greater than the plate half-width, the correction factor is approximately 1.0 which means that the resistance-limited formula is a good approximation. When the plate half-width gets to be several times the skin

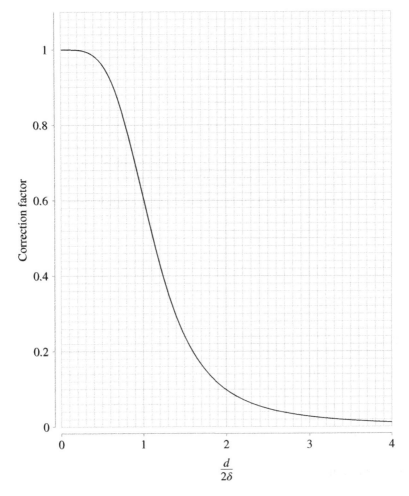

Figure 2.7 Correction factor in Equation (2.44).

depth, the loss is much lower than the resistance-limited formula would predict. This is due to the cancelation of the flux due to the eddy current field. We can also comment that the resistance-limited formula will always over-predict the losses in the plate. It is often used in design due to its simplicity and the knowledge that the results will be *conservative*. The loss formula is then

$$W'' = K \frac{2\sigma (\omega B_0 d)^2}{3} \tag{2.45}$$

We have found the power dissipated in a plate is

$$P = \frac{H_0^2}{\sigma \delta} \left(\frac{\sinh (2d/\delta) - \sin (2d/\delta)}{\cosh (2d/\delta) + \cos (2d/\delta)} \right) \tag{2.46}$$

Another way of writing this is

$$P = \frac{H_0^2}{\sigma \delta} \left(\frac{e^\alpha - e^{-\alpha} - 2 \sin \alpha}{e^\alpha + e^{-\alpha} + 2 \cos \alpha} \right) \tag{2.47}$$

where we define $\alpha = 2d/\delta$.

As we can see from Equation (2.47), if α is very large, or the plate width is much larger than δ this simplifies to

$$P = \frac{H_0^2}{\sigma \delta} \tag{2.48}$$

This is just two times the result that we obtained for the semi-infinite plate since we now consider both sides of the plate. We can also analyze the equation in the limit in which the plate thickness is much smaller that the skin depth. Replacing the exponential and trigonometric functions by series expansions we have

$$P = \frac{H_0^2}{\sigma \delta} \left[\frac{\left(1 + \alpha + \frac{\alpha^2}{2!} + \frac{\alpha^3}{3!} \cdots \right) - \left(1 - \alpha + \frac{\alpha^2}{2!} - \frac{\alpha^3}{3!} \cdots \right) - 2 \left(\alpha - \frac{\alpha^3}{3!} \cdots \right)}{\left(1 + \alpha + \frac{\alpha^2}{2!} + \frac{\alpha^3}{3!} \cdots \right) + \left(1 - \alpha + \frac{\alpha^2}{2!} - \frac{\alpha^3}{3!} \cdots \right) + 2 \left(\alpha \frac{\alpha^3}{3!} \cdots \right)} \right] \tag{2.49}$$

This simplifies to

$$P = \frac{H_0^2}{\sigma \delta} \left[\frac{4 \frac{\alpha^3}{3!} + \cdots}{4 + 4 \frac{\alpha^4}{4!} + \cdots} \right] \tag{2.50}$$

Ignoring terms higher than third order we get

$$P = \frac{H_0^2}{\sigma \delta} \frac{\alpha^3}{6} = \frac{1}{3} (\mu H_0)^2 \sigma \omega^2 d^3 \tag{2.51}$$

We can find the loss per unit volume is then

$$P = \frac{\omega^2 B_0^2 \sigma d^2}{3} \tag{2.52}$$

If we express the loss in terms of the lamination thickness, $b = 2d$, then Equation (2.52) becomes

$$P = \frac{\omega^2 B_0^2 \sigma b^2}{12} \tag{2.53}$$

This result agrees with that found in Equation (2.10).

2.4 Superposition of Eddy Losses in a Conductor

We have seen that the current distribution in a conductor is modified by the self-field of the applied current and also by any external fields produced by outside sources. We can say that the load current distribution depends on the material properties and geometry of the conductor as well as the frequency of the applied source. The eddy currents produced by external fields will also result in a redistribution of the current in the conductor. Let us assume that this external field is constant over the conductor and can be resolved into two perpendicular components which we will call B_x and B_y as shown in Figure 2.8. We will call the eddy current density produced by these two components J_x and J_y, respectively. The subscripts here refer to the flux that is producing the eddy currents. All currents are in the z direction. We also recall from previous discussions that these eddy currents are circulating currents and therefore sum to zero over the surface of the conductor. Calling the current distribution produced by the self field J_0, we have

$$J(x, y) = J_0(x, y) + J_x(x, y) + J_y(x, y) \tag{2.54}$$

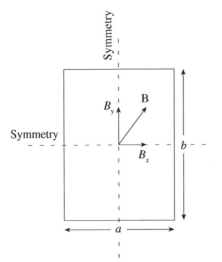

Figure 2.8 Superposition of eddy losses in a rectangular conductor.

The eddy loss is

$$W = \rho \int_0^a \int_0^b J^2 dx dy \tag{2.55}$$

Multiplying out the terms, we find

$$W = \rho \int_0^a \int_0^b \left(J_0^2 + J_x^2 + J_y^2 + 2J_0 J_x + 2J_0 J_y + 2J_x J_y \right) dx dy \tag{2.56}$$

Combining terms

$$W = \rho \int_0^a \int_0^b J_0^2 \, dx \, dy + \rho b \int_0^a J_x^2 \, dx + \rho a \int_0^b J_y^2 \, dy$$
$$+ 2\rho b \int_0^a J_0 J_x \, dx + 2\rho a \int_0^b J_0 J_y \, dy + 2\rho \int_0^a J_x \, dx \int_0^b J_y \, dy \tag{2.57}$$

The first three terms on the right are the self-loss, the loss due to the x component of the flux density and the loss due to the y component of the flux density respectively. Since J_x and J_y are circulating currents, their integrals over the cross section of the conductor vanish. This means that the last term is zero. It remains to consider the fourth and fifth terms which have the product of the load current density and eddy current density. These generally will not be zero as we saw in the discussion of proximity effect. Let us consider the symmetry involved in this problem. Referring to Figure 2.8 again, we have a line of symmetry along the vertical center and horizontal center. For a homogeneous conductor, we, therefore, expect that the load current distribution will have the same symmetries. There may be exceptions to this. For example, if the conductor is near a magnetic body, the symmetry may be modified. In this example, we are assuming a conductor in empty space. With the external fields B_x and B_y assumed constant, we also will have that the eddy current density, J_x will be symmetric around the horizontal axis of symmetry (symmetric top to bottom), and the eddy current density, J_y will be symmetric around the vertical symmetry axis (symmetric right to left). If these conditions are met, then the integrals of the products do indeed vanish over the conductor surface. This leaves

$$W = \rho \int_0^a \int_0^b J_0^2 \, dx \, dy + \rho b \int_0^a J_x^2 \, dx + \rho a \int_0^b J_y^2 \, dy \tag{2.58}$$

We therefore conclude that, under these circumstances, the losses due to each component can be separately computed and the results added together.

2.5 Discussion of Losses in Rectangular Plates

With the theory we have developed, we are in a position to better understand the variation of losses in plates and can consider the one-dimensional solution in a

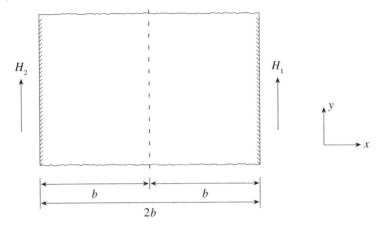

Figure 2.9 Plate with tangential excitation.

number of practical applications [30, 47]. The geometry under consideration is illustrated in Figure 2.9. We begin with the one-dimensional diffusion equation written for the magnetic field. The magnetic field has only a y component and the current density has only a z component.

$$\frac{\partial^2 H_y}{\partial x^2} = j\omega\sigma\mu H_y \tag{2.59}$$

With

$$\beta = \sqrt{j\omega\sigma\mu} \tag{2.60}$$

The solution is

$$H_y = Ce^{\beta x} + De^{-\beta x} \tag{2.61}$$

The constants are found by applying boundary conditions. If the surface field is H_1 at $x = b$ and H_2 at $x = -b$, we can solve for the constants.

$$C = \frac{H_1 e^{\beta b} - H_2 e^{-\beta b}}{e^{2\beta b} - e^{-2\beta b}} \tag{2.62}$$

$$D = -\frac{H_1 e^{-\beta b} - H_2 e^{\beta b}}{e^{2\beta b} - e^{-2\beta b}} \tag{2.63}$$

By taking the curl of Equation (2.61), we find the electric field as

$$E_z = \frac{\beta}{\sigma}\left(Ce^{\beta x} + De^{-\beta x}\right) \tag{2.64}$$

In comparing the examples below, it will be useful to find the loss per unit surface area (yz). It is convenient to use the Poynting vector, which gives the real and reactive power transferred across a surface. Since the directions of the magnetic

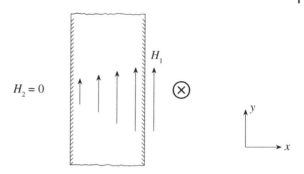

Figure 2.10 Case 1: plate with current parallel.

field and electric field are perpendicular, we can evaluate the Poynting vector at the surface of the plate as

$$S = P + jQ = \frac{1}{2} \left(E_z H_y^* |_{x=b} - E_z H_y^* |_{x=-b} \right) \tag{2.65}$$

The factor of $1/2$ comes from the fact that we are using peak values of the field vectors and are looking for the time average. Using this result, we can analyze several different situations by applying the appropriate values of H_1 and H_2. We will consider three practical examples that illustrate some of the physics of eddy current loss in plates. The first is a current-carrying conductor or current sheet parallel to the surface of the plate as shown in Figure 2.10.

The flux inside the plate is parallel to the surface, in the y direction. If we set the tangential field on the side opposite the current source to zero, we are imposing the condition that no flux fully penetrates the plate and the current in the plate is the *reflection* of the current source. Applying the boundary conditions $H_1 = 1.0$ and $H_2 = 0.0$ A m^{-1}, we find the constants and evaluate the magnetic field and the electric field at the surface. Then we evaluate the Poynting vector to find the power. As an example, consider a copper plate, ($\sigma = 5.8 \times 10^7$ S m^{-1}, $\mu_0 = 4\pi \times 10^{-7}$ H m^{-1}) with 60 Hz excitation. To make the results more general, we plot the normalized surface loss density, i.e. dividing the loss by $P_0 = H_1^2/(2\sigma\delta)$ vs. the normalized depth (b/δ). The results for this example are shown in Figure 2.11.

We notice that initially, as the plate gets thicker, the losses go down. This is because our boundary conditions force a certain current to flow in the plate, and with a thicker plate, there is more area for the current to pass through. As the plate gets thicker than around two skin depths, the results asymptotically approach unity. This is consistent with the results that we found for finite plates with thickness much greater than the skin depth. The loss is the same as if all of the current were uniformly distributed in one skin depth. Increasing the thickness of the plate has no effect on the total losses in this case.

For the second example, we consider a conductor penetrating a conducting plate, such as would be the case for a bushing. In this example, the magnetic field

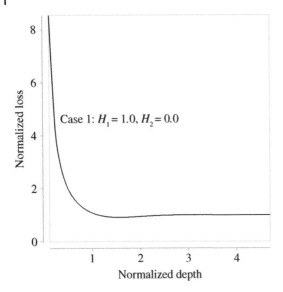

Figure 2.11 Normalized loss density vs. normalized depth for parallel conductor case.

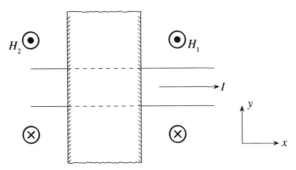

Figure 2.12 Case 2: current perpendicular to and through the surface.

on both sides of the plate are equal. This is illustrated in Figure 2.12. We then evaluate the constants and the Poynting vector as before. The results for this case are displayed in Figure 2.13.

We see that the trend is different compared to Figure 2.11. In the this case, the losses initially go up as the plate thickness grows. This is because, with the field excitation constant, more eddy currents will flow in the plate as it gets thicker. We are in the resistance-limited regime. Decreasing the resistance increases the loss. As the plate continues to get thicker, however, the loss density peaks and approaches a normalized value of 2 (since both sides of the plate are involved), and there is relatively little current in the center of the plate.

In the third example, we have current flowing along a plate as in a bus bar. The tangential magnetic field in this case is equal and opposite on the two sides of the plate. This is illustrated in Figure 2.14.

Figure 2.13 Normalized loss density vs. normalized depth for perpendicular conductor.

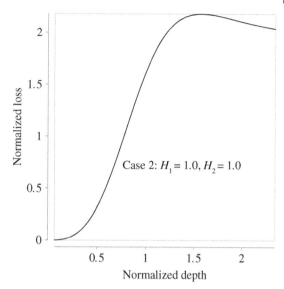

Figure 2.14 Case 3: current flowing along plate.

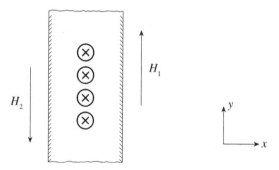

Using Equation (2.65), we find the result shown in Figure 2.15. The normalized losses are very high for a thin plate since we are forcing a constant current through the area. As the plate increases in thickness, the loss comes down quickly and approaches a value of 2 as the skin depth becomes much smaller than the thickness. We have the same asymptote as in the second example since both sides of the plate are involved.

These examples, although idealized, are of practical importance and the analysis here is useful in the design of electric machines and transformers as well as other application in which conductors pass near or through conducting structures. The application of the theory that we developed aids in the understanding of how these losses vary as the geometry changes.

In order to emphasize the usefulness of the formulations we have discussed, let us apply these results to other practical problems. We have analyzed the case of

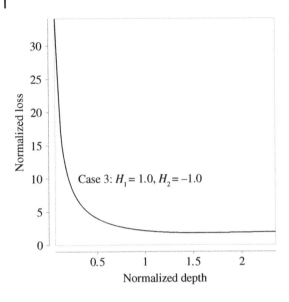

Figure 2.15 Loss density for a current-carrying plate vs. normalized thickness.

rectangular bus bar conductors with go and return currents in Chapter 1 using a coupled circuit model. The problem there was two-dimensional due to the geometry in which the conductors were narrow and were in the same plane. We, therefore, had both x and y components of the field. There are practical geometries for these bus bars in which the fields and eddy currents are approximately one-dimensional, and we can use our solutions to the diffusion equation to calculate current distribution and losses in these cases. Referring to Figure 2.16, we consider two parallel conducting plates separated by a distance $2d$. Each plate has thickness $2b$ and height $2a$.

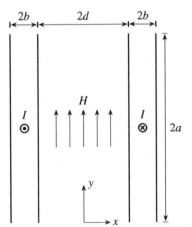

Figure 2.16 Parallel conducting plates carrying go-and-return current.

In this example, current is only in the z direction and we assume the conductors extend to infinity in this direction. We will also assume that the height, $2a$, is much greater than the thickness, $2b$, of the conductors. In this case, the field is solenoidal and the field is negligible outside the two plates and uniform in the y direction between the plates with a value of

$$H_y = \frac{I}{2a} \tag{2.66}$$

We have already seen this problem and can write the general solution as

$$H_y = Ce^{\beta x} + De^{-\beta x} \tag{2.67}$$

where

$$\beta = \sqrt{j\omega\mu\sigma} \tag{2.68}$$

Applying the boundary conditions, we see that

$$Ce^{\beta d} + De^{-\beta d} = \frac{I}{2a} \tag{2.69}$$

We also have

$$Ce^{\beta(d+2b)} + De^{-\beta(d+2b)} = 0 \tag{2.70}$$

Solving Equations (2.69) and (2.70) for C and D, we obtain

$$H_y = \frac{I}{2a} \frac{\sinh \beta(x - d - 2b)}{\sinh 2\beta b} \tag{2.71}$$

Using

$$\nabla \times H = J = \sigma E \tag{2.72}$$

we obtain

$$J = \frac{\beta I}{2a} \frac{\cosh \beta(x - d - 2b)}{\sinh 2\beta b} \tag{2.73}$$

Integrating the square of Equation (2.73) over the conductor, we obtain the losses in one of the conductors [18].

$$P = \frac{I^2}{2\delta\sigma a} \frac{\sinh(4b/\delta) + \sin(4b/\delta)}{\cosh(4b/\delta) - \cos(4b/\delta)} \tag{2.74}$$

We note here that this problem, for each conductor separately, is the case that we treated earlier as example 1, with magnetic field applied tangential to one surface of the conductor and no magnetic field on the opposite side. This is the situation that will occur in the long solenoid. In that case, we were able to find and plot the normalized loss as a function of the normalized width of the plate. We can compare this to the result that we get by evaluating Equation (2.74) and let the width ($2b$) of the plate vary. The result of (2.74) is divided by the coefficient $\frac{I^2}{2\delta\sigma a}$

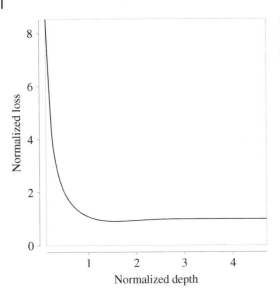

Figure 2.17 Normalized loss vs. normalized width for parallel rectangular conductors.

to get the normalized loss. The result is plotted in Figure 2.17. The results are the same as the results of Figure 2.11.

As another example, let us reconsider the lamination problem in which a plate is placed in a uniform field. This problem was solved in Section 2.3. In this case, we applied a field parallel to the surface of the conductor, and the field was the same on both sided of the conductor. This is the same situation as our Case 2 as shown in Figure 2.12. We have plotted the results of normalized loss vs. normalized width in Figure 2.13. In that Section 2.3, we found the loss in the plate as

$$P = \frac{2H_0^2}{\delta\sigma} \frac{\sinh(4b/\delta) - \sin(4b/\delta)}{\cosh(4b/\delta) + \cos(4b/\delta)} \tag{2.75}$$

We can now find the normalized loss by dividing Equation (2.75) by the coefficient $\frac{H_0^2}{\sigma\delta}$. If we plot this formula against the normalized width of the plate, we obtain the graph in Figure 2.18. As we can see, the results are the same.

As a third example, let us consider the problem of an isolated plate carrying current I as shown in Figure 2.19. Since all currents must form closed paths, we can look at this isolated conductor as the superposition of the two problems as shown in the figure. If the conductors are tall compared to the width and conductor spacing, we can continue to use the one-dimensional analysis. We see that each of the problems with go-and-return currents equal to $I/2$ are the parallel plate conductor problem that we discussed above. Recall, in this case, there is no magnetic field outside the conducting sheet, so the two problems are decoupled.

We now can look at the normalized loss in the plate vs. the normalized depth. To find this, we can use Equation (2.74) and make adjustments to account for the

Figure 2.18 Normalized loss in a finite plate with uniform field applied vs. normalized width.

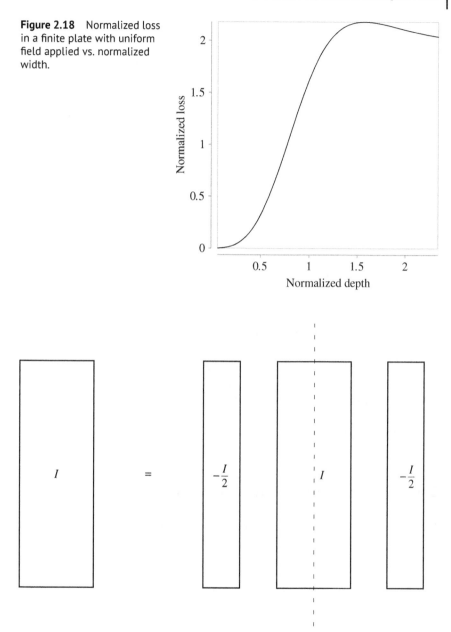

Figure 2.19 Single conductor as the superposition of two sets of parallel conductors.

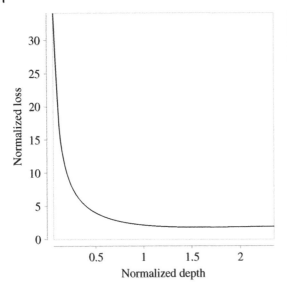

fact that we have two identical problems, each with current $I/2$. Then we adjust the normalized thickness to $2d/\delta$. This is plotted in Figure 2.20.

There are two curves plotted in Figure 2.20, but as they are identical they lie on top of each other. One is the evaluation of Equation (2.74), modified as discussed above and the other is the Poynting vector solution.

While this discussion was limited to one-dimensional analysis, the results illustrate some of the practical implications of conductors in configurations that are commonly found in design.

2.6 Eddy Currents in a Nonlinear Plate

We have seen in the one-dimensional case [2, 42, 62] that

$$\frac{\partial^2 B_x}{\partial z^2} = \mu\sigma\frac{\partial B_x}{\partial t} \tag{2.76}$$

and the flux density at any depth can be found as

$$B_x = \Re\left\{ B_0 e^{(j\omega t - z\sqrt{j\omega\mu\sigma})} \right\} \tag{2.77}$$

where B_0 is the flux density at the surface. The total flux is then

$$\psi_m(t) = \int_0^\infty B_x\, dz = \frac{B_0}{\sqrt{j\omega\mu\sigma}} e^{j\omega t} \tag{2.78}$$

Figure 2.21 Limiting B – H curve of steel.

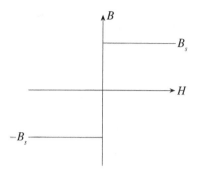

In terms of the total flux, the flux density can be written

$$B_x = \Re e\left\{ \sqrt{j\omega\mu\sigma}\,\psi_{m\,\text{max}}e^{(j\omega t - z\sqrt{j\omega\mu\sigma})} \right\} \tag{2.79}$$

Now let us consider a nonlinear material characteristic as shown in Figure 2.21. The flux density has only two states $\pm B_s$, the saturation flux density. The material can switch only when **H** = 0. A sinusoidal flux, $\psi_m(t) = \Re\{\psi_m e^{j\omega t}\}$, can be supported in a material like this by a series of square or rectangular waves.

Since $\psi_m(t)$ is periodic, we can construct it from square waves of the same period. This is illustrated in Figure 2.22. This leaves two possibilities for the magnetized

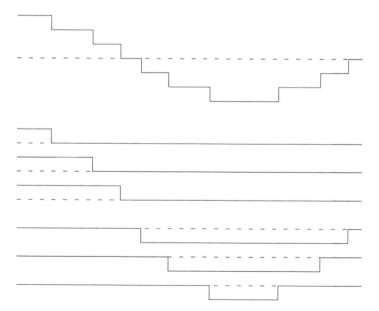

Figure 2.22 Making a sinusoidal wave from square waves.

iron. It is magnetized $\pm B_s$ periodically at a frequency ω, or it is constant at $\pm B_s$ being left at that state from a previous excitation.

At some instant of time, the material above a certain level, z_1, is magnetized to $-B_s$, while below this level it is $+B_s$. At some later time, this layer of separation has moved to z_2. The movement of the surface took place in a time Δt. The phase shift between the rectangular waves at z_1 and z_2 is $\omega \Delta t$ radians. The surface of separation will move to a certain depth and no further. At this depth, the total flux above the surface of separation is $-\psi_{m\,max}$. If we call this depth δ, then

$$\delta = \frac{\psi_{m\,max}}{B_s} \tag{2.80}$$

We note several things. First, the depth of penetration, δ, is interpreted differently than in the linear case. In the linear case, the depth of penetration is the depth at which the flux density is $1/e$ of its value at the surface. The depth of penetration depends only on the frequency and the material properties (conductivity and permeability). In the nonlinear limiting case, the depth of penetration is the maximum distance at which the material switches from $+B_s$ to $-B_s$ and vice versa. This distance depends on the total flux and the saturation flux density B_s.

The surface of separation has moved from $z = 0$ to $z = \delta$ in one half-cycle since the flux has changed from $\psi_{m\,max}$ to $-\psi_{m\,max}$ in that time. This means that the phase shift between the rectangular waves at $z = 0$ and $z = \delta$ is π radians.

Consider now that the surface of separation is at location z'. As the surface moves from z' to $z' + \Delta z'$, the change in flux is

$$\Delta \psi_m = -2B_s \Delta z' \tag{2.81}$$

In differential form, we have

$$\frac{dz'}{d\psi_m} = -\frac{1}{2B_s} \tag{2.82}$$

We can integrate this equation starting from $\psi_{m\,max}$ and $z = 0$

$$z' = \frac{1}{2B_s} \int_{\psi_{m\,max}}^{\psi_m} -1 \, d\psi \tag{2.83}$$

So that

$$z' = \frac{\psi_{m\,max} - \psi_m(t)}{2B_s} \tag{2.84}$$

With this model, we can find the eddy current distribution as

$$\mathbf{J} = \sigma \mathbf{E} \tag{2.85}$$

where the electric field is found as

$$\nabla \times \mathbf{E} = -\frac{\partial \mathbf{B}}{\partial t} \tag{2.86}$$

Figure 2.23 Separation surface and coordinate system.

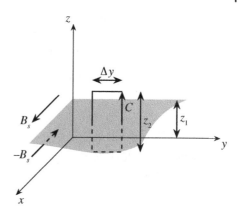

In integral form, this becomes

$$\oint_C \mathbf{E} \cdot d\ell = -\frac{\partial}{\partial t} \iint_S \mathbf{B} \cdot dS \tag{2.87}$$

where the curve C bounds the surface S. If the curve C encloses a region which does not contain the surface of separation between $+B_s$ and $-B_s$, then the integral will be zero. That is to say, the flux density in the region is constant. If, however, the contour intersects the surface of separation, then there will be an induced electric field which will depend on the velocity at this surface. For this one-dimensional case, we can evaluate the integral directly (Figure 2.23).

The flux density has only an x component. We choose the integration path to be a closed contour in the yz plane. At some instant, the surface of separation is at a depth z' and is moving in the z direction. The intersection of the line integral and the surface of separation is a line segment of length Δy. The electric field, $\mathbf{E}$, has only a y component. It is constant above the surface of separation and below the surface of separation, with a discontinuity at the surface. At some time, t, the velocity of the surface is $\frac{\partial z'}{\partial t}$ and the area with $-B_s$ is S_2, where $S_1 + S_2 = S$. At some later time $t + \Delta t$, the surface has moved a distance $\frac{\partial z'}{\partial t}\Delta t$ and an area with $-B_s$ is $S_1 + \frac{\partial z'}{\partial t}\Delta t\Delta y$ and the area with $+B_s$ is $S_2 - \frac{\partial z'}{\partial t}\Delta t\Delta y$. The integral can then be written as

$$E_y|_{z'-\Delta z'}\Delta y - E_y|_{z'+\Delta z'}\Delta y = \frac{2B_s \frac{\partial z'}{\partial t}\Delta t\Delta y}{\Delta t} \tag{2.88}$$

so that

$$E_y|_{z'-\Delta z'}\Delta y - E_y|_{z'+\Delta z'}\Delta y = 2B_s \frac{\partial z'}{\partial t} \tag{2.89}$$

which describes the discontinuity in the electric field at the surface of separation.

From a physical standpoint, the electric field below the surface of separation must be zero. If this were not true, uniform current would flow from the surface to infinity which would require an infinite amount of energy. Therefore,

$$E_y = 2B_s \frac{\partial z'}{\partial t}, \ (0 \leq z \leq z') \tag{2.90}$$

and

$$E_y = 0, \ (z > z') \tag{2.91}$$

During the next half-cycle, the sign of **E** will be reversed since the flux density will be changing from $-B_s$ to $+B_s$, but the velocity remains positive. In this model then, at any instant of time, the eddy current per unit depth is then

$$\mathbf{J} = 2B_s \sigma z' \frac{\partial z'}{\partial t} \tag{2.92}$$

In terms of the applied sinusoidal flux,

$$\mathbf{J} = -2B_s \sigma \left(\frac{\psi_{m \max} - \psi_m(t)}{2B_s} \right) \left(-\frac{1}{2B_s} \frac{\partial \psi_m}{\partial t} \right) \tag{2.93}$$

or

$$\mathbf{J} = \frac{\sigma}{2B_s} \left(\psi_{m \max} - \psi_m(t) \right) \frac{\partial \psi_m}{\partial t} \tag{2.94}$$

Using $\psi_m(t) = \psi_{m \max} \cos(\omega t)$,

$$\mathbf{J} = \frac{\omega \sigma}{2B_s} \psi_{m \max}^2 \left(1 - \cos(\omega t) \right) \sin(\omega t) \tag{2.95}$$

As an example of the use of this model, consider the current-carrying steel cylinder shown in Figure 2.24. We will assume that the radius is larger than

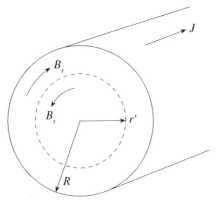

Figure 2.24 Steel cylinder with axial current and skin depth.

the skin depth. The separation boundary between the material magnetized $+B_s$ and $-B_s$ is located at $r = r'$. As the boundary moves it causes a change in flux linkage in the region $r < r'$. There is no change in the region $r' < r < R$. The flux linkage of a small element in webers per unit length at time, t, is

$$\lambda(t) = -B_s \left(r' - r_s \right) + B_s \left(R - r' \right) \tag{2.96}$$

At time $t + \Delta t$, the surface of separation has moved to $r' + \Delta r'$ and the flux linkage has changed to

$$\lambda(t + \Delta t) = -B_s \left(r' + \Delta r' - r_s \right) + B_s \left(R - r' - \Delta r' \right) \tag{2.97}$$

The rate of change of flux linkage is therefore

$$\frac{d\lambda}{dt} = -2B_s \frac{dr'}{dt} \tag{2.98}$$

This expression is independent of r_s so that everywhere in the region $0 < r < r'$ has the same induced electric field,

$$E_s = -2B_s \frac{dr'}{dt} \tag{2.99}$$

The current density in the conducting annulus is

$$J = -2B_s \sigma \frac{dr'}{dt} \tag{2.100}$$

and the total current in the conductor is

$$I = \pi \left(R^2 - r'^2 \right) J \tag{2.101}$$

or

$$I = -2B_s \pi \sigma \left(R^2 - r'^2 \right) \frac{dr'}{dt} \tag{2.102}$$

For the case in which I is sinusoidal, using the geometry in the figure, the flux is increasing, and the current I is positive. Integrating

$$-\int_R^{r'} 2\pi B_s \sigma \left(R^2 - r'^2 \right) dr' = \int_0^t I\, dt \tag{2.103}$$

$$\frac{2}{3}\pi B_s \sigma \left[r'^3 - R^2 r' + 2R^3 \right] = \frac{I_{peak}}{\omega} \left(1 - \cos(\omega t)\right) \tag{2.104}$$

If the current is not sufficient to cause the surface of separation to reach the center, i.e. $\delta < R$, we define the depth of penetration, δ, as

$$\delta = R - r'_{min} \tag{2.105}$$

This gives

$$\frac{2}{3}\pi B_s \sigma \delta^2 (3R - \delta) = \frac{2I_{peak}}{\omega} \tag{2.106}$$

If the depth of penetration is very small compared to the cylinder, ($\delta \ll R$), then,

$$\delta = \sqrt{\frac{I_{peak}}{\psi_m R B_s \omega \sigma}} \tag{2.107}$$

In terms of the peak surface current density

$$J_{peak} = \frac{I_{peak}}{2\pi R} \tag{2.108}$$

$$\delta = \sqrt{\frac{2J_{peak}}{\omega \sigma B_s}} \tag{2.109}$$

If the current is such that the surface of separation just reaches the center of the cylinder at the end of each half-cycle, then $r' = 0$ and

$$I_{peak} = \frac{2}{3}\pi \omega \sigma B_s R^2 \tag{2.110}$$

If the current is great enough that it reaches the center of the cylinder before the end of the half-cycle, then for the rest of the cycle the conductor looks like a dc conductor with uniform current density throughout.

We can now use these concepts to find the losses in a nonlinear conductor and with this, find the effective impedance. For the case of a small depth of penetration, ($\delta \ll R$), the loss density (per unit volume) can be found as

$$P(t) = \frac{J^2}{\sigma} = \frac{i(t)^2}{\pi^2 \left(R^2 - r'^2\right)^2} \tag{2.111}$$

and the loss per unit length is then

$$P(t) = \frac{i(t)^2}{\pi \sigma \left(R^2 - r'^2\right)} \tag{2.112}$$

The radius of the surface of separation, r', is a function of time and for small δ we have

$$P(t) = \sqrt{\frac{B_s}{2\pi R \sigma}} \left(i(t)^2\right) \left(\int_0^t i(t)\,dt\right)^{-1/2} \tag{2.113}$$

Assuming that the current is sinusoidal,

$$P(t) = I_{peak}^{\frac{3}{2}} \sqrt{\frac{B_s \omega}{2\pi R \sigma}} \left(\frac{\sin^2(\omega t)}{\sqrt{1 - \cos(\omega t)}}\right) \tag{2.114}$$

The average power over a half-cycle is

$$P_{avg} = I_{peak}^{\frac{3}{2}} \sqrt{\frac{B_s \omega}{2\pi R \sigma}} \int_0^\pi \frac{\sin^2\theta}{\sqrt{1 - \cos\theta}} d\theta \tag{2.115}$$

$$P_{avg} = I_{peak}^{\frac{3}{2}} \sqrt{\frac{B_s \omega}{2\pi R \sigma}} \frac{4\sqrt{2}}{3\pi} \tag{2.116}$$

In terms of the depth of penetration, δ, the loss in watts/unit length is

$$P_{avg} = I_{peak}^2 \frac{8}{3\pi} \frac{1}{2\pi R \delta \sigma} \tag{2.117}$$

If we now define the effective resistance of the cylinder as

$$R_{eff} = \frac{P_{avg}}{I_{rms}^2} \tag{2.118}$$

then

$$R_{eff} = \frac{16}{3\pi} \left(\frac{1}{2\pi R \delta \sigma} \right) \tag{2.119}$$

Assume now that the applied field is sinusoidal. In this case,

$$2B_s \sigma z' \frac{\partial z'}{\partial t} = J_{s_{peak}} \sin(\omega t) \tag{2.120}$$

The surface of separation is located at

$$z' = \sqrt{\frac{1}{B_s \sigma} \int_0^t J_{s_{peak}} \sin(\omega t)} \tag{2.121}$$

for each half-cycle and the depth of penetration is

$$\delta = z_{max} = \sqrt{\frac{2J_{s_{peak}}}{B_s \omega \sigma}} \tag{2.122}$$

In the conducting layer, the current density is

$$J = \frac{J_s}{z'} = \sqrt{B_s \omega \sigma J_{s_{peak}}} \frac{\sin(\omega t)}{\sqrt{1 - \cos(\omega t)}} \tag{2.123}$$

and the loss dissipated per unit of surface area is

$$P = z' \frac{J^2}{\sigma} = J_{s_{peak}}^{\frac{3}{2}} \sqrt{\frac{B_s \omega}{\sigma}} \frac{\sin^2(\omega t)}{\sqrt{1 - \cos(\omega t)}} \tag{2.124}$$

The average power is

$$P_{avg} = \frac{4\sqrt{2}}{3\pi} J_{s_{peak}} \sqrt{\frac{B_s \omega}{\sigma}} \tag{2.125}$$

We can now find the real and reactive components of the electric field due to the eddy currents. This will be useful in determining the power factor of the load. The electric field due to the flux in the magnetic material is

$$E = \sqrt{\frac{B_s \omega J_{S_{peak}}}{\sigma}} \left(\frac{\sin(\omega t)}{\sqrt{1 - \cos(\omega t)}} \right) \tag{2.126}$$

in each half-cycle. This can be written as

$$E = \sqrt{\frac{B_s \omega J_{S_{peak}}}{\sigma}} \cos\left(\frac{\omega t}{2} \right) \tag{2.127}$$

The fundamental component of the electric field is

$$E = \sqrt{\frac{B_s \omega J_{S_{peak}}}{\sigma}} \left(\frac{8}{3\pi} \sin(\omega t) + \frac{4}{3\pi} \cos(\omega t) \right) \tag{2.128}$$

The phase angle between the electric field and the exciting current is

$$\theta = \tan^{-1}(0.5) = 26.6° \tag{2.129}$$

and the power factor is

$$\cos \theta = 0.895 \tag{2.130}$$

We note that the phase angle of the impedance is different than in the case of the linear conductor where we found that the angle is 45°. In the linear case, the real power and reactive power are the same and the resistance is equal to the reactance, while in the limiting nonlinear case, the resistance is just twice the reactance. We also note that the losses in a linear material are proportional to the current squared, while in the limiting nonlinear case, the losses are proportional to the current to the 3/2 power.

2.6.1 Numerical Example

As an example, we will consider the case of a current sheet at the surface of a non-linear steel conductor. In Section 6.1, we will introduce the finite element method, which is a numerical technique that can be used to solve eddy current problems. As an illustration of the eddy currents in nonlinear steel, we created a finite element model for a deep steel plate. The problem is therefore one-dimensional. The magnetization characteristic used in the model is shown in Figure 2.25.

The saturation flux density is 2.0T. The curve is not exactly the square wave non-linear characteristic of Figure 2.21. This would be numerically unstable, with an infinite slope at the beginning and a discontinuous slope at the saturation point.

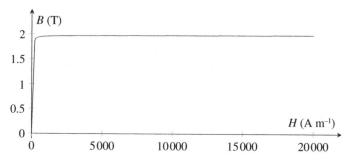

Figure 2.25 *B – H* curve for finite element example.

The curve is a fairly good approximation of the idealized characteristic of the nonlinear theory. The conductivity used is $\sigma = 1.0 \times 10^7$ S m^{-1}. This model was solved in the time domain and run over a number of cycles until steady state was reached.

The theory above tells us that the current density above the surface of separation is the same at all locations, while the current density below the surface is zero. In the finite element problem, we do not have the ideal square-wave $B - H$ characteristic, but if we look at the current density over a cycle at various depths from the finite element analysis, we obtain the results of Figure 2.26.

In analyzing Figure 2.26, we see that at a particular depth, the current density is essentially zero, until a certain point in the cycle, at which it jumps to a value very close to the current density at all points above it. The further down in the material

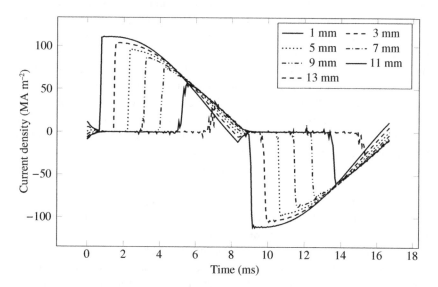

Figure 2.26 Finite element results of current density at different depths.

we go, the longer the time at zero current. It is also the case that the total current in the steel is sinusoidal in time. In other words, these distorted current densities, when integrated over the conductor, produce a sine wave.

In Figure 2.26, we see that as we go further into the material, the current stays at zero longer into the cycle. The peak of the applied current sheet was approximately 0.833×10^6 A m^{-1}. From Equation (2.107), we can find the penetration depth. This is the furthest point at which we will have current.

Using the values in this example, we obtain

$$\delta = 0.015 \text{ m} \tag{2.131}$$

This agrees quite well with the finite element results. So our conclusion is that while the square wave saturation curve is an idealized characteristic that does not exist in nature, the approximation is well-justified and the results for a smoother curve, show the same behavior as in the limiting nonlinear case. We note that there are some irregularities in the current waveform found with the finite element analysis. This is numerical noise and not physical. These occur at the points where the current is discontinuous and the flux density switches. In these regions, the flux density is changing very quickly and the numerical methods used have difficulty converging. In these regions, we have relaxed the convergence requirements and this gives rise to the numerical noise.

2.6.2 Effective Permeability

Many problems involving losses in nonlinear steel can be solved using a complex phasor formulation with a degree of approximation. In phasor analysis, we assume that all quantities are sinusoidally time varying and all are at the same frequency. As pointed out above, the current and flux density are not sinusoidal in the nonlinear steel. By considering the $B - H$ loop in Figure 2.27, we can see that if B varies sinusoidally, then H will have harmonics included. Similarly, if H is sinusoidal, then B will have harmonics. In practice, for periodic problems in nonlinear steel, both B and H will normally include harmonics. In a transformer, for example, we may excite the primary with a sinusoidal voltage. This means, by Faraday's law, that the total flux is sinusoidal. But, if we look at the local flux density in the steel, we find it is non-sinusoidal. An approximation that is often used and has been verified by test results [13, 66, 67] involves finding an effective permeability over a cycle.

There are many computational advantages in treating the problem using steady-state sinusoidal (complex) mathematics. In this case, the problem is formulated in terms of algebraic rather than differential equations. The problem can be solved once without having to time-step through several cycles to get to a periodic result. Further, quantities such as loss, resistance, and reactance come directly from the phasor solution. These are more difficult to find from the time domain solution. We do lose some information however, especially the effect of the harmonics.

Figure 2.27 Determination of effective permeability.

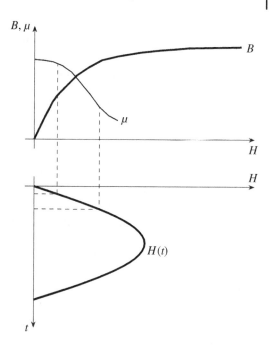

The question is, how can we find an effective permeability? One method that has been used and is well-verified by measurements, is to create an artificial saturation curve, each point of which corresponds to the value of permeability that gives the same stored energy as the normal $B - H$ curve.

The process is illustrated in Figure 2.27. Let us assume a sinusoidal H as shown. We do this for each peak value of H. We then use the normal magnetization curve and find the energy stored in a cycle. (In practice, we need only consider 1/4 cycle.)

We evaluate

$$W_m = \frac{1}{T} \int_0^T \frac{1}{2} B \cdot H \, dt \tag{2.132}$$

where T is the period. If the material were linear we could use

$$W_m = \frac{1}{T} \int_0^T \frac{\mu}{2} H^2 \, dt \tag{2.133}$$

We can now evaluate the *effective permeability* as

$$\mu_{\text{eff}} = \frac{\int_0^T B \cdot H \, dt}{\int_0^T H^2 \, dt} \tag{2.134}$$

This process is done for a number of peak values of H and an artificial curve of permeability vs. H or B vs. H is created. Then a standard iteration process is used as if this curve was the normal magnetization curve. There are a number of variations on the method described above. For example, we can assume that B is sinusoidal instead of H. This will give a slightly different curve. We may also use both sinusoidal B and H and take different weighted averages to produce the final curve. The results, although approximate, have been found to match experience and measurements quite well. The skin depth will be a function of the current as we found in the limiting nonlinear theory. The effective permeability model gives losses that vary between the 1.5 power and the square of the current. As we have seen, the losses in the limiting nonlinear model vary as the 1.5 power of the current and in linear materials we get the second power.

The numerical example above was repeated using the effective permeability method. A modified saturation curve was created using the data of Figure 2.25 by assuming a sinusoidal variation of the magnetic field, H. The problem was then solved using phasor analysis. This involved an iterative process since the permeability of the nonlinear material changes depending on the local value of the field. Comparing the results, we find the loss for a 1.0 cm slice of the deep plate as 39.2 W. The limiting nonlinear theory given by Equation (2.116) gives $P = 39.3$ W. The phase angle of the current should be 26.5° in the limiting nonlinear theory and from the effective permeability calculation we find 34.2°. In the linear case, we would expect 45 °. So we see that the effective permeability method produces results between the linear and the limiting nonlinear theory.

2.7 Plate with Hysteresis and Complex Permeability

O'Kelly [25] has introduced the idea of a complex permeability as a convenient way to account for hysteresis loss. In this analysis, we ignore the harmonics so that both the flux density and magnetic field are sinusoidal. The hysteresis is represented by introducing a phase shift between the complex B and H. The magnetization curve is then an ellipse. The angle of the major axis of the ellipse depends on the phase shift introduced by the complex permeability and is illustrated in Figure 2.28.

We define the permeability as

$$\mu = \frac{B}{H} = |\mu| e^{-j\theta} \tag{2.135}$$

We have seen that the one-dimensional partial differential equation for the magnetic field is

$$\frac{\partial^2 H}{\partial x^2} = \alpha^2 H \tag{2.136}$$

We will change the notation and write

$$\frac{\partial^2 H}{\partial x^2} = K^2 H \tag{2.137}$$

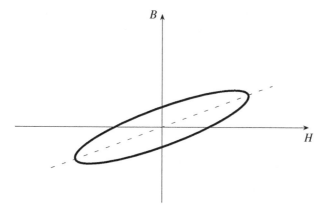

Figure 2.28 Elliptical $B - H$ curve.

We see that α has been replaced by K, where

$$K = \alpha e^{-j\theta/2} = k_r + jk_i \tag{2.138}$$

and

$$k_r = \frac{\sqrt{2}}{\delta} \cos \phi$$

$$k_i = \frac{\sqrt{2}}{\delta} \sin \phi \tag{2.139}$$

$$\phi = \frac{\pi}{4} - \frac{\theta}{2}$$

The mathematics is the same as in Section 2.3 of losses in a plate. We consider a plate of width $2b$ and with conductivity σ. We replace the parameter α with the new parameter K. We find that the current density is given as

$$J = KH_s \frac{\sinh Kx}{\cosh Kx} \tag{2.140}$$

The magnitude of the current density is

$$|J| = \frac{\sqrt{2}H_s}{\delta} \left(\frac{\cosh \frac{2k_r x}{\delta} - \cos \frac{2k_i x}{\delta}}{\cosh \frac{2k_r b}{\delta} + \cos \frac{2k_i x}{\delta}} \right)^{1/2} \tag{2.141}$$

The eddy current loss per unit total surface area is then

$$P_e = \frac{H_s^2}{\sqrt{2}\sigma\delta} \left(\frac{\sinh 2k_r b/\cos \phi - \sin 2k_i b/\sin \phi}{\cosh 2k_r b + \cos 2k_i b} \right) \tag{2.142}$$

We note that for the case of $\theta = 0$ or no hysteresis, Equation (2.142) reverts to the result we found in Section 2.3.

To add the hysteresis component of the loss, we use the Poynting vector. We have seen that the complex power flow crossing the surface can be found as

$$P = \frac{1}{2}\Re e\left[EH^*\right]_{x=b} - \frac{1}{2}\Re e\left[EH^*\right]_{x=-b} = \Re e\left[\frac{1}{\sigma}J_{y=b}H_s\right] \qquad (2.143)$$

Substituting for J we have

$$P = \frac{H_s^2}{\sigma}\left(K\frac{\sinh Kb}{\cosh Kb}\right) \qquad (2.144)$$

or

$$P = \frac{\sqrt{2}H_s^2}{\sigma\delta}\left(\frac{\cos\phi\sinh 2k_r b - \sin\phi\sin 2k_r b}{\cosh 2k_r b + \cos 2k_r b}\right) \qquad (2.145)$$

Stoll [26] presents the results as loss per unit surface area divided by the coefficient $H_s^2/\sigma\delta$, which results in a per unit loss, vs. the thickness of the plate divided by the skin depth. Following this example, we see Equation (2.145) evaluated for $\theta = 10°$ and $\theta = 0°$ in Figure 2.29.

While this process does not accurately represent the physical process, in which the flux density and magnetic fields are non-sinusoidal, it does allow us to approximate the hysteresis loss in the material.

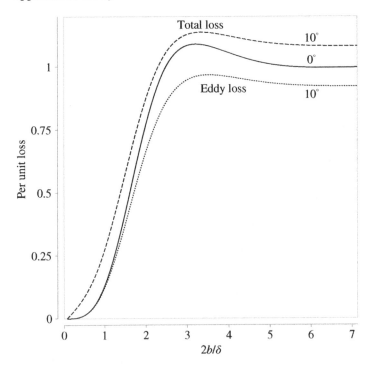

Figure 2.29 Loss at different angles including hysteresis.

2.8 Conducting Plates with Sinusoidal Space Variation of Field

First, we consider the problem of a conducting plate with a sinusoidally time varying applied flux density that is sinusoidally distributed in space. Such a situation is encountered when computing eddy currents in electric machines. To begin, consider the plate in Figure 2.30, with a standing wave of magnetic flux density in the z direction given by

$$B_z(x) = B_0 \cos\left(\frac{\pi x}{\tau}\right) e^{j\omega t} \tag{2.146}$$

Here B_0 is the peak flux density and τ is the pole pitch or one-half the wavelength of the standing wave. The plate has dimensions $\beta\tau$ in the x direction (corresponding to the peripheral direction in the machine), ℓ in the y direction (corresponding to the radial direction in the machine), and depth d in the z direction (axial direction in the machine). The fraction of a pole pitch that the plate spans is denoted as β. The assumption is that there is only one component of the magnetic field (z) and two components of the induced current (x, y), as shown.

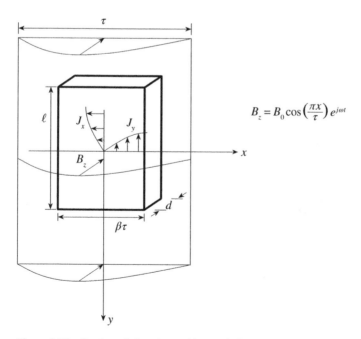

$$B_z = B_0 \cos\left(\frac{\pi x}{\tau}\right) e^{j\omega t}$$

Figure 2.30 Portion of electric machine end plate subject to axial flux density that varies sinusoidally in time and in space.

We begin with the low-frequency Maxwell equations

$$\nabla \times \mathbf{E} = -\frac{\partial \mathbf{B}}{\partial t} \tag{2.147}$$

$$\mathbf{J} = \sigma \mathbf{E}. \tag{2.148}$$

Substituting, we get for the current density

$$\nabla \times \mathbf{J} = -\sigma \frac{\partial \mathbf{B}}{\partial t} \tag{2.149}$$

In our case, the magnetic field has only a z component and the current has only x and y components. This gives

$$\frac{\partial J_y}{\partial x} - \frac{\partial J_x}{\partial y} = -\sigma \frac{\partial B_z}{\partial t} \tag{2.150}$$

We now differentiate with respect to x and y to get

$$\frac{\partial^2 J_y}{\partial x^2} - \frac{\partial^2 J_x}{\partial y \, \partial x} = -\sigma \frac{\partial^2 B_z}{\partial x \, \partial t} \tag{2.151}$$

and

$$\frac{\partial^2 J_y}{\partial x \, \partial y} - \frac{\partial^2 J_x}{\partial y^2} = -\sigma \frac{\partial^2 B_z}{\partial y \, \partial t} \tag{2.152}$$

From the current continuity equation

$$\nabla \cdot \mathbf{J} = 0 \tag{2.153}$$

we can differentiate to find

$$\frac{\partial^2 J_x}{\partial x^2} + \frac{\partial^2 J_y}{\partial x \, \partial y} = 0 \tag{2.154}$$

$$\frac{\partial^2 J_x}{\partial x \, \partial y} + \frac{\partial^2 J_y}{\partial y^2} = 0 \tag{2.155}$$

Now substituting

$$\frac{\partial^2 J_x}{\partial x^2} + \frac{\partial^2 J_x}{\partial y^2} = \sigma \frac{\partial^2 B_z}{\partial y \, \partial t} \tag{2.156}$$

$$\frac{\partial^2 J_y}{\partial x^2} + \frac{\partial^2 J_y}{\partial y^2} = -\sigma \frac{\partial^2 B_z}{\partial x \, \partial t} \tag{2.157}$$

Since the magnetic flux density is independent of y, we have

$$\frac{\partial^2 J_x}{\partial x^2} + \frac{\partial^2 J_x}{\partial y^2} = 0 \tag{2.158}$$

$$\frac{\partial^2 J_y}{\partial x^2} + \frac{\partial^2 J_y}{\partial y^2} = -\sigma \frac{\partial^2 B_z}{\partial x \, \partial t} \tag{2.159}$$

The first differential equation is the homogeneous equation

$$\nabla^2 J_x = 0 \tag{2.160}$$

which can be solved using separation of variables. We first assume a solution of the form

$$J_x(x, y) = J_x(x) \cdot J_x(y) \tag{2.161}$$

Then we have, by substitution

$$J_x(y)\frac{\partial^2 J_x(x)}{\partial x^2} + J_x(x)\frac{\partial^2 J_x(y)}{\partial y^2} = 0 \tag{2.162}$$

Dividing (2.161) and (2.162), we get

$$\frac{1}{J_x(x)}\frac{\partial^2 J_x(x)}{\partial x^2} = -\frac{1}{J_x(y)}\frac{\partial^2 J_x(y)}{\partial y^2} \tag{2.163}$$

Since the left hand side of the equation is a function only of x and the right hand side only a function of y, the equality implies that each side is a constant.

$$\frac{\partial^2 J_x(x)}{\partial x^2} + \alpha^2 J_x(x) = 0 \tag{2.164}$$

$$\frac{\partial^2 J_x(y)}{\partial y^2} - \alpha^2 J_x(y) = 0 \tag{2.165}$$

The solution is

$$J_x(x) = C_1 e^{-j\alpha x} + C_2 e^{j\alpha x} \tag{2.166}$$

$$J_x(y) = C_3 e^{-\alpha y} + C_4 e^{\alpha y} \tag{2.167}$$

or

$$J_x(x) = D_1 \cos(\alpha x) + D_2 \sin(\alpha x) \tag{2.168}$$

$$J_x(y) = D_3 \cosh(\alpha y) + D_4 \sinh(\alpha y) \tag{2.169}$$

Now that we have the homogeneous solution, we focus next on the particular solution. The magnetic field has the form

$$B_z = B_0 \cos\left(\frac{\pi x}{\tau}\right) e^{j\omega t} \tag{2.170}$$

Recalling (2.159):

$$\frac{\partial^2 J_y}{\partial x^2} + \frac{\partial^2 J_y}{\partial y^2} = -\sigma \frac{\partial^2 B_z}{\partial x \partial t}$$

it can be shown by substitution that the particular solution is

$$J_y(x) = -\frac{j\omega\sigma\tau}{\pi} B_0 \sin\left(\frac{\pi x}{\tau}\right) e^{j\omega t} \tag{2.171}$$

Since J_y is not a function of y,

$$\frac{\partial^2 J_y}{\partial y^2} = 0 \tag{2.172}$$

and we only consider the differentiation with respect to x:

$$\frac{\partial^2 J_y}{\partial x^2} = \frac{j\omega\sigma\pi}{\tau}\sin\left(\frac{\pi x}{\tau}\right)e^{j\omega t} \tag{2.173}$$

Therefore

$$-\sigma\frac{\partial^2 B_z}{\partial x\,\partial t} = -\sigma B_0\left(\frac{-\pi}{\tau}\right)\sin\left(\frac{\pi x}{\tau}\right)je^{j\omega t} = \frac{j\omega\sigma\pi B_0}{\tau} \tag{2.174}$$

The expressions for the components of J are now written as

$$J_x = \sum_n \left(A_{x1n}\cosh\left(\alpha_n y\right) + A_{x2n}\sinh\left(\alpha_n y\right)\right)$$
$$\times \left(A_{x3n}\cos\left(\alpha_n x\right) + A_{x4n}\sin\left(\alpha_n x\right)\right) \tag{2.175}$$

$$J_y = \sum_n \left(A_{y1n}\cosh\left(\alpha_n y\right) + A_{y2n}\sinh\left(\alpha_n y\right)\right)$$
$$\times \left(A_{y3n}\cos\left(\alpha_n x\right) + A_{y4n}\sin\left(\alpha_n x\right) - \frac{j\omega\sigma\tau B_0}{\pi}\sin\left(\frac{\pi x}{\tau}\right)e^{j\omega t}\right) \tag{2.176}$$

As shown in Figure 2.31, along the plane $y = 0$, we must have $J_x = 0$ and along the plane $x = 0$ we must have $J_y = 0$. From these boundary conditions, we conclude that

$$A_{x1n} = 0 \tag{2.177}$$

$$A_{y3n} = 0 \tag{2.178}$$

Since we must have

$$\frac{\partial J_x}{\partial x} = -\frac{\partial J_y}{\partial y} \tag{2.179}$$

This continuity condition requires that

$$\frac{\partial J_x}{\partial x} = \sum_n \alpha_n A_{x2n}\sinh\left(\alpha_n y\right)\left(-A_{x3n}\sin\left(\alpha_x\right) + A_{x4n}\cos\left(\alpha_n x\right)\right) \tag{2.180}$$

$$\frac{\partial J_y}{\partial y} = \sum_n \alpha_n A_{y4n}\sin\left(\alpha_n x\right)\left(A_{y1n}\sinh\left(\alpha_n y\right) + A_{y2n}\cosh\left(\alpha_n y\right)\right) \tag{2.181}$$

This is satisfied if

$$A_{x4n} = 0 \tag{2.182}$$

$$A_{y2n} = 0 \tag{2.183}$$

The current densities now become

$$J_x = \sum_n A_{x2n}A_{x3n}\sinh\left(\alpha_n y\right)\cos\left(\alpha_n x\right) \tag{2.184}$$

Figure 2.31 Eddy current contours at edges of end plate.

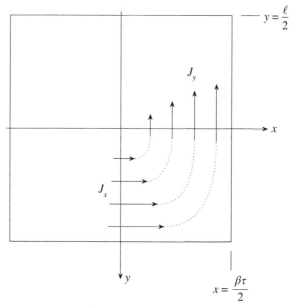

Boundary conditions:

$$\text{(i)} \ x = 0, J_y = 0$$
$$\text{(ii)} \ y = 0, J_x = 0$$
$$\text{(iii)} \ x = \pm\frac{\beta\tau}{2}, J_x = 0$$
$$\text{(iv)} \ y = \pm\frac{\ell}{2}, J_y = 0$$

$$J_y = \sum_n A_{y1n}A_{y4n}\cosh\left(\alpha_n y\right)\sin\left(\alpha_n x\right) - \frac{j\omega\sigma B_0}{\pi}\sin\left(\frac{\pi x}{\tau}\right)e^{j\omega t} \tag{2.185}$$

From the condition that the x component of current vanishes on the right and left edges of the plate ($x = \pm\beta\tau/2$), we have

$$\sum_n A_{x2n}A_{x3n}\sinh\left(\alpha_n y\right)\cos\left(\alpha_n\frac{\beta\tau}{2}\right) = 0 \tag{2.186}$$

which is satisfied if

$$\alpha_n\frac{\beta\tau}{2} = \frac{\pi}{2}, \frac{3\pi}{2}, \frac{5\pi}{2}, \ldots, \frac{(2n-1)\pi}{2} \quad \text{for } n = 1, 2, 3, \ldots \tag{2.187}$$

This means that

$$\alpha_1 = \frac{\pi}{\beta\tau} \tag{2.188}$$

$$\alpha_2 = \frac{3\pi}{2}\cdot\frac{2}{\beta\tau} = \frac{3\pi}{\beta\tau} \tag{2.189}$$

$$\alpha_3 = \frac{5\pi}{2}\cdot\frac{2}{\beta\tau} = \frac{5\pi}{\beta\tau}, \text{etc.} \tag{2.190}$$

From the condition that the y component of the current vanishes at the top and bottom edges of the plate ($y = \pm\ell/2$), we require that

$$\sum_n A_{y1n}A_{y4n} \cosh\left(\alpha_n \frac{\ell}{2}\right) \sin(\alpha_n x) - \frac{j\omega\tau B_0}{\pi} \sin\left(\frac{\pi x}{\tau}\right) e^{j\omega t} = 0 \qquad (2.191)$$

or

$$\sum_n A_{y1n}A_{y4n} \cosh\left(\alpha_n \frac{\ell}{2}\right) \sin(\alpha_n x) = \frac{j\omega\tau B_0}{\pi} \sin\left(\frac{\pi x}{\tau}\right) e^{j\omega t} \qquad (2.192)$$

Since the sine is an odd function, the Fourier decomposition is

$$A_{y1n}A_{y4n} \cosh\left(\alpha_n \frac{\ell}{2}\right) = \frac{2}{\beta\tau/2} \int_0^{\beta\tau/2} \frac{j\omega\tau B_0}{\pi} \sin\left(\frac{\pi x}{\tau}\right) \sin(\alpha_n x) e^{j\omega t} dx$$

$$= \frac{j4\omega\sigma B_0}{\beta\pi} \int_0^{\beta\tau/2} \sin\left(\frac{\pi x}{\tau}\right) \sin(\alpha_n x) e^{j\omega t} dx \qquad (2.193)$$

We integrate to find the first four coefficients. The result is

$$A_{y11}A_{y41} = A_{x21}A_{x31} = \frac{2j\omega\sigma B_0 \tau e^{j\omega t}}{\pi^2 \cosh\left(\dfrac{\pi\ell}{2\beta\tau}\right)} \qquad (2.194)$$

$$A_{y12}A_{y42} = A_{x22}A_{x32} = \frac{2j\omega\sigma B_0 \tau e^{j\omega t}}{\pi^2 \cosh\left(\dfrac{3\pi\ell}{2\beta\tau}\right)} \qquad (2.195)$$

$$A_{y13}A_{y43} = A_{x23}A_{x33} = \frac{2j\omega\sigma B_0 \tau e^{j\omega t}}{\pi^2 \cosh\left(\dfrac{5\pi\ell}{2\beta\tau}\right)} \qquad (2.196)$$

$$A_{y14}A_{y44} = A_{x24}A_{x34} = \frac{2j\omega\sigma B_0 \tau e^{j\omega t}}{\pi^2 \cosh\left(\dfrac{7\pi\ell}{2\beta\tau}\right)} \qquad (2.197)$$

The current densities can now be found by using these coefficients in (2.184) and (2.185). For example, the fundamental components are

$$J_{x1} = A_1 \sinh(\alpha_1 y)\cos(\alpha_1 x) \qquad (2.198)$$

$$J_{y1} = A_1 \cosh(\alpha_1 y)\sin(\alpha_1 x) - \frac{j\omega\tau B_0}{\pi} \sin\left(\frac{\pi x}{\tau}\right) e^{j\omega t} \qquad (2.199)$$

where A_1 is found from (2.194)

$$A_1 = A_{y11}A_{y41} = A_{x21}A_{x31} = \frac{2j\omega\sigma B_0 \tau e^{j\omega t}}{\pi^2 \cosh\left(\dfrac{\pi\ell}{2\beta\tau}\right)} \qquad (2.200)$$

and α_1 is as given in (2.188)

$$\alpha_1 = \frac{\pi}{\beta\tau}$$

The Joule losses in the plate are given by

$$W = \int_V \frac{1}{\sigma} J^2 \, dV \tag{2.201}$$

Since we have assumed no variation in the z direction, we can integrate over the surface to find the loss per meter depth into the plate.

$$W' = \int_S \frac{1}{\sigma} J^2 \, dx \, dy \tag{2.202}$$

We now use

$$J^2 = J_x^2 + J_y^2 \tag{2.203}$$

and therefore

$$W' = \int_S \frac{1}{\sigma} \left(J_x^2 + J_y^2 \right) \, dx \, dy \tag{2.204}$$

We will compute the x and y components separately.

$$W_x' = \int_S \frac{1}{\sigma} J_x^2 \, dx \, dy \tag{2.205}$$

$$W_y' = \int_S \frac{1}{\sigma} J_y^2 \, dx \, dy \tag{2.206}$$

Substituting (2.198) into (2.205) and taking advantage of the symmetry of the problem to integrate over a quarter of the plate, we obtain[1]

$$
\begin{aligned}
W_x' &= \frac{A_1^2}{\sigma} \int_0^{\beta\tau/2} \cos^2 \left(\alpha_1 x \right) \, dx \int_0^{\ell/2} \sinh^2 \left(\alpha_1 y \right) \, dy \\
&= \frac{A_1^2}{\sigma} \int_0^{\beta\tau/2} \cos^2 \left(\frac{\pi x}{\beta\tau} \right) \, dx \int_0^{\ell/2} \sinh^2 \left(\frac{\pi y}{\beta\tau} \right) \, dy \\
&= \frac{A_1^2}{\sigma} \left\{ \frac{\beta\tau}{4} \right\} \left\{ \frac{\beta\tau}{4} \left[\frac{1}{\pi} \sinh \left(\frac{\pi\ell}{\beta\tau} \right) - \frac{\ell}{\beta\tau} \right] \right\} \\
&= \frac{(A_1\beta\tau)^2}{16\sigma} \left[\frac{1}{\pi} \sinh \left(\frac{\pi\ell}{\beta\tau} \right) - \frac{\ell}{\beta\tau} \right]
\end{aligned}
\tag{2.207}
$$

Similarly, J_{y1} is used to obtain W_y'. However, since the expression for J_{y1} in (2.199) contains two terms, its square contains three terms and solving for W_y' involves three integrations:

$$
\begin{aligned}
W_y' &= \frac{A_1^2}{\sigma} \int_0^{\beta\tau/2} \sin^2 \left(\alpha_1 x \right) \, dx \int_0^{\ell/2} \cosh^2 \left(\alpha_1 y \right) \, dy \\
&\quad - \frac{2j\omega\tau B_0}{\pi} \left(\frac{A_1 e^{j\omega t}}{\sigma} \right) \int_0^{\beta\tau/2} \sin \left(\frac{\pi x}{\beta\tau} \right)
\end{aligned}
$$

---

[1] The integrations are carried out with the aid of the identities $\cos^2 x = \dfrac{1 + \cos(2x)}{2}$ and $\sinh^2 x = \dfrac{\cosh(2x) - 1}{2}$.

$$\times \sin\left(\frac{\pi x}{\tau}\right) e^{j\omega t} dx \int_0^{\ell/2} \cosh\left(\frac{\pi y}{\beta \tau}\right) dy$$

$$- \left(\frac{\omega \tau B_0}{\pi}\right)^2 (\sigma e^{2j\omega t}) \int_0^{\beta\tau/2} \sin^2\left(\frac{\pi x}{\tau}\right) dx \int_0^{\ell/2} dy$$

Carrying out the integrations yields

$$W_y' = \frac{(A_1 \beta \tau)^2}{16\sigma} \left[\frac{1}{\pi} \sinh\left(\frac{\pi \ell}{\beta \tau}\right) + \frac{\ell}{\beta \tau}\right]$$

$$- \frac{A_1^2}{\sigma} \left[\frac{\beta \tau}{2} \cosh\left(\frac{\pi \ell}{2\beta \tau}\right)\right] \left[\frac{\beta \tau}{\pi} \sinh\left(\frac{\pi \ell}{2\beta \tau}\right)\right]$$

$$- \left(\frac{\omega \tau B_0}{\pi}\right)^2 (\sigma e^{2j\omega t}) \frac{\ell}{2} \left[\frac{-\beta \tau^2}{16\pi} \sin(\beta \tau)\right] \qquad (2.208)$$

Finally the total power is found by combining (2.207) and (2.208)

$$W' = W_x' + W_y'$$

$$= \frac{(A_1 \beta \tau)^2}{8\pi\sigma} \sinh\left(\frac{\pi \ell}{\beta \tau}\right)$$

$$- \frac{(A_1 \beta \tau)^2}{2\pi\sigma} \cosh\left(\frac{\pi \ell}{2\beta \tau}\right) \sinh\left(\frac{\pi \ell}{2\beta \tau}\right)$$

$$- \left(\frac{\omega \tau B_0}{\pi}\right)^2 (\sigma e^{2j\omega t}) \frac{\beta \tau \ell}{8} \left[1 - \frac{1}{\beta \pi} \sin(\beta \pi)\right] \qquad (2.209)$$

Using the identity $\cosh(2x) = 2\cosh x \sinh x$, this becomes

$$W' = \frac{(A_1 \beta \tau)^2}{8\pi\sigma} \sinh\left(\frac{\pi \ell}{\beta \tau}\right) - \left(\frac{\omega \tau B_0}{\pi}\right)^2 (\sigma e^{2j\omega t}) \frac{\beta \tau \ell}{8} \left[1 - \frac{1}{\beta \pi} \sin(\beta \pi)\right]$$

Next the time-varying factor is replaced by $1/2$ (because the rms value of the square of current density is half the square of the peak value) and A_1^2 is expanded using (2.200), yielding

$$W' = \frac{2\sigma}{\pi} \left(\frac{\omega B_0 \beta \tau^2}{2\pi^2}\right)^2 \tanh\left(\frac{\pi \ell}{2\beta \tau}\right) \left\{\frac{\sin\left[(1-\beta)\pi/2\right]}{1-\beta} - \frac{\sin\left[(1+\beta)\pi/2\right]}{1+\beta}\right\}^2$$

$$- \left(\frac{\omega \tau B_0}{4\pi}\right)^2 (\beta \tau \ell \sigma) \left[1 - \frac{1}{\beta \pi} \sin(\beta \pi)\right] \qquad (2.210)$$

Combining common coefficients, this becomes

$$W' = (\omega \tau B_0)^2 \left[\frac{8\beta^2 \tau^2 \sigma}{\pi} \tanh\left(\frac{\pi \ell}{2\beta \tau}\right) \left\{\frac{\sin\left[(1-\beta)\pi/2\right]}{1-\beta} - \frac{\sin\left[(1+\beta)\pi/2\right]}{1+\beta}\right\}^2\right.$$

$$- \beta \tau \ell \pi^2 \sigma \left[1 - \frac{1}{\beta \pi} \sin (\beta \pi) \right] \Bigg) \tag{2.211}$$

Finally, we introduce a parameter λ for the ratio of the radial dimension of the plate ℓ to the peripheral dimension $\beta \tau$

$$\lambda = \frac{\ell}{\beta \tau} \tag{2.212}$$

and then (2.211) becomes

$$W' = \left(\omega \tau B_0 \right)^2 \Bigg(\frac{8 \ell^2 \sigma}{\pi \lambda^2} \tanh \left(\frac{\pi \lambda}{2} \right) \left\{ \frac{\sin \left[(1 - \beta) \pi / 2 \right]}{1 - \beta} - \frac{\sin \left[(1 + \beta) \pi / 2 \right]}{1 + \beta} \right\}^2$$

$$- \pi^2 \sigma \frac{\ell^2}{\lambda} \left[1 - \frac{1}{\beta \pi} \sin (\beta \pi) \right] \Bigg) \tag{2.213}$$

or

$$W' = \left(\omega \tau B_0 \right)^2 \frac{\ell^2}{\lambda} \Bigg(\frac{8 \sigma}{\pi \lambda} \tanh \left(\frac{\pi \lambda}{2} \right) \left\{ \frac{\sin \left[(1 - \beta) \pi / 2 \right]}{1 - \beta} - \frac{\sin \left[(1 + \beta) \pi / 2 \right]}{1 + \beta} \right\}^2$$

$$- \pi^2 \sigma \left[1 - \frac{1}{\beta \pi} \sin (\beta \pi) \right] \Bigg) \tag{2.214}$$

We will now examine this expression further by splitting it into two parts; the first may be considered a "field" term and the second a correction factor:

$$W' = W'_{\text{field}} \Lambda \tag{2.215}$$

where

$$W'_{\text{field}} = \left(\omega \tau B_0 \right)^2 \frac{\ell^2}{\lambda} \tag{2.216}$$

$$\Lambda = \frac{8 \sigma}{\pi \lambda} \tanh \left(\frac{\pi \lambda}{2} \right) \left\{ \frac{\sin \left[(1 - \beta) \pi / 2 \right]}{1 - \beta} - \frac{\sin \left[(1 + \beta) \pi / 2 \right]}{1 + \beta} \right\}^2$$

$$- \pi^2 \sigma \left[1 - \frac{1}{\beta \pi} \sin (\beta \pi) \right] \tag{2.217}$$

A plot of the correction factor Λ vs. the end plate radial length to peripheral width ratio λ, for various fractional pitch values β, is given in Figure 2.32. Similarly, a plot of Λ vs. β for various λ is given in Figure 2.33. These figures show the relative

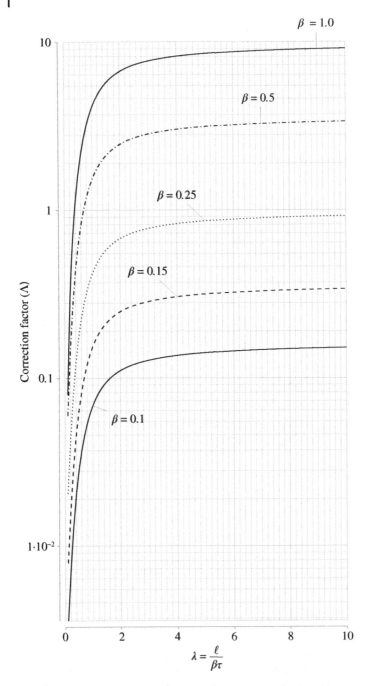

Figure 2.32 Loss correction factor Λ as a function of $\dfrac{\ell}{\beta\tau}$ for various β.

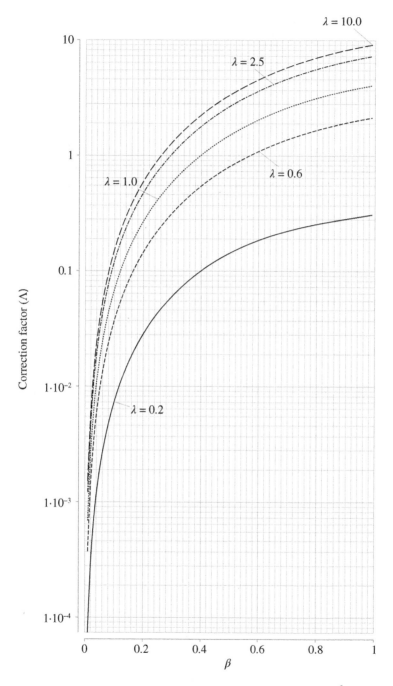

Figure 2.33 Loss correction factor Λ as a function of β for various $\dfrac{\ell}{\beta\tau}$.

influence of the geometrical parameters on the loss solution and indicate regimes where approximations may be made for (2.214).

2.9 Eddy Currents in Multi-Layered Plate Geometries

In this section, we look at the problem of a multi-layered geometry with a sinusoidally time varying and sinusoidally space-varying current sheet as the source. The problem is fairly general in that other types of excitation can be produced by decomposing that source into its Fourier coefficients and then superimposing the results. In the example that follows, we use two layers of different material properties and dimensions as well as free space regions. This method can be extended to larger numbers of layers by following the same analytic process.

We begin with a two-dimensional Cartesian coordinate system. The source is a sinusoidally distributed current sheet carrying a sinusoidally time-varying current. Material 1 has thickness d_1, permeability μ_1, and conductivity σ_1. Material 2 has thickness d_2, permeability μ_2, and conductivity σ_2. The material properties of both regions are linear and isotropic. The problem, being two-dimensional, has no variation in the z direction which we assume extends to infinity. The geometry is illustrated in Figure 2.34.

Let the source current be

$$J(x) = J_0 e^{jnkx} e^{jm\omega t} \tag{2.218}$$

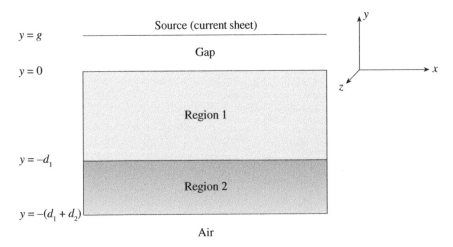

Figure 2.34 Two-dimensional layered geometry and source.

where J_0 is the peak value of the current sheet, n is the space harmonic order, k is the wave number, m is the time harmonic order, and ω is the fundamental angular frequency.

For the air-gap region, we have the homogeneous Laplace equation for the z component of the magnetic vector potential.

$$\nabla^2 A_g = 0 \tag{2.219}$$

The general solution of Equation (2.219) is

$$A_g(x,y) = \left(C_1^g e^{nky} + C_2^g e^{-nky} \right) e^{j(nkx+m\omega t)} \tag{2.220}$$

In region 1, we have

$$\nabla^2 A_1 - jm\omega\mu_1\sigma_1 A_1 = 0 \tag{2.221}$$

The general solution of Equation (2.221) is

$$A_1(x,y) = \left(C_1^1 e^{a_1 y} + C_2^1 e^{-a_1 y} \right) e^{j(nkx+m\omega t)} \tag{2.222}$$

where

$$a_1 = \sqrt{(nk)^2 + jm\omega\mu_1\sigma_1} \tag{2.223}$$

In region 2, we have

$$\nabla^2 A_2 - jm\omega\mu_2\sigma_2 A_2 = 0 \tag{2.224}$$

The general solution of Equation (2.224) is

$$A_2(x,y) = \left(C_1^2 e^{a_2 y} + C_2^2 e^{-a_2 y} \right) e^{j(nkx+m\omega t)} \tag{2.225}$$

where

$$a_2 = \sqrt{(nk)^2 + jm\omega\mu_2\sigma_2} \tag{2.226}$$

In the air region, which extends to infinity, we have

$$\nabla^2 A_a = 0 \tag{2.227}$$

The general solution of Equation (2.227) is

$$A_a(x,y) = \left(C_1^a e^{nky} \right) e^{j(nkx+m\omega t)} \tag{2.228}$$

We now must evaluate the constants. There are seven complex numbers to evaluate, and there are seven interface conditions to enforce, these being the normal flux density and tangential field continuity conditions. Note that at the current source, $H_t = J$. We have the following interface conditions.

At the source,

$$H_t|_{y=g} = \left. \frac{1}{\mu_0} \frac{\partial A_g}{\partial y} \right|_{y=g} = J_0 e^{j(nkx+m\omega t)} \tag{2.229}$$

which leads to

$$\frac{nk}{\mu_0} \left(C_1^g e^{nkg} - C_2^g e^{-nkg} \right) e^{j(nkx+m\omega t)} = J_0 e^{j(nkx+m\omega t)} \qquad (2.230)$$

Therefore,

$$e^{nkg} C_1^g - e^{-nkg} C_2^g = \frac{\mu_0 J_0}{nk} \qquad (2.231)$$

At the gap – region 1 interface, B normal is continuous so that

$$-\frac{\partial A_g}{\partial x}\bigg|_{y=0} = -\frac{\partial A_1}{\partial x}\bigg|_{y=0} \qquad (2.232)$$

which gives

$$-jnk \left(C_1^g + C_2^g \right) e^{j(nkx+m\omega t)} = -jnk \left(C_1^1 + C_2^1 \right) e^{j(nkx+m\omega t)} \qquad (2.233)$$

Therefore,

$$C_1^g + C_2^g - C_1^1 - C_2^1 = 0 \qquad (2.234)$$

Also, the tangential component of H is continuous so

$$\frac{1}{\mu_0} \frac{\partial A_g}{\partial y}\bigg|_{y=0} = \frac{1}{\mu_1} \frac{\partial A_1}{\partial y}\bigg|_{y=0} \qquad (2.235)$$

This results in

$$\frac{nk}{\mu_0} \left(C_1^g - C_2^g \right) e^{j(nkx+m\omega t)} = \frac{a_1}{\mu_1} \left(C_1^1 - C_2^1 \right) e^{j(nkx+m\omega t)} \qquad (2.236)$$

We multiply both sides of Equation (2.236) by the permeability of free space, μ_0, so that we can use the relative permeability of region 1. Rearrange the terms to get

$$nkC_1^g - nkC_2^g - \frac{a_1}{\mu_{r1}} C_1^1 + \frac{a_1}{\mu_{r1}} C_2^1 = 0 \qquad (2.237)$$

At the region 1–2 interface, we must have the normal component B continuous, so that

$$-\frac{\partial A_1}{\partial x}\bigg|_{y=-d_1} = -\frac{\partial A_2}{\partial x}\bigg|_{y=-d_1} \qquad (2.238)$$

This gives

$$-jnk \left(C_1^1 e^{-a_1 d_1} + C_2^1 e^{a_1 d_1} \right) e^{j(nkx+m\omega t)} = \qquad (2.239)$$
$$-jnk \left(C_1^2 e^{-a_2 d_1} + C_2^2 e^{a_2 d_1} \right) e^{j(nkx+m\omega t)}$$

Therefore,

$$e^{-a_1 d_1} C_1^1 + e^{a_1 d_1} C_2^1 - e^{-a_2 d_1} C_1^2 - e^{a_2 d_1} C_2^2 = 0 \qquad (2.240)$$

The tangential H condition gives

$$\frac{1}{\mu_1} \frac{\partial A_1}{\partial y}\bigg|_{y=-d_1} = \frac{1}{\mu_2} \frac{\partial A_2}{\partial y}\bigg|_{y=-d_1} \tag{2.241}$$

which results in

$$\frac{a_1}{\mu_1} \left(C_1^1 e^{-a_1 d_1} - C_2^1 e^{a_1 d_1} \right) e^{j(nkx+m\omega t)} = \frac{a_2}{\mu_2} \left(C_1^2 e^{-a_2 d_1} - C_2^2 e^{a_2 d_1} \right) e^{j(nkx+m\omega t)} \tag{2.242}$$

Similar to Equation (2.237), we multiply by μ_0 to use the relative permeabilities of regions 1 and 2. Therefore,

$$\frac{a_1}{\mu_{r1}} e^{-a_1 d_1} C_1^1 - \frac{a_1}{\mu_{r1}} e^{a_1 d_1} C_2^1 - \frac{a_2}{\mu_{r2}} e^{-a_2 d_1} C_1^2 + \frac{a_2}{\mu_{r2}} e^{a_2 d_1} C_2^2 = 0 \tag{2.243}$$

At the region 2–air interface, for the B normal condition, we have

$$-\frac{\partial A_2}{\partial x}\bigg|_{y=-(d_1+d_2)} = -\frac{\partial A_a}{\partial x}\bigg|_{y=-(d_1+d_2)} \tag{2.244}$$

$$-jnk \left(C_1^2 e^{-a_2(d_1+d_2)} + C_2^2 e^{a_2(d_1+d_2)} \right) e^{j(nkx+m\omega t)} =$$
$$-jnk C_1^a e^{-nk(d_1+d_2)} e^{j(nkx+m\omega t)} \tag{2.245}$$

Therefore,

$$e^{-a_2(d_1+d_2)} C_1^2 + e^{a_2(d_1+d_2)} C_2^2 - e^{-nk(d_1+d_2)} C_1^a = 0 \tag{2.246}$$

For the tangential H condition

$$\frac{1}{\mu_2} \frac{\partial A_2}{\partial y}\bigg|_{y=-(d_1+d_2)} = \frac{1}{\mu_0} \frac{\partial A_a}{\partial y}\bigg|_{y=-(d_1+d_2)} \tag{2.247}$$

so

$$\frac{a_2}{\mu_2} \left(C_1^2 e^{-a_2(d_1+d_2)} - C_2^2 e^{a_2(d_1+d_2)} \right) e^{j(nkx+m\omega t)} = \frac{nk}{\mu_0} C_1^a e^{-nk(d_1+d_2)} e^{j(nkx+m\omega t)} \tag{2.248}$$

Similar to Equations (2.237) and (2.243), we multiply by μ_0 to use the relative permeability of region 2. Therefore,

$$\frac{a_2}{\mu_{r2}} e^{-a_2(d_1+d_2)} C_1^2 - \frac{a_2}{\mu_{r2}} e^{a_2(d_1+d_2)_1} C_2^2 - nk e^{-nk(d_1+d_2)} C_1^a = 0 \tag{2.249}$$

This gives seven equations and seven unknown coefficients. To evaluate the unknowns, we solve the simultaneous linear set of equations shown below

$$
\begin{pmatrix}
e^{nkg} & -e^{-nkg} & 0 & 0 & 0 & 0 & 0 \\
1 & 1 & -1 & -1 & 0 & 0 & 0 \\
nk & -nk & \frac{-a_1}{\mu_{r1}} & \frac{a_1}{\mu_{r1}} & 0 & 0 & 0 \\
0 & 0 & e^{-a_1 d_1} & e^{a_1 d_1} & -e^{-a_2 d_1} & -e^{a_2 d_1} & 0 \\
0 & 0 & \frac{a_1}{\mu_{r1}}e^{-a_1 d_1} & -\frac{a_1}{\mu_{r1}}e^{a_1 d_1} & -\frac{a_2}{\mu_{r2}}e^{-a_2 d_1} & \frac{a_2}{\mu_{r2}}e^{a_2 d_1} & 0 \\
0 & 0 & 0 & 0 & e^{-a_2(d_1+d_2)} & e^{a_2(d_1+d_2)} & -e^{-nk(d_1+d_2)} \\
0 & 0 & 0 & 0 & \frac{a_2}{\mu_{r2}}e^{-a_2(d_1+d_2)} & -\frac{a_2}{\mu_{r2}}e^{a_2(d_1+d_2)} & -nke^{-nk(d_1+d_2)}
\end{pmatrix}
$$

$$
\times
\begin{pmatrix}
C_1^g \\
C_2^g \\
C_1^1 \\
C_2^1 \\
C_1^2 \\
C_2^2 \\
C_1^a
\end{pmatrix}
=
\begin{pmatrix}
\frac{\mu_0 J_0}{nk} \\
0 \\
0 \\
0 \\
0 \\
0 \\
0
\end{pmatrix}
$$

With these seven (complex) coefficients, we can completely describe the field in the entire domain. The reader can see that the method can be extended to more layers in the same way, by equating the normal flux density and tangential magnetic field at each boundary.

Once the coefficients are found, we find the eddy current density as

$$
J = -\sigma \frac{dA}{dt} = -jm\omega\sigma A \tag{2.250}
$$

The loss density is then

$$
q = \frac{JgJ^*}{2\sigma} = \frac{(m\omega)^2 AgA^*}{2} \tag{2.251}
$$

These expressions give the watts per cubic meter, but it is convenient to find the loss for a unit surface area in regions 1 and 2.

The surface loss density is

$$
q_1 = \int_{-d_1}^{0} q\,dy = \frac{(m\omega)^2 \sigma_1}{2} \int_{-d_1}^{0} A_1 g A_1^*\,dy \tag{2.252}
$$

$$
A_1 = \left(C_1^1 e^{a_1 y} + C_2^1 e^{-a_1 y} \right) e^{j(nkx+m\omega t)}
$$

Let

$$C_1^1 = |C_1^1|e^{j\angle C_1^1}, \ C_2^1 = |C_2^1|e^{j\angle C_2^1}, a_1 = a_{1r} + ja_{1i} \tag{2.253}$$

$$A_1 = \left(|C_1^1|e^{a_{1r}y}e^{j(\angle C_1^1 + a_{1i}y)} + |C_2^1|e^{-a_{1r}y}e^{j(\angle C_2^1 - a_{1i}y)}\right)e^{j(nkx + m\omega t)}$$

$$A_1 g A_1^* = |C_1^1|^2 e^{2a_{1r}y} + |C_2^1|^2 e^{-2a_{1r}y} + 2|C_1^1||C_2^1|\cos\left(\angle C_1^1 - \angle C_2^1 + 2a_{1i}y\right)$$

$$q_1 = \frac{(m\omega)^2 \sigma_1}{2}\left(|C_1^1|^2 \int_{-d_1}^{0} e^{2a_{1r}y}dy \right. \tag{2.254}$$

$$\left. + |C_2^1|^2 \int_{-d_1}^{0} e^{-2a_{1r}y}dy + 2|C_1^1||C_2^1|\int_{d_1}^{0}\cos\left(\angle C_1^1 - \angle C_2^1 + 2a_{1i}y\right)dy\right)$$

Performing the integration we get

$$q_1 = \frac{(m\omega)^2 \sigma_1}{4}\left(\frac{|C_1^1|^2\left(1 - e^{-2a_{1r}d_1}\right) + |C_2^1|^2\left(e^{2a_{1r}d_1} - 1\right)}{a_{1r}} + \right.$$

$$\left. \frac{2|C_1^1||C_2^1|\left(\sin\left(\angle C_1^1 - \angle C_2^1\right) - \sin\right)\left(\angle C_1^1 - \angle C_2^1\right) - 2a_{1i}d_1}{a_{1i}}\right) \tag{2.255}$$

For region 2, we use the same process

$$q_2 = \int_{-(d_1+d_2)}^{-d_1} qdy = \frac{(m\omega)^2 \sigma_2}{2}\int_{-(d_1+d_2)_1}^{-d_1} A_2 g A_2^* dy \tag{2.256}$$

$$A_2 = \left(C_1^2 e^{a_2 y} + C_2^2 e^{-a_2 y}\right)e^{j(nkx+m\omega t)}$$

Let

$$C_1^2 = |C_1^2|e^{j\angle C_1^2} \tag{2.257}$$

$$C_2^2 = |C_2^2|e^{j\angle C_2^2}$$

$$a_2 = a_{2r} + ja_{2i}$$

$$q_2 = \frac{(m\omega)^2 \sigma_2}{4}\left(\frac{|C_1^2|^2 e^{-2a_{2r}d_1}\left(1 - e^{-2a_{2r}d_1}\right) + |C_2^2|^2 e^{2a_{2r}d_1}\left(e^{2a_{2r}d_2} - 1\right)}{a_{2r}}\right.$$

$$\left. + \frac{2|C_1^2||C_2^2|\left(\sin\left(\angle C_1^2 - \angle C_2^1 - 2a_{2i}d_1\right) - \sin\right)\left(\angle C_1^2 - \angle C_2^2\right) - 2a_{2i}\left(d_1 + d_2\right)}{a_{1i}}\right) \tag{2.258}$$

This process has been applied to a problem with two conducting layers, but the same procedure can be used if there are more than two layer. In this case, the coefficient matrix will be larger to account for the extra interface conditions.

2.10 Thin Wire Carrying Current Above Conducting Plates

In many engineering applications, eddy current losses in metallic structures due to the proximity of current-carrying conductors must be evaluated. Where possible, these losses are reduced by, for example, laminating the cores in transformers and rotating machines. However, many regions of solid conducting material remain exposed to alternating fields. Among these are end iron structures of electrical machines, terminal equipment, transformer tanks, reinforcing rods, isolated phase bus bars, and others. Eddy currents induced in these structures also impose an additional burden on the source, resulting in reduced efficiency and higher temperatures.

It is therefore necessary to analyze this problem rigorously to obtain results accurate enough for engineering design. The work was pioneered by Poritsky and Jerrard [27], and later by Jain and Ray [28, 29]. Lawrenson et al. [30] carried out a two-dimensional traveling wave solution of the field problem in solid poles of electrical machines. Stoll and Hammond [31, 32] and Sen and Adkins [33] presented a solution with a semi-infinite slab with a single-valued permeability. Agarwal [19], McConnell [20, 34], Winchester [35], Lawrenson and coworker [36], and Stoll [26] presented a nonlinear solution based on a rectangular approximation of the $B - H$ curve (see Section 2.6).

In this section, a two-dimensional Cartesian analytical solution is presented using the Fourier integral method applicable to a variety of problems.

The following assumptions are made:

- The source field is a harmonic function of time.
- The properties of each material are constant and single-valued.
- Temperature effects on the material properties are neglected.
- The field problem is solved in two-dimensions, end effects are ignored.
- Currents are in the z direction and are not a function of z.

From Maxwell's equations we have

$$\nabla \times H = J \tag{2.259}$$

$$\nabla \times E = -\frac{\partial B}{\partial t} \tag{2.260}$$

$$\nabla \cdot B = 0 \tag{2.261}$$

and constitutive equations

$$J = \sigma E \tag{2.262}$$

$$B = \mu H \tag{2.263}$$

Assuming that the electric field, magnetic field, and current density are all time-varying steady-state sinusoidal quantities, we obtain from Equations (2.259)–(2.263)

$$\frac{\partial^2 B_x}{\partial x^2} + \frac{\partial^2 B_x}{\partial y^2} = j\omega\sigma\mu B_x \tag{2.264}$$

$$\frac{\partial^2 B_y}{\partial x^2} + \frac{\partial^2 B_y}{\partial y^2} = j\omega\sigma\mu B_y \tag{2.265}$$

In the nonconducting regions, Laplace's equation holds so

$$\frac{\partial^2 B_x}{\partial x^2} + \frac{\partial^2 B_x}{\partial y^2} = 0 \tag{2.266}$$

$$\frac{\partial^2 B_y}{\partial x^2} + \frac{\partial^2 B_y}{\partial y^2} = 0 \tag{2.267}$$

We solve Equations (2.264)–(2.267) by the separation of variables method and match the interface conditions at the boundaries.

2.10.1 A Semi-Infinite Slab Excited by a Filamentary Current

Referring to Figure 2.35, for region 1 ($h < y < \infty$) we have

$$B_{x1} = C_1 \cos mx \, e^{-my}$$
$$B_{y1} = -C_1 \sin mx \, e^{-my} \tag{2.268}$$

For region 2 ($0 < y < h$)

$$B_{x2} = \left(C_2 e^{my} + C_3 e^{-my}\right) \cos mx$$
$$B_{y2} = \left(C_2 e^{my} - C_3 e^{-my}\right) \sin mx \tag{2.269}$$

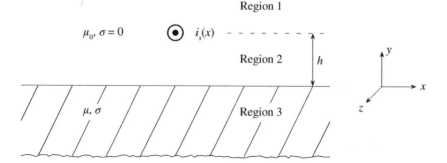

Figure 2.35 Semi-infinite slab with filament conductor.

For region 3 $(-\infty < y < 0)$

$$B_{x3} = C_4 e^{\beta y} \cos mx$$

$$B_{y3} = \frac{m}{\beta} C_4 e^{\beta y} \sin mx \tag{2.270}$$

where

$$\beta = \sqrt{m^2 + j\omega\mu\sigma} \tag{2.271}$$

Equating the tangential components of the magnetic field and the normal components of the flux density, respectively, we have for the interface between regions 1 and 2, at $y = h$

$$\frac{B_{x1} - B_{x2}}{\mu_0} = i_s(x) \tag{2.272}$$

$$B_{y1} = B_{y2} \tag{2.273}$$

Between regions 2 and 3 $(y = 0)$

$$H_{x2} = \frac{B_{x3}}{\mu_r} = H_{x3} \tag{2.274}$$

$$B_{y2} = B_{y3} \tag{2.275}$$

The point source of ac current can be expanded in a Fourier series and the solution can be found by superposition of the solutions of a series of harmonics. To avoid singularities, we will represent the source current as a surface current with a very small but finite width ξ.

$$i_s(x) = i_0 \cos(mx) = \begin{cases} J, & -\xi/2 > x > \xi/2 \\ 0, & \text{elsewhere} \end{cases} \tag{2.276}$$

The current filament can be expressed by the Fourier integral

$$i_s(x) = \int_0^\infty g(m) \cos(mx)\, dm \tag{2.277}$$

From Equation (2.277) above

$$g(m) = \frac{1}{\pi} \int_0^\infty i_s(x) \cos(mx)\, dx = \frac{1}{\pi} \int_{-\xi/2}^{\xi/2} J \cos(mx)\, dx = 2J \frac{\sin \frac{m\xi}{2}}{m\pi} \tag{2.278}$$

For small values of ξ, $g(m)$ approaches the value $\frac{J\xi}{\pi}$. Since J is the surface current density, $J\xi$ will be the total conductor current. Hence we have

$$i_0 = |i_s(x)| = \int_0^\infty \frac{1}{\pi}\, dm \tag{2.279}$$

Using the interface conditions from Equations (2.272)–(2.275) and substituting for $i_s(x)$ from Equation (2.279) and after considerable algebra, we find the flux density in the different regions as

Region 1

$$B_{x1} = \frac{\mu_0 I}{\pi} \int_0^\infty \left(\sinh(mh)e^{-my} + \frac{me^{-m(h+y)}}{m + \frac{\beta}{\mu_r}} \right) \cos(mx) \, dm$$

$$B_{y1} = \frac{\mu_0 I}{\pi} \int_0^\infty \left(\sinh(mh)e^{-my} + \frac{me^{m(h+y)}}{m + \frac{\beta}{\mu_r}} \right) \sin(mx) \, dm \qquad (2.280)$$

Region 2

$$B_{x2} = \frac{-\mu_0 I}{\pi} \int_0^\infty \left(\cosh(my)e^{-my} - \frac{me^{-m(h+y)}}{m + \frac{\beta}{\mu_r}} \right) \cos(mx) \, dm$$

$$B_{y2} = \frac{-\mu_0 I}{\pi} \int_0^\infty \left(\sinh(mh)e^{my} + \frac{me^{m(h+y)}}{m + \frac{\beta}{\mu_r}} \right) \sin(mx) \, dm \qquad (2.281)$$

Region 3

$$B_{x3} = \frac{-\mu_0 I}{\pi} \int_0^\infty \frac{\beta e^{-mh} e^{\beta y}}{m + \frac{\beta}{\mu_r}} \cos(mx) \, dm$$

$$B_{y3} = \frac{-\mu_0 I}{\pi} \int_0^\infty \frac{\beta e^{-mh} e^{\beta y}}{m + \frac{\beta}{\mu_r}} \sin(mx) \, dm \qquad (2.282)$$

The electric field in the different regions are obtained from Maxwell's equation for the two-dimensional case

$$\nabla \times E = \frac{\partial E}{\partial y}\hat{u}_x - \frac{\partial E}{\partial x}\hat{u}_y = -\frac{\partial B}{\partial t} = -j\omega B = -j\omega B_x \hat{u}_x - j\omega B_y \hat{u}_y \qquad (2.283)$$

Equating the Cartesian components and integrating we obtain

$$E = \int -j\omega B_x \, dy - \int j\omega B_y \, dx \qquad (2.284)$$

Substituting for B_x and B_y for the regions from Equations (2.280)–(2.282) into Equation (2.284)

$$E_1 = \frac{j\omega \mu_0 I}{\pi} \int_0^\infty \left(\frac{\sinh(mh)e^{-my}}{m} + \frac{e^{-m(h+y)}}{m + \frac{\beta}{\mu_r}} \right) \cos(mx) \, dm \qquad (2.285)$$

$$E_2 = \frac{j\omega\mu_0 I}{\pi} \int_0^\infty \left(\frac{\sinh(my)e^{-mh}}{m} + \frac{e^{-m(h+y)}}{m + \frac{\beta}{\mu_r}} \right) \cos(mx)\, dm \qquad (2.286)$$

$$E_3 = \frac{j\omega\mu_0 I}{\pi} \int_0^\infty \left(\frac{e^{-mh}e^{\beta y}}{m + \frac{\beta}{\mu_r}} \right) \cos(mx)\, dm \qquad (2.287)$$

As an example, we show a finite element solution to the problem with a small circular wire carrying 1.0 *A* at 60 Hz above a thick copper plate with $\sigma = 5.8 \times 10^7 \, \mathrm{S\,m^{-1}}$ and $\mu = 4\pi \times 10^{-7} \, \mathrm{H\,m^{-1}}$.

Figure 2.36 shows the imaginary part of the flux distribution. The applied current is $1.0 + j0.0 \, A$, so that the current in the wire is zero at this instant in the cycle. All of the flux is then due to the eddy currents in the plate. We can see in the figure that there appears to be a source in the copper plate directly below the filamentary wire. In Figure 2.37, we see the real part of the flux distribution. This is the component that is in-phase with the excitation current. We note that the flux does not penetrate very deeply into the plate.

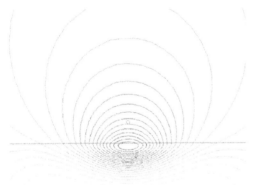

Figure 2.36 Imaginary component of the flux.

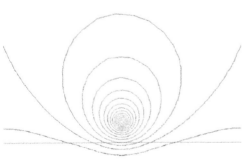

Figure 2.37 Real component of the flux.

Figure 2.38 Losses in the plate.

In Figure 2.38, we see the loss density in the plate. As expected, the losses are highest just below the current-carrying wire. The losses drop off quickly as we go deep into the plate and as we move in the horizontal direction. This behavior agrees with the formulation.

2.10.2 A Finite Slab Excited by a Filamentary Current

A conducting slab of finite thickness is shown in Figure 2.39. The equations for the magnetic flux density in the four regions are

Region 1 $(h < y < \infty)$

$$B_{x1} = C_1 e^{-my} \cos mx$$
$$B_{y1} = -C_1 e^{-my} \sin mx \quad (2.288)$$

Region 2 $(0 < y < h)$

$$B_{x2} = \left(C_2 e^{my} + D_2 e^{-my}\right) \cos mx$$
$$B_{y2} = \left(C_2 e^{my} - D_2 e^{-my}\right) \sin mx \quad (2.289)$$

Region 3 $(-t < y < 0)$

$$B_{x3} = \left(C_3 e^{\beta y} + D_3 e^{-\beta y}\right) \cos mx$$
$$B_{y3} = \frac{m}{\beta} \left(C_3 e^{\beta y} - D_3 e^{-\beta y}\right) \sin mx \quad (2.290)$$

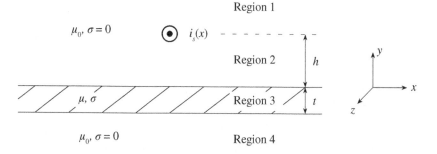

Figure 2.39 Finite thickness slab with filament conductor.

Region 4 $(-\infty < y < -t)$

$$B_{x4} = C_4 e^{my} \cos mx$$

$$B_{y4} = C_4 e^{my} \sin mx \qquad (2.291)$$

The following interface conditions apply: At the interface between regions 1 and 2 $(y = h)$

$$\frac{B_{x1} - B_{x2}}{\mu_0} = i_0 \cos mx$$

$$B_{y1} = B_{y2} \qquad (2.292)$$

At the interface between regions 2 and 3 $(y = 0)$

$$B_{x2} = \frac{B_{x3}}{\mu_r}$$

$$B_{y2} = B_{y3} \qquad (2.293)$$

At the interface between regions 3 and 4 $(y = -t)$

$$\frac{B_{x3}}{\mu_r} = B_{x4}$$

$$B_{y3} = B_{y4} \qquad (2.294)$$

Solving Equations (2.288)–(2.294) and substituting the expression for $i_s(x)$

$$B_{x1} = \frac{\mu_0 I}{\pi} \int_0^\infty \left(m e^{-m(y+h)} \frac{S_4 e^{-2t\beta} - S_3}{S_3^2 - S_4^2 e^{-2t\beta}} - e^{-my} \sinh(mh) \right) \cos(mx)\, dm \qquad (2.295)$$

$$B_{y1} = -\frac{\mu_0 I}{\pi} \int_0^\infty \left(m e^{-m(y+h)} \frac{S_4 e^{-2t\beta} - S_3}{S_3^2 - S_4^2 e^{-2t\beta}} - e^{-my} \sinh(mh) \right) \sin(mx)\, dm \qquad (2.296)$$

$$E_1 = \frac{-j\omega\mu_0 I}{\pi} \int_0^\infty \left(e^{-m(y+h)} \frac{S_4 e^{-2t\beta} - S_3}{S_3^2 - S_4^2 e^{-2t\beta}} - e^{-my} \frac{\sinh(mh)}{m} \right) \cos(mx)\, dm \qquad (2.297)$$

Where

$$S_3 = m + \frac{\beta_3}{\mu_{r3}}$$

$$S_4 = m - \frac{\beta_3}{\mu_{r3}} \qquad (2.298)$$

2.10.3 A Slab of Finite Thickness Near a Semi-Infinite Slab Excited by a Filamentary Current

In several practical cases, a conducting shield is inserted between the source and another conducting body to shield that body from stray fields and produce an

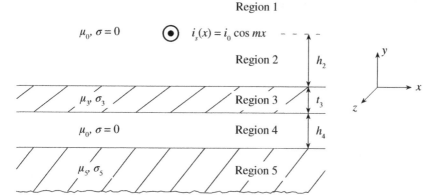

Figure 2.40 Finite thickness shield in front of a semi-infinite slab excited by a filamentary current.

overall reduction of loss in the protected object. We see in Figure 2.40 a filament current parallel to a shielding plate and a semi-infinite body below.

Using the same process described above, we have for the various regions

Region 1: $(h_2 < y < \infty)$

$$B_{x1} = D_1 e^{-my} \cos mx$$
$$B_{y1} = -D_1 e^{-my} \sin mx \tag{2.299}$$

Region 2: $(0 < y < h)$

$$B_{x2} = \left(C_2 e^{my} + D_2 e^{-my}\right) \cos mx$$
$$B_{y2} = \left(C_2 e^{my} - D_2 e^{-my}\right) \sin mx \tag{2.300}$$

Region 3: $(-t_3 < y < 0)$

$$B_{x3} = \left(C_3 e^{\beta_3 y} + D_3 e^{-\beta_3 y}\right) \cos mx$$
$$B_{y3} = \frac{m}{\beta_3} \left(C_3 e^{\beta_3 y} - D_3 e^{-\beta_3 y}\right) \sin mx \tag{2.301}$$

Region 4: $(-(t_3 + h_4) < y < -t_3)$

$$B_{x4} = \left(C_4 e^{my} + D_4 e^{-my}\right) \cos mx$$
$$B_{y4} = \left(C_4 e^{my} - D_4 e^{-my}\right) \sin mx \tag{2.302}$$

Region 5: $(-\infty < y < -(t_3 + h_4))$

$$B_{x5} = C_5 e^{-\beta_5 y} \cos mx$$
$$B_{y5} = \frac{m}{\beta_5} C_5 e^{-\beta_5 y} \sin mx \tag{2.303}$$

The integration constants in Equations (2.299)–(2.303) are evaluated using the interface conditions below.

Regions 1–2 interface ($y = h_2$)

$$i_0 = \frac{B_{x1} - B_{x2}}{\mu_0} = \int_0^\infty \frac{I}{\pi}\, dm$$

$$B_{y1} = B_{y2} \tag{2.304}$$

Regions 2–3 interface ($y = 0$)

$$B_{x2} = \frac{B_{x3}}{\mu_{r3}}$$

$$B_{y2} = B_{y3} \tag{2.305}$$

Regions 3–4 interface ($y = -t$)

$$B_{x4} = \frac{B_{x3}}{\mu_{r3}}$$

$$B_{y3} = B_{y4} \tag{2.306}$$

Regions 4–5 interface ($y = -(t_3 + h_4)$)

$$B_{x4} = \frac{B_{x5}}{\mu_{r5}}$$

$$B_{y4} = B_{y5} \tag{2.307}$$

We shall define two expressions in addition to S_3 and S_4 of Equation (2.298).

$$S_5 = m + \frac{\beta_5}{\mu_{r5}}$$

$$S_6 = m - \frac{\beta_5}{\mu_{r5}} \tag{2.308}$$

The field quantities in terms of S_3, S_4, S_5, and S_6 are

$$B_{x1} = \frac{\mu_0 I}{\pi} \int_0^\infty e^{-my} \left(\sinh(mh_2) \right. \tag{2.309}$$

$$\left. + \frac{me^{-mh_2} \left(S_5 e^{mh_4}(S_3 - S_4 e^{-2\beta_3 t_3}) - S_6 e^{-mh_4}(S_4 - S_3 e^{-2\beta_3 t_3}) \right)}{S_5 e^{mh_4}(S_3^2 - S_4^2 e^{-2\beta_3 t_3}) - S_3 S_4 S_6 e^{-mh_4}(1 - e^{-2\beta_3 t_3})} \right) \cos mx\, dm$$

$$B_{y1} = \frac{-\mu_0 I}{\pi} \int_0^\infty e^{-my} \left(\sinh(mh_2) \right. \tag{2.310}$$

$$\left. + \frac{me^{-mh_2} \left(S_5 e^{mh_4}(S_3 - S_4 e^{-2\beta_3 t_3}) - S_6 e^{-mh_4}(S_4 - S_3 e^{-2\beta_3 t_3}) \right)}{S_5 e^{mh_4}(S_3^2 - S_4^2 e^{-2\beta_3 t_3}) - S_3 S_4 S_6 e^{-mh_4}(1 - e^{-2\beta_3 t_3})} \right) \sin mx\, dm$$

$$E_1 = \frac{j\omega\mu_0 I}{\pi} \int_0^\infty \left(\frac{e^{-my}\sinh(mh_2)}{m} \right.$$

$$\left. + \frac{e^{-m(y+h_2)}\left(S_5 e^{mh_4}(S_3 - S_4 e^{-2\beta_3 l_3}) - S_6 e^{-mh_4}(S_4 - S_3 e^{-2\beta_3 l_3})\right)}{S_5 e^{mh_4}(S_3^2 - S_4^2 e^{-2\beta_3 l_3}) - S_3 S_4 S_6 e^{-mh_4}(1 - e^{-2\beta_3 l_3})} \right) \cos mx \, dm \quad (2.311)$$

There are in general two ways in which surface ohmic losses due to the eddy currents can be determined. The first is to determine the current density vector and then obtain the losses by evaluating the volume integral

$$W = \int \frac{J \cdot J^* \rho}{2} \, dv \quad (2.312)$$

where J^* is the complex conjugate of the current density J.

The second method is to find the source impedance and then evaluate the loss as the product of the resistive part of the source impedance and the square of the RMS source current. This second method is more convenient as it eliminates the volume integral and the computation of the conjugate of the current density. The expression for the source impedance is the ratio of the surface value of the electric field to the source current.

So

$$Z = \frac{E(0,h)}{I} \quad (2.313)$$

The following are the expressions for the surface impedance for the different cases we have discussed.

For the semi-infinite conducting slab, we use Equation (2.285) to find the surface impedance as

$$Z_1 = \frac{E_1(0,y)}{I} = \frac{-j\omega\mu_0}{\pi} \int_0^\infty \left(\frac{\sinh(mh)e^{-my}}{m} + \frac{e^{-m(h+y)}}{m + \frac{\beta}{\mu_r}} \right) dm \quad (2.314)$$

The first term in the definite integral of Equation (2.314) can be written as

$$\int_0^\infty -\frac{e^{-m(y-h)} - e^{-m(y+h)}}{2m} \, dm \quad (2.315)$$

which can be integrated to find

$$-\frac{1}{2}\ln\left(\frac{y-h}{y+h}\right) \quad (2.316)$$

If $y = h$, the definite integral will tend to infinity. To overcome this difficulty, we will assume the conductor to be very small with a finite radius r so that $y - h = r$ and $y + h \approx 2h$. Therefore

$$Z_1 = \frac{-j\omega\mu_0}{\pi} \left(-\frac{1}{2}\ln\frac{r}{2h} + \int_0^\infty \frac{e^{-2mh}}{m + \frac{\beta}{\mu_r}} \, dm \right) \quad (2.317)$$

When the slab is of finite thickness we obtain the impedance

$$Z_1 = \frac{E(0,h)}{I} = \frac{j\omega\mu_0}{\pi}\left(\frac{-1}{2}\ln\frac{r}{2h} + \int_0^\infty \frac{e^{-2mh}\left(S_4 e^{-2\beta t} - S_3\right)}{S_3^2 - S_4^2 e^{-2\beta t}}\, dm\right) \quad (2.318)$$

For the case of the finite thickness conducting shield interposed between the source and the semi-infinite conducting slab, we expand Equation (2.312)

$$Z_1 = \frac{j\omega\mu_0}{\pi}\left(\frac{1}{2}\ln\frac{r}{2h}\right. \quad (2.319)$$

$$+ \int_0^\infty e^{-2mh_2}\left(S_5 e^{mh_4}(S_3 - S_4 e^{-2\beta_3 t_3})\right.$$

$$\left.\left. - \frac{S_6 e^{-mh_4}(S_4 - S_3 e^{-2\beta_3 t_3})}{S_5 e^{mh_4}(S_3^2 - S_4^2 e^{-2\beta_3 t_3}) - S_3 S_4 S_6 e^{-mh_4}(1 - e^{-2\beta_3 t_3})}\right) dm\right)$$

In the above expressions, if $\mu_r = 1$, $\sigma_5 = 0$, and $h_4 = 0$ we have

$$\mu_{r5}\mu_0\sigma_5 = 0 \quad (2.320)$$

$$S_6 = m - \frac{\beta_5}{\mu_{r5}} = 0$$

Substituting Equation (2.321) into (2.320) we obtain

$$Z_1 = \frac{E(0,h)}{I} = \frac{j\omega\mu_0}{\pi}\left(\frac{1}{2}\ln\frac{r}{2h} + \int_0^\infty \frac{e^{-2mh}\left(S_4 e^{-2\beta t} - S_3\right)}{S_3^2 - S_4^2 e^{-2\beta t}}\, dm\right) \quad (2.321)$$

Which, as a check, is the same as Equation (2.318).

For these cases, the power loss per unit length of source current is

$$W = I^2\Re e(Z_1) \quad (2.322)$$

We can find a condition for minimum loss. The minimum power loss (for a nontrivial case) occurs at a specific value of the shield thickness, this thickness being independent of the distance of the shield from the source current. Thus for every shield type, magnetic or nonmagnetic, this minimum power loss occurs at a unique value of the thickness of the slab, or more precisely, at a definite value of the thickness-to-skin depth ratio. We demonstrate this by considering the loss equation for a single finite-thickness slab.

The power loss is

$$P = I^2\Re e\left(\frac{j\omega\mu_0}{\pi}\left(\frac{1}{2}\ln\frac{r}{2h} + \int_0^\infty \frac{e^{-2mh}\left(S_4 e^{-2\beta t} - S_3\right)}{S_3^2 - S_4^2 e^{-2\beta t}}\, dm\right)\right) \quad (2.323)$$

Since for a given conducting shield type, the losses are only a function of the thickness as seen in Equation (2.318), we can obtain the minimum loss condition by differentiating Equation (2.323) with respect to the thickness t, assuming $m = 0$.

This last assumption means that for each value of m, the loss contribution is a minimum.

Substituting $(1+j)/\delta$ for $\sqrt{j\omega\mu\sigma}$ in the expressions for S_3 and S_4, and setting $m = 0$, we obtain

$$S_3 = m + \frac{\beta}{\mu_r} = \frac{1+j}{\delta} \tag{2.324}$$

$$S_4 = m - \frac{\beta}{\mu_r} = -\frac{1+j}{\delta} \tag{2.325}$$

If we now express

$$e^{-2\beta t} = e^{-\alpha(1+j)} \tag{2.326}$$

where $\alpha = 2t/\delta$, then

$$P = I^2 \Re e \left(\frac{j\omega\mu_0}{\pi} \left(\frac{-1}{2} \ln \frac{r}{2h} + \int_0^\infty \frac{\frac{1+j}{\delta} + \frac{1+j}{\delta} e^{-\alpha(1+j)}}{\frac{(1+j)^2}{\delta} - \frac{(1+j)^2}{\delta} e^{-\alpha(1+j)}} \, dm \right) \right) \tag{2.327}$$

Expanding $e^{-j\alpha}$ as $\cos\alpha - j\sin\alpha$ the expression for power loss becomes

$$P = I^2 \Re e \left(\frac{j\omega\mu_0}{\pi} \left(\frac{-1}{2} \ln \frac{r}{2h} + \int_0^\infty \frac{\mu_r \delta}{(1+j)} \frac{1 + e^{-\alpha} \cos\alpha - j e^{-\alpha} \sin\alpha}{1 - e^{-\alpha} \cos\alpha + j e^{-\alpha} \sin\alpha} \, dm \right) \right)$$

$$= \frac{-I^2 \omega\mu_0}{\pi} \int_0^\infty \frac{e^{2\alpha} - 2\sin\alpha e^{-\alpha} - 1}{1 + e^{-2\alpha} - 2e^{-\alpha}\cos\alpha} \, dm \tag{2.328}$$

Note that Equations (2.327) and (2.328), based on the assumption that $m = 0$, is valid for all plate thickness except zero thickness. Hence $\alpha = 2t/\delta$ can not be zero.

Differentiating Equation (2.328) with respect to α and setting the result to zero

$$\frac{dP}{d\alpha} = \frac{d}{d\alpha} \left(\frac{e^{2\alpha} - 2\sin\alpha e^{-\alpha}}{1 + e^{-2\alpha} - 2e^{-\alpha}\cos\alpha} \right) = 0 \tag{2.329}$$

Which gives

$$e^{-\alpha} \sin(1 - e^{-2\alpha}) = 0 \tag{2.330}$$

For nontrivial solutions, we consider the values of α which yield $\sin\alpha = 0$ so that $\alpha = \pi, 2\pi, 3\pi, \dots, n\pi$.

Assuming $\alpha = \pi$ we obtain

$$t = \frac{\alpha\delta}{2} = \frac{\pi\delta}{2} \tag{2.331}$$

If we had substituted a value of m not equal to zero, it can be shown that the differential of P with respect to α yields the same solution for t. Hence the assumptions of $m = 0$ and $\alpha \neq 0$ do not affect the generality of the solution.

For a copper shield at 60 HZ and $\sigma = 5.8 \times 10^7$ Sm^{-1}, we have $\delta = 0.0085$ m, and the theoretical minimum loss occurs at a value of 0.0134 m.

The solution for the power loss as well as the electric and magnetic fields in the different regions involves the evaluation of infinite integrals. Poritsky and Jerrard [27] suggested closed-form solutions, by separating the integrals by parts and dealing with the individual components by means of summation of infinite series. In this analysis, much care had to be taken to avoid singularities in the kernels of the definite integrals making the analysis cumbersome. We suggest that numerical integration using quadrature formulae be used to evaluate these terms. The evaluation of these integrals has been done with a three-point Newton–Coates formula commonly known as Simpson's rule. In each subdivision, the quadrature formula is used to evaluate the function, and this value is then summed over the total number of subdivisions. The final result is multiplied by the interval length.

For a given interval, the integrated value of any function is

$$f(O) = \frac{f_1 + 4f_2 + f_3}{3} \times h \tag{2.332}$$

where h is the interval length. If the total number of intervals is n the value of the integral is given by

$$S = \frac{h}{3} \left(f_1 + 4f_2 + 2f_3 + 4f_4 + \cdots + f_n \right) \tag{2.333}$$

where $f_1, f_2, f_3, \ldots f_n$ are the function values at the ends of the respective intervals starting with zero. With 3000 intervals, convergence of the definite integral has been found to be within 1% of the analytical solution. As an example, we consider the solution for the semi-infinite slab of permeable steel with a filament current at height 10 cm. The material had properties $\rho = 80 \times 10^{-8}$ Ω m and $\mu_r = 100$. The surface resistance found was $R = 6.46474 \times 10^{-5}$ $\Omega\,\mathrm{m}^{-1}$. Poritsky's analytical solution gives $R = 6.433 \times 10^{-5}$ $\Omega\,\mathrm{m}^{-1}$. The difference is 0.497%.

2.11 Eddy Currents in Materials with Anisotropic Permeability

A problem that is common in electrical equipment such as power transformers, motors, and generators, is that there is a component of magnetic flux that impinges on the laminations on the wide surface. We have discussed eddy currents produced by ac flux which is parallel to the surface of the lamination (flux is normal to the small dimension) and we saw that for cases in which δ, the skin depth, is greater than the lamination thickness we could use the resistance-limited formula. For cases where δ is equivalent to or larger than the width, we would need the reactance-limited analysis. These formulations do not apply in cases that have the flux normal to the stack on laminations since the effective permeability in the direction of the flux (across laminations) is much lower than the permeability along the laminations. This is due to the "air" layers in series with the flux. A typical stacking factor is 0.95, which means that for the axial dimension of the stack, we have 95% steel and 5% space. This reduces the axial permeability substantially as the large reluctance of the air is in series with the reluctance of the steel. Typical permeabilities in that direction are in the range of 10 – 20 times μ_0. The effective permeability along the laminations is only slightly lowered due to the stacking, as the air reluctance is in parallel with the steel reluctance. The analysis which follows allows for the anisotropy of the permeability.

The formulation is based on the following assumptions.

- The eddy currents are invariant in the z (radial) direction. See Figure 2.41.
- The permeabilities $\mu_x \neq \mu_y$, but the conductivities $\sigma_x = \sigma_y$.
- Edge effects are ignored.
- Each tooth is a semi-infinite region in the z direction.
- The flux density is decomposed into its components and the losses are found as the sum of the contributions of each component.

Considering the geometry in Figure 2.41, we can write Maxwell's equations as

$$\nabla \times H = J \tag{2.334}$$

$$\nabla \times E = -\frac{\partial B}{\partial t} \tag{2.335}$$

$$\nabla \cdot B = 0 \tag{2.336}$$

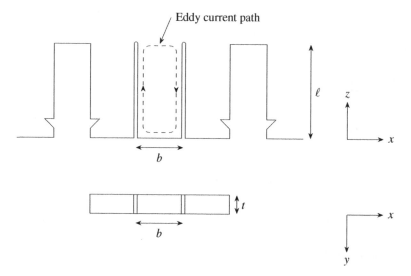

Figure 2.41 End region laminations: geometry and coordinate system.

and

$$J = \sigma E \tag{2.337}$$

Assuming all of the field quantities are sinusoidally time varying, we obtain

$$\nabla^2 B = j\omega\sigma\mu B \tag{2.338}$$

We will allow for the case in which $\mu_x \neq \mu_y$.
In Cartesian coordinates

$$\mu_x \frac{\partial H_x}{\partial x} + \mu_y \frac{\partial H_y}{\partial y} = 0 \tag{2.339}$$

$$\frac{\partial H_y}{\partial x} - \frac{\partial H_x}{\partial y} = J \tag{2.340}$$

$$\frac{\partial E_z}{\partial y} = -j\omega B_x \tag{2.341}$$

$$\frac{\partial E_z}{\partial x} = -j\omega B_y \tag{2.342}$$

After some manipulation we have

$$\frac{1}{\alpha^2} \left\{ \frac{\partial^2 B_x}{\partial x^2} + \frac{\partial^2 B_x}{\partial y^2} \right\} = j\omega\sigma\mu_x B_x \tag{2.343}$$

$$\frac{1}{\alpha^2} \left\{ \frac{\partial^2 B_y}{\partial x^2} + \frac{\partial^2 B_y}{\partial y^2} \right\} = j\omega\sigma\mu_x B_y \tag{2.344}$$

where $\alpha^2 = \frac{\mu_y}{\mu_x}$, $k_x = \mu_x\sigma$ and $k_y = \mu_y\sigma$.
The solution can be written as a summation

$$B_x = \sum_{n=1,3,5,\dots}^{\infty} C\sin(nmx)e^{-\beta y} \tag{2.345}$$

$$B_y = \sum_{n=1,3,5,\dots}^{\infty} \frac{nmC}{\beta} \cos(nmx)e^{-\beta y} \tag{2.346}$$

Here b is the width of the lamination, $m = \pi/b$, and $\beta = \sqrt{n^2 m^2 + j\omega\mu_y\mu_0\sigma}$.
At $y = 0$ we have $B_y = B_0$, so we can find the integration constant C as

$$\int_{-b/2}^{b/2} \frac{nmC}{\beta}\cos^2(nmx)\,dx = \int_{-b/2}^{b/2} B_0\cos(nmx)\,dx \tag{2.347}$$

The value of C is then

$$C = \left(\frac{4B_0\beta}{n^2 m^2 b}\right)\sin\left(\frac{n\pi}{2}\right) \tag{2.348}$$

Using this value of C, we get

$$B_x = \sum_{n=1,3,5,\ldots} \frac{4B_0\beta}{n^2m^2b} \sin(\frac{n\pi}{2}) \sin(nmx)e^{-\beta y} \tag{2.349}$$

$$B_y = \sum_{n=1,3,5,\ldots} \frac{4B_0}{nmb} \sin(\frac{n\pi}{2}) \cos(nmx)e^{-\beta y} \tag{2.350}$$

As we see, the axial flux varies exponentially. We now obtain the expression for the eddy current density from the relation

$$\nabla \times H = J = \frac{1}{\mu_0} \left(\frac{1}{\mu_y} \frac{\partial B_y}{\partial x} - \frac{1}{\mu_x} \frac{\partial B_x}{\partial y} \right) \tag{2.351}$$

This gives

$$J = -\frac{1}{\mu_y\mu_0} \sum_{n=1,3,5,\ldots} \frac{4B_0}{w} \sin \frac{n\pi}{2} \left(1 - \frac{\beta^2\alpha^2}{n^2m^2} \right) \sin(nmx)e^{-\beta y} \tag{2.352}$$

The power loss can now be found as

$$P = \iiint \frac{J \cdot J^*}{2\sigma} \, dx \, dy \, dz \tag{2.353}$$

$$P = \frac{b\ell}{\mu_0^2\mu_y^2} \left(\frac{8B_0^2}{b^2\sigma} \right) \sum_{n=1,3,5,\ldots}^{\infty} \sin^2 \frac{n\pi}{2} \left(1 - \frac{\beta^2\alpha^2}{n^2m^2} \right) \left(1 - \frac{\beta^{*2}\alpha^2}{n^2m^2} \right)$$
$$\times \int_{\frac{-b}{2}}^{\frac{b}{2}} \sin^2(nmx) \, dx \int_0^t e^{-(\beta+\beta^*)} \, dy \tag{2.354}$$

or

$$P = \frac{b\ell}{\mu_0^2\mu_y^2} \left(\frac{8B_0^2}{b^2\sigma} \right) \sum_{n=1,3,5,\ldots}^{\infty} \sin^2 \frac{n\pi}{2} \left(1 - \frac{\beta^2\alpha^2}{n^2m^2} \right)$$
$$\times \left(1 - \frac{\beta^{*2}\alpha^2}{n^2m^2} \right) \left(\frac{1}{\beta+\beta^*} \right) \left(1 - e^{-(\beta+\beta^*)t} \right) \tag{2.355}$$

We have now extended the solution of eddy currents in laminations to include the effect of anisotropic permeability. This formulation is useful in cases in which magnetic flux crosses laminations and the small gaps between them.

2.12 Isolated Rectangular Conductor with Axial Current Applied

The solution for the current density in a long-isolated rectangular conductor with axial field was first solved by Press [37]. If we consider the case of a long isolated rectangular conductor, with an applied axial ac current, we can obtain a

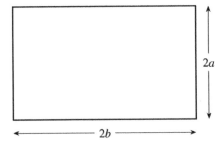

Figure 2.42 Rectangular conductor with ac axial current applied.

$2a$

$2b$

two-dimensional solution for the current distribution using an infinite series. In Section 2.5, we have seen the one-dimensional distribution for the reactance limited case. This is now generalized to two-dimensions and the effect of the eddy currents on the current density are included. The rectangular conductor is shown in Figure 2.42. The conductor has dimensions $2b \times 2a$. The current density has only a z component and the current density can vary in magnitude and phase in the x and y directions.

The current density is described by the equation

$$\frac{\partial^2 J_z}{\partial x^2} + \frac{\partial^2 J_z}{\partial y^2} = \beta^2 J_z \tag{2.356}$$

where

$$\beta = \frac{1+j}{\delta} \tag{2.357}$$

The solution to Equation (2.356) will be written as an infinite series. By symmetry, we can deduce that only even terms are possible solutions. In other words, the solution at points $\pm x$ or $\pm y$ must be the same. In this case, with the origin of the coordinate system at the center of the conductor, the summation must be only over even functions like the cosine function.

We will make the assumption that due to the isolated nature of this conductor, the magnetic field follows the contour of the boundary and the current density at the surface is uniform around the conductor perimeter. The assumption that the boundary of the conductor is a flux line is an approximation. Near the center lines of the rectangle, the flux is nearly parallel to the conductor boundary, but at the corners, the flux lines do not follow the contour of the conductor. Using the assumption that the contour of the conductor is a flux line, that the current density is constant along that boundary, can be seen by considering the magnetic vector potential. If the contour is a flux line, then it is a contour of equi-magnetic vector potential. This comes from the curl relationship between the flux density and the vector potential. A constant vector potential contour means that the derivative is zero as we move along the contour. Using the curl relationship, there is no flux perpendicular to this path, therefore all flux is along the path. We now use the relationship that the current density is found by $J = j\omega\sigma A$, and can see that if A is constant, then the current density is constant as well.

We then have

$$J(x,y) = \frac{4}{\pi}J_0 \sum_{n=1,3,5...}^{\infty} \frac{j^{n+1}}{n}\left(\left(\frac{\cosh\sqrt{\beta^2 + \frac{n^2\pi^2}{4a^2}}x}{\cosh\sqrt{\beta^2 + \frac{n^2\pi^2}{4a^2}}b}\right)\cos\frac{n\pi}{2a}y\right.$$

$$\left. + \left(\frac{\cosh\sqrt{\beta^2 + \frac{n^2\pi^2}{4a^2}}y}{\cosh\sqrt{\beta^2 + \frac{n^2\pi^2}{4a^2}}a}\right)\cos\frac{n\pi}{2b}x\right)\sin(\omega t) \quad (2.358)$$

In Figure 2.43, we see an example of Equation (2.358) evaluated for a copper conductor with $a = 0.01$ m and $b = 0.02$ m excited with an axial current of $J_0 = 1.0$ Am^{-2}. In the figure, we see the magnitude of the current density plotted for the upper right quadrant of the conductor. We note, as expected, the exponential-like decay of the current density as we move toward the center of the conductor.

As a further example, let us consider a case in which the conductor is very tall with respect to the width. In this case, we have $a = 0.1$ m, $b = 0.01$ m, and copper at 60 HZ, with $J_0 = 1.0$ Am^{-2}. We would expect that on the axes of symmetry, we would get a result similar to the one-dimensional solutions. In Figures 2.44 and 2.45, we see the solutions for the magnitude of the current density along the

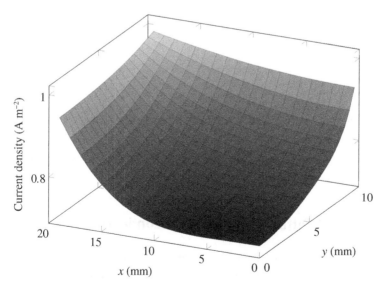

Figure 2.43 Magnitude of the current density using copper at 60 HZ for $a = 0.01$ m, $b = 0.02$ m.

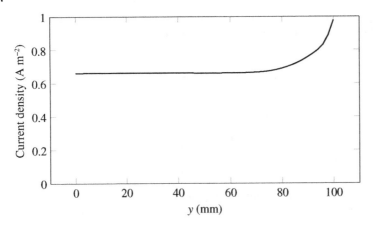

Figure 2.44 Magnitude of the current density along the *y* axis for copper at 60 HZ for the long narrow conductor.

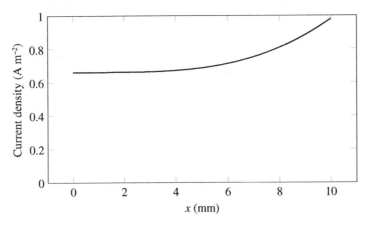

Figure 2.45 Magnitude of the current density along the *x* axis for copper at 60 HZ for the long narrow conductor.

x and *y* directions. These solutions are very much like those we have seen for the isolated plate in one-dimension.

2.13 Transient Diffusion Into a Solid Conducting Block

We will now consider problems that include transient or time-domain diffusion of flux and current into a conductor. Consider the problem described in Figure 2.46.

We have a uniform, homogeneous, linear conductor which extends far in the *y* and *z* directions. By symmetry then, the magnetic field *H*, will have only a *y*

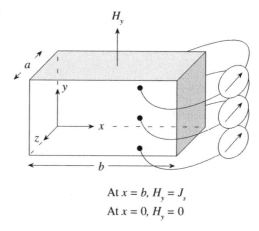

Figure 2.46 Uniform conducting block with step function of current applied to terminals.

At $x = b$, $H_y = J_s$

At $x = 0$, $H_y = 0$

component and the current density J, will have only a z component. We will assume that there is no variation of the field or current density in the y or z directions. These variables then are only a function of x and t. A current sheet source is connected to the conducting block in the (x, y) plane by means of highly conducting plates, so we can assume an equi-potential surface in the back and front of the block.

This problem is described by the one-dimensional diffusion equation in rectangular coordinates. We have

$$\frac{\partial^2 H}{\partial x^2} = \mu\sigma \frac{\partial H}{\partial t} \tag{2.359}$$

From Ampere's law, we can write

$$J_z = -\frac{\partial H_y}{\partial x} \tag{2.360}$$

Using the separation of variables technique, (see Appendix B), we can assume that the solution can be expressed as the product of two functions, one a function of x and one a function of time.

$$H(x, t) = X(x)T(t) \tag{2.361}$$

Since each term is a function of only one variable, then each term must be a constant.

Let

$$\frac{1}{X}\frac{\partial^2 X}{\partial x^2} = -\gamma^2 \tag{2.362}$$

or

$$\frac{\partial^2 X}{\partial x^2} + \gamma^2 X = 0 \tag{2.363}$$

and

$$-\frac{\mu\sigma}{T}\frac{\partial T}{\partial t} = \gamma^2 \tag{2.364}$$

or

$$\frac{\partial T}{\partial t} - \frac{\gamma^2}{\mu\sigma}T = 0 \tag{2.365}$$

We now consider the boundary conditions. We can find constraints at $x = 0$ and $x = b$. At $x = 0$, we must have $H_y = J_s$. At $x = b$, we must have $H_y = 0$.

The solution is composed of a particular part, which at steady state or $t = \infty$, is a constant, and a transient part or the homogeneous solution.

For the steady-state solution, we have $\frac{d}{dt} = 0$, so that

$$\frac{\partial^2 H_y}{\partial x^2} = 0 \tag{2.366}$$

The steady-state solution is then

$$H_p = -J_s\frac{x}{b} \tag{2.367}$$

Using the boundary conditions at the edges, we can state that the solution is of the form

$$X = \sin\beta x \tag{2.368}$$

where $\beta = \frac{n\pi}{b}$.

The time-dependent part of the solution is an exponential with time constant

$$\tau_n = \frac{\mu\sigma b^2}{(n\pi)^2} \tag{2.369}$$

$$H = \sum_{n=1}^{\infty} C_n \sin\left(\frac{n\pi}{b}x\right) e^{-\frac{t}{\tau_n}} \tag{2.370}$$

We now multiply each side by $\sin\left(\frac{m\pi x}{b}\right) dx$ and integrate from 0 to b.

$$\int_0^b J_s\frac{x}{b}\sin\left(\frac{m\pi x}{b}\right) dx = \sum_{m=1}^{\infty} C_m \int_0^b \sin\left(\frac{n\pi x}{b}\right)\sin\left(\frac{m\pi x}{b}\right) dx \tag{2.371}$$

Due to the orthogonality of the sine function, the only non-zero term in the summation is the $n = m$ term. We therefore find that[2]

$$C_n = -2J_s\frac{(-1)^n}{n\pi} \tag{2.372}$$

The expression for the magnetic field is then

$$H(x, t) = -J_s\frac{x}{b} - \sum_{n=1}^{\infty} 2J_s\frac{(-1)^n}{n\pi}\sin\left(\frac{n\pi x}{b}\right) e^{-\frac{t}{\tau_n}} \tag{2.373}$$

---

2 We are using the result $\int \theta\sin\theta\, d\theta = \sin\theta - \theta\cos\theta$.

The current density is found by taking the curl of Equation (2.373).

$$J(x,t) = \frac{J_s}{b} + \sum_{n=1}^{\infty} 2J_s \frac{(-1)^n}{b} \cos\left(\frac{n\pi x}{b}\right) e^{-\frac{t}{\tau_n}} \tag{2.374}$$

As an example, consider the case of a copper block ($\sigma = 5.8 \times 10^7$ S m^{-1}, $\mu_0 = 4\pi \times 10^{-7}$ H m^{-1}) and a width of $b = 0.1$ m.

We switch on a current sheet of 1.0 A m^{-1} at $t = 0$. Figure 2.47 shows the field across the conducting region at several instants of time. Initially, we see that there is a field only very close to the current sheet side. After a long time, we expect the steady, state solution, with a linear drop to zero at the edge of the conductor. The longest diffusion time constant, ($n = 1$), in this case is $\tau = \mu \sigma b^2 / \pi^2 = 0.074$ s.

In Figure 2.48, we see the current density in the block at different times. As expected, there is current only at the surface close to the current sheet at short times and in steady state we have uniform current density.

As we found in Section 1.10, for the transient case, as the field penetrates into the conducting region, *skin depth* takes on a different meaning and depends on $\sqrt{t}$. Using the result, we just found for transient diffusion, we can verify this behavior. In Figure 2.49, we have a block of copper, 1 m wide, and a current sheet of 1.0 A m^{-1} applied on one face of the conductor with the current returning through the conductor. The magnetic field at the surface will then be 1.0A m^{-1} at all times greater than $t = 0$. If we now look at the location where the field is $1/e$ of the surface value at different times, we find that as time increases, the location moves

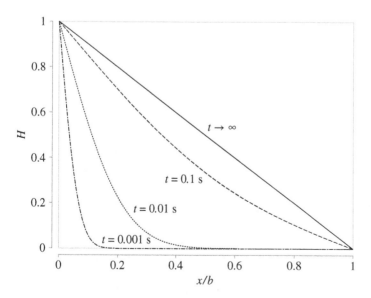

Figure 2.47 Field diffusion into copper block.

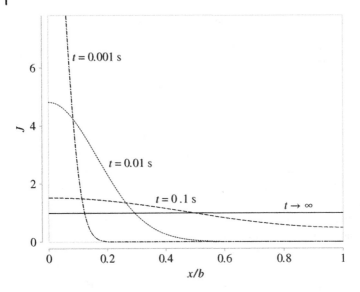

Figure 2.48 Current diffusion into copper block.

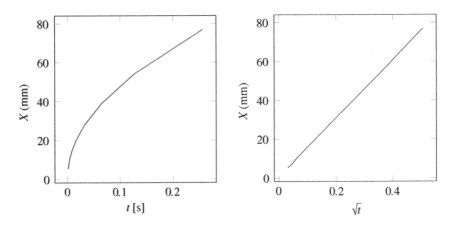

Figure 2.49 Location of a particular value of the field as a function of time.

in the positive direction as the square root of time. For example, at $t = 0.004$ s, the location is approximately 10 mm. At $t = 0.016$ s the location is approximately 20 mm. Figure 2.49 shows the location at which the field is $1/e$ as a function of time. We plot the location vs. time and vs. the square root of time, in which case we see a linear relationship.

Another example was suggested by Silvester [38]. Referring to Figure 2.50, we are interested in the turn-off transient in the conducting region between the two current sheets.

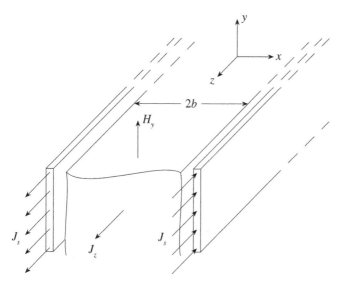

Figure 2.50 Conducting region in solenoidal field: turn-off transient.

In this problem, the symmetry is different and the sine functions are replaced by cosine functions. The time constant is now

$$\tau_n = \frac{4\mu\sigma b^2}{(n\pi)^2} \tag{2.375}$$

due to the width of the conducting region being twice as large.

The solution is then

$$H(x, t) = \sum_{n=1}^{\infty} 4J_s \frac{\sin(n\pi/2)}{n\pi} \cos\left(\frac{n\pi x}{2b}\right) e^{-\frac{t}{\tau_n}} \tag{2.376}$$

For the turn on transient, we obtain

$$H(x, t) = \sum_{n=1}^{\infty} 4J_s \frac{\sin(n\pi/2)}{n\pi} \cos\left(\frac{n\pi x}{2b}\right) \left(1 - e^{-\frac{t}{\tau_n}}\right) \tag{2.377}$$

The current density is

$$J(x, t) = \sum_{n=1}^{\infty} -2J_s \frac{\sin(n\pi/2)}{b} \sin\left(\frac{n\pi x}{2b}\right) e^{-\frac{t}{\tau_n}} \tag{2.378}$$

Using a uniform copper conducting region of width $2b$ and an initial magnetic field of $1.0\,\text{A}\,\text{m}^{-1}$, we find the field decay as shown in Figure 2.51. We note that the field starts out as the initial magnetic field and decays to zero.

In Figure 2.52, we see the turn-off transient current density at various times. As expected, the current density goes to zero after several time constants.

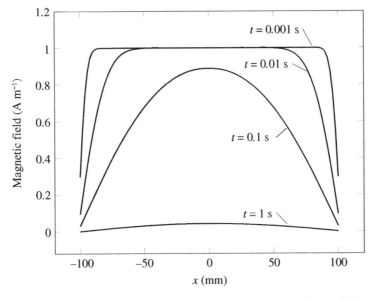

Figure 2.51 Turn-off magnetic field transient in conducting region at different times.

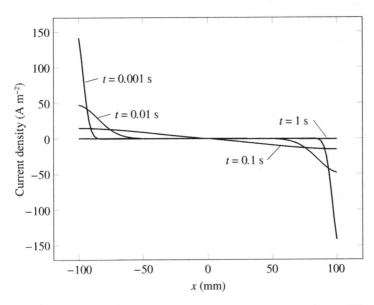

Figure 2.52 Turn-off current density transient in conducting region at different times.

2.14 Eddy Current Modes in a Rectangular Core

We have looked at one-dimensional transient diffusion into a conducting block in Section 2.13. We will now look at transient diffusion into a rectangular bar as shown in Figure 2.53 [38]. The dimensions of the bar are $2b \times 2h$. We will assume that the material properties are linear and homogeneous.

We will also assume that the bar is very long compared to its width and height so that variation in the z direction may be neglected. We will employ the separation of variables technique (see Appendix B) that was used in Section 2.13. The difference is that we now have two space variables, x and y. Considering the space variables together, we assume a solution of the form

$$S = X(x)Y(y) \tag{2.379}$$

Then the Laplacian part of the diffusion equation becomes

$$\frac{\partial^2 S}{\partial x^2} + \frac{\partial^2 S}{\partial y^2} = -\gamma^2 S \tag{2.380}$$

Combining these we get

$$\frac{X''}{X} + \frac{Y''}{Y} = -\gamma^2 \tag{2.381}$$

Let

$$\gamma^2 = \alpha^2 + \beta^2 \tag{2.382}$$

Figure 2.53 Rectangular conductor with applied magnetic field.

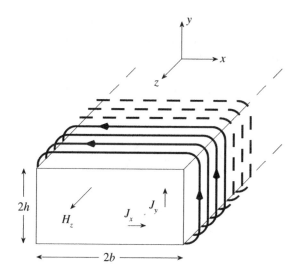

From the geometry, we deduce that the solution must have even symmetry, so that sine functions cannot be a solution. We, therefore, obtain a solution of the form

$$X'' = K_x \cos(\alpha x)$$
$$Y'' = K_y \cos(\beta y) \tag{2.383}$$

The total solution then has the form

$$H = K \cos(\alpha x) \cos(\beta y) e^{-t/\tau} \tag{2.384}$$

where

$$\tau = \frac{\mu\sigma}{\alpha^2 + \beta^2} \tag{2.385}$$

While these functions satisfy the diffusion equation, they do not satisfy the boundary conditions. To make sure that the field vanishes at the boundaries $(x = \pm b)$ and $(y = \pm h)$, we require that the parameters α and β are limited to

$$\alpha = \frac{m\pi}{2b}$$

$$\beta = \frac{n\pi}{2h} \tag{2.386}$$

for any odd integers m and n.

This general solution satisfies the boundary conditions at the conductor's edges, but not the initial conditions or steady-state distribution for the turn-on or turn-off transients. For example, in the turn-off transient, the initial state of the conductor is uniform magnetization. For the turn-on transient, the steady-state solution is a uniform field. Referring to the general solution of Equation (2.384), we see that a single term cannot produce either of these results. We now multiply each term (in x and y) by a cosine function, and use the orthogonality property, where only the $i = j$ term survives the integration. We are left with a Fourier series, in which the summation will give us the desired result. As an example, Figure 2.54 shows the $m = 5$, $n = 7$ solution.

As in the previous example, each term in the series decays with a different time constant, the time constants getting shorter as the spatial orders get larger. The solution is then

$$H(x,y,t) = \sum_{m=1,3,5,\dots}^{\infty} \sum_{n=1,3,5,\dots}^{\infty} K_{mn} \cos\left(\frac{m\pi}{2b}x\right) \cos\left(\frac{n\pi}{2h}y\right) \left(1 - e^{-t/\tau_{mn}}\right) \tag{2.387}$$

where

$$\tau_{mn} = \frac{\mu\sigma}{\left(\frac{\pi m}{2b}\right)^2 + \left(\frac{\pi n}{2h}\right)^2} \tag{2.388}$$

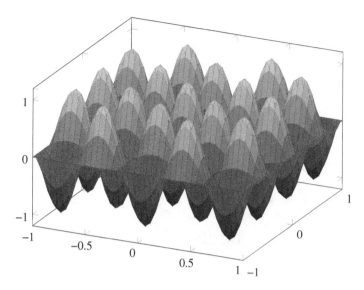

Figure 2.54 The 5,7 mode in a rectangular conductor.

and

$$K_{mn} = \frac{16J_s}{\pi^2 mn} \sin \frac{m\pi}{2} \sin \frac{n\pi}{2} \tag{2.389}$$

To find the current density, we can use Ampere's law and take the curl of Equation (2.387).

$$J_x(x,y,t) = -\sum_{m=1,3,5...}^{\infty} \sum_{n=1,3,5,...}^{\infty} K_{mn} \frac{n\pi}{2h} \cos\left(\frac{m\pi}{2h}x\right) \sin\left(\frac{n\pi}{2h}y\right) e^{-t/\tau_{mn}} \tag{2.390}$$

and

$$J_y(x,y,t) = \sum_{m=1,3,5...}^{\infty} \sum_{n=1,3,5,...}^{\infty} K_{mn} \frac{n\pi}{2b} \sin\left(\frac{m\pi}{2b}x\right) \cos\left(\frac{n\pi}{2h}y\right) e^{-t/\tau_{mn}} \tag{2.391}$$

As an example, in Figures 2.55–2.57 we see the magnetic field for the turn-on transient across the rectangular surface of the conducting bar. The bar is initially unmagnetized when, at $t = 0$, a current sheet of 1.0 A m^{-1} is applied. The figure shows the field distribution three times during the transient. Shortly after the current is applied, the edges of the bar become magnetized, some time later the field has diffused further into the center. After a longer time interval, we see that steady state is nearly achieved.

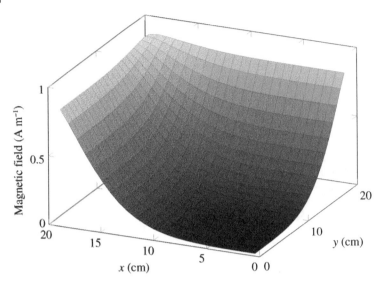

Figure 2.55 Magnetic field distribution in bar soon after current is applied at $t = 0.1$ s.

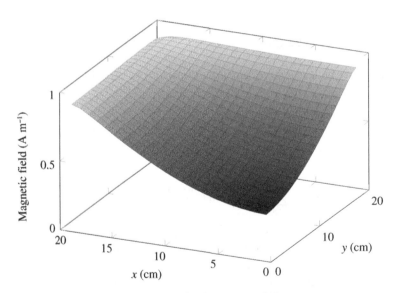

Figure 2.56 Magnetic field distribution in bar at $t = 0.5$ s.

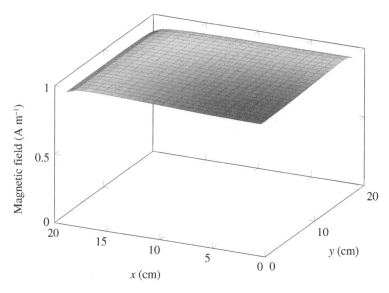

Figure 2.57 Magnetic field distribution in bar near steady state at $t = 2$ s.

2.15 Summary

In this chapter, we have presented many solutions for conductors with rectangular cross section. The theory is based on the principles developed in Chapter 1. We discussed the lamination problem covering both the resistance and reactance-limited cases. In the later example, the skin depth for steady-state ac fields was developed from a first principles argument from Ampere's law and Faraday's law. This is an extremely practical example and the results presented here are still in use in the design of many electrical devices. We then presented an extensive discussion of the conducting plate problem and compared different techniques to find losses. These examples were all for linear, homogeneous, and isotropic materials. We then introduced a limiting nonlinear analysis for eddy currents in saturating steel. This analysis used a square wave saturation curve. The analysis allows us to find the current density and loss in a saturating material and this analysis is still commonly used in design. We also introduced the concepts of *effective permeability* as a way of dealing with saturation in steady-state ac analysis and *complex permeability* as a way of accounting for hysteresis loss. These method are approximations but give useful results for losses in plates and laminations. We then looked at problems in which the magnetic field has both space and time variations in one-dimension and two-dimensions. This analysis is applicable to find losses in structural members, especially in electrical machines. We then analyzed a number of problems involving layered geometries with sinusoidal space-varying excitation and with filamentary excitation. These included single plates and multiple layered plates with or without gaps. Formulas for eddy currents and losses were derived for these cases. We then turned our attention to problems in which the material was anisotropic. When designing magnetic circuits with laminations, most of the flux should travel perpendicular to the thin edge of the lamination and the losses are usually resistance limited. At the end of these structures, however, some of the flux enters through one of the wide sides, wide being compared to the skin depth. In this case, the flux crossed the lamination stack and encounters a number of small air gaps. This has a significant effect on the permeability in that particular direction. Since the permeability is different for the different directions of flux, we found a formulation that will account for the eddy currents and losses in this particular application. We then introduced a closed-form solution for eddy currents in an isolated rectangular conductor, in which the eddy currents vary in magnitude and phase in two-dimensions. We concluded Chapter 2 with a number of problems in the time domain. These included two examples of magnetic diffusion into a solid conductor and an example that introduced the concept of eddy current modes. This formulation was used to find the transient eddy current distribution on the face of a two-dimensional conductor responding to a step function of field excitation. We now turn our attention to problems in which the conductor cross section is circular.

3

Conductors with Circular Cross Sections

In Chapter 2, we studied a number of problems involving eddy currents in conductors with rectangular cross sections. These problems were formulated in Cartesian or rectangular coordinates. As we have seen, the solution of the diffusion equation in rectangular coordinates usually results in exponential functions which, depending on boundary conditions, we can express as sines and cosines or hyperbolic functions. In this section, we will consider problems involving conductors with circular cross sections. There are, of course, many applications for circular conductors. Many motors and transformers are wound with circular wire, as are many electromagnets and inductors. Conductors with circular cross section are universally used in overhead power transmission and for underground or underwater cables. In the field of instrumentation, co-axial conductors are in widespread use. These problems are best described in polar or cylindrical coordinates. The solution of the diffusion equation in polar coordinates often results in Bessel functions. Since these functions are often less familiar, we refer the reader to Appendix A for more information. There are also a number of excellent works on the subject [1, 34, 39, 74]. The process we have followed in Chapter 2 is continued here so that we consider circular conductors with axial fields and axial current, and circular conductors with transversely directed fields applied. We consider both the resistance limited case, in which the reaction field of the conductor can be neglected, and the reactance-limited case, in which the field produced by eddy currents redistributes the field.

3.1 Axial Current in a Conductor with Circular Cross Section: Reactance-Limited Case

This example deals with the case of skin effect for a long straight wire of circular cross-section carrying steady-state sinusoidal current in the axial direction. The geometry is illustrated in Figure 3.1. We assume that the wire is made of linear

Eddy Currents: Theory, Modeling, and Applications, First Edition.
Sheppard J. Salon, M. V. K. Chari, Lale T. Ergene, David Burow, and Mark DeBortoli.

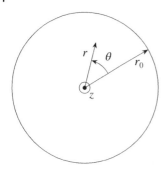

Figure 3.1 Coordinate system defined for the long straight wire.

homogeneous material and we ignore any perturbation of the field from external sources or magnetic material. In this case, the current has only one component, along the wire. The magnetic field has only a θ component. The current and the magnetic field are only a function of the radial coordinate.

We have from Maxwell's equations

$$\nabla \times J = -\sigma \mu \frac{\partial H}{\partial t} \tag{3.1}$$

Taking the curl of both sides of Equation (3.1) and assuming steady-state time-harmonic behavior, we have

$$\nabla \times \nabla \times J = -\sigma \mu \nabla \times \frac{\partial H}{\partial t} = -\sigma \mu \frac{\partial J}{\partial t} = j\omega \mu \sigma J \tag{3.2}$$

Expanding Equation (3.2) in cylindrical coordinates

$$\frac{\partial^2 J}{\partial r^2} + \frac{1}{r}\frac{\partial J}{\partial r} = j\omega\mu\sigma J \tag{3.3}$$

Equation (3.3) is a Bessel equation of zero order whose solution is found in [39]

$$J = C\left(\operatorname{ber}(\alpha r) + j\operatorname{bei}(\alpha r)\right) \tag{3.4}$$

where $\alpha = \sqrt{\omega \mu \sigma}$.

We can find the constant of integration, C, by setting the current density at the outer radius of the conductor r_0 to J_0. This value, J_0, is the uniform current density which would exist at dc conditions. We find then that

$$C = \frac{J_0}{\operatorname{ber}(\alpha r_0) + j\operatorname{bei}(\alpha r_0)} \tag{3.5}$$

The current density as a function of r is then

$$J = J_0 \frac{\left(\operatorname{ber}(\alpha r) + j\operatorname{bei}(\alpha r)\right)}{\operatorname{ber}(\alpha r_0) + j\operatorname{bei}(\alpha r_0)} \tag{3.6}$$

As an example, let us look at the solution for the case of a circular copper wire with 60 HZ current. In this example, the wire has a radius of $r_0 = 0.05$ m and is

Figure 3.2 Current density magnitude as a function of r for 60 HZ.

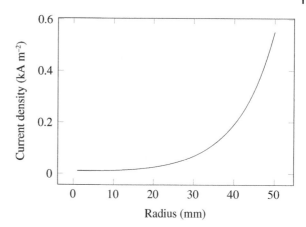

Figure 3.3 Real and imaginary components of current density as a function of r for 60 HZ.

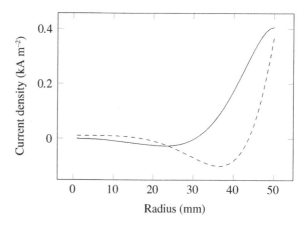

carrying $(1.0 + j0.0)$ A. Evaluating Equation (3.6), the magnitude of the current density is shown in Figure 3.2. In Figure 3.3, we see the real and imaginary components of the current density. Here the radius is larger than the skin depth and we expect that the current density at the surface is much higher than that in the center.

We can find the current in the conductor up to a given radius r by integrating the current density over the cross section of the wire.

$$I_r = \int_0^{2\pi} \int_0^r Jr \, dr \, d\theta = \int_0^r 2\pi Jr \, dr = \int_0^r \frac{J_0 \left(\text{ber}(\alpha r) + j\,\text{bei}(\alpha r)\right)}{\text{ber}(\alpha r_0) + j\,\text{bei}(\alpha r_0)} \, dr \quad (3.7)$$

We have the relations

$$\int_0^r \text{ber}(\alpha r)r \, dr = \frac{r}{\alpha}\text{bei}'(\alpha r) \quad (3.8)$$

$$\int_0^r \text{bei}(\alpha r) r \, dr = \frac{r}{\alpha} \text{ber}'(\alpha r) \tag{3.9}$$

The current in the entire wire is then

$$I_r = \int_0^{r_0} 2\pi J r \, dr = \int_0^{r_0} 2\pi J r \, dr = 2\pi J_0 \frac{r_0}{\alpha} \frac{\text{bei}'(\alpha r_0) - j\,\text{ber}'(\alpha r_0)}{\text{ber}(\alpha r_0) + j\,\text{bei}(\alpha r_0)} \tag{3.10}$$

and

$$I_r = I_0 \frac{r}{r_0} \frac{\text{bei}'(\alpha r_0) - j\,\text{ber}'(\alpha r_0)}{\text{ber}(\alpha r_0) + j\,\text{bei}(\alpha r_0)} \tag{3.11}$$

We may now find the impedance per unit length as

$$Z = \frac{E_0}{I} = \frac{\rho J_0}{2\pi J_0 \frac{r_0}{\alpha}} \frac{\text{ber}(\alpha r_0) + j\,\text{bei}(\alpha r_0)}{\text{bei}'(\alpha r_0) - j\,\text{ber}'(\alpha r_0)} \tag{3.12}$$

Rewriting Equation (3.12), we have

$$R + j\omega L = \frac{\rho \alpha}{2\pi r_0} \frac{\text{ber}(\alpha r_0) + j\,\text{bei}(\alpha r_0)}{\text{bei}'(\alpha r_0) - j\,\text{ber}'(\alpha r_0)} \tag{3.13}$$

Rationalizing, by multiplying numerator and denominator by the complex conjugate of the denominator, then taking the real and imaginary parts, we find the resistance and reactance per unit length as

$$R = \frac{\rho \alpha}{2\pi r_0} \frac{\text{ber}(\alpha r_0)\,\text{bei}'(\alpha r_0) - \text{bei}(\alpha r_0)\,\text{ber}'(\alpha r_0)}{\text{ber}'^2(\alpha r_0) + \text{bei}'^2(\alpha r_0)} \tag{3.14}$$

$$X = \omega L = \frac{\rho \alpha}{2\pi r_0} \frac{\text{bei}(\alpha r_0)\,\text{bei}'(\alpha r_0) + \text{ber}(\alpha r_0)\,\text{ber}'(\alpha r_0)}{\text{ber}'^2(\alpha r_0) + \text{bei}'^2(\alpha r_0)} \tag{3.15}$$

The power loss per unit length is then

$$P = I^2 \frac{\rho \alpha}{2\pi r_0} \frac{\text{ber}(\alpha r_0)\,\text{bei}'(\alpha r_0) - \text{bei}(\alpha r_0)\,\text{ber}'(\alpha r_0)}{\text{ber}'^2(\alpha r_0) + \text{bei}'^2(\alpha r_0)} \tag{3.16}$$

As an example, let us consider a long circular wire, carrying 1.0 A at various frequencies: 60, 200, and 500 HZ. The radius of the wire is $a = 0.01$ m and the material is copper ($\sigma = 5.8 \times 10^7 \, \text{S}\,\text{m}^{-1}$).

The magnitude of the current density in the wire vs. radius is illustrated in Figure 3.4. The dc resistance per meter depth is

$$R_{dc} = \frac{1}{\sigma \pi a^2} = 5.488 \times 10^{-5} \, \Omega\,\text{m}^{-1} \tag{3.17}$$

Using Equation (3.14) for 60 HZ, we get $R_{ac} = 5.697 \times 10^{-5} \, \Omega$. This is somewhat higher than the dc resistance as expected. Stoll [26] uses asymptotic expansions for the Bessel functions to obtain approximate solutions for the resistance and reactance. This is convenient since it is no longer necessary to evaluate the

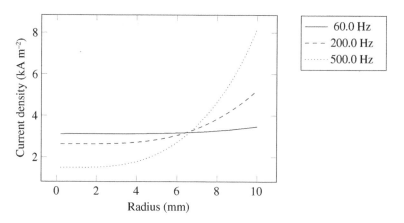

Figure 3.4 Magnitude of current density in copper wire at three frequencies vs. radius.

Bessel functions, only a simple polynomial expression. These approximations are valid either in the low frequency regime ($a < \delta$) or in the high frequency regime ($a > \delta$). In the low frequency regime, when the skin depth is comparable or larger than the radius, we obtain

$$R_{ac} \approx R_{dc}\left(1 + \frac{a^4}{48\delta^4}\right) \tag{3.18}$$

Evaluating this expression for 60 Hz, we obtain the result $R_{ac} = 5.704 \times 10^{-5}\,\Omega\,\mathrm{m}^{-1}$. There is a slight difference compared to the exact solution since the radius is slightly larger than the 60 Hz skin depth. We now consider the same conductor but with current at 10 HZ ($\delta = 0.021\,\mathrm{m}$). The exact solution gives $R_{ac} = 5.494 \times 10^{-5}\,\Omega\,\mathrm{m}^{-1}$ while the approximate result is $R_{ac} = 5.494 \times 10^{-5}\,\Omega\,\mathrm{m}^{-1}$. The solutions agree to four significant figures.

Stoll [26] also uses asymptotic expansions to obtain a high frequency ($a \gg \delta$) approximation for the resistance of the conductor.

$$R_{ac} \approx R_{dc}\left(\frac{a}{2\delta} + \frac{1}{4} + \frac{3\delta}{32a}\right) \tag{3.19}$$

If we now consider, the case of 500 HZ ($\delta = 0.002955\,\mathrm{m}$), the exact solution is $R_{ac} = 1.0809 \times 10^{-4}\,\Omega\,\mathrm{m}^{-1}$ and the approximation in Equation (3.19) gives $R_{ac} = 1.0789 \times 10^{-4}\,\Omega\,\mathrm{m}^{-1}$. The value here is considerably greater than the dc resistance as expected. In Figure 3.5, we see the exact solution for the ac resistance divided by the dc resistance and the high frequency approximation plotted as a function of the variable $\xi = r_0/\delta$. We can see from the figure that at high frequencies, where $\delta < r_0$, the approximation is valid for $\xi > 2$.

In Figure 3.6, we have plotted the normalized resistance vs. ξ for the low frequency approximation, $\delta > r_0$. We see from the figure that the approximation is very good for values of $\xi < 2$.

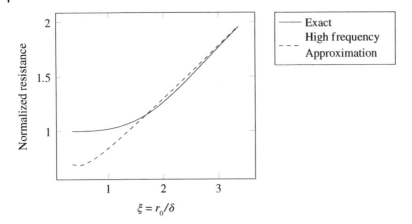

Figure 3.5 Normalized resistance vs. $\xi = \frac{r_0}{\delta}$, closed-form and high-frequency approximation.

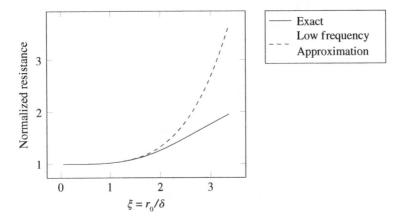

Figure 3.6 Normalized resistance vs. $\xi = \frac{r_0}{\delta}$, closed-form and low-frequency approximation.

3.2 Axial Current in Composite Circular Conductors

There are several applications in which we have composite conductors with circular cross sections as illustrated in Figure 3.7. One of these is overhead power transmission lines. These lines are typically made with an inner layer of steel, which supplies the strength, and an outer layer of aluminum which carries most of the current. Aluminum is used instead of copper due to it's light weight and mechanical strength. These conductors are referred to as ACSR or Aluminum

Figure 3.7 Two layer conductor with circular cross section.

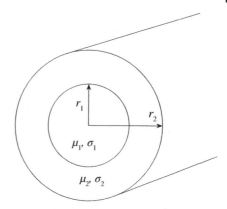

Conductor Steel Reinforced. The center steel layer does not carry much of the load current due to its higher resistance and higher inductance. In Section 3.1, we solved for the magnetic field and eddy current density directly. In this section, we will use the magnetic vector potential to solve the coupled set of equations in the two regions.

We will first consider the case of a solid homogeneous conductor to introduce the vector potential formulation, and then extend this method to the case of multiple layers. The magnetic vector potential is defined by

$$\nabla \times A = B \tag{3.20}$$

We can write the electric field in terms of the flux density by using Faraday's law,

$$\nabla \times E = -j\omega B \tag{3.21}$$

As shown in Chapter 1, the electric field has two sources, one the induced field and the other the field produced by an electric charge distribution. Therefore

$$E = -j\omega A + \nabla V \tag{3.22}$$

Since we are only interested in the curl of the vector potential, and since we can arbitrarily set the divergence, we choose to set the divergence to zero.

$$\nabla \cdot A = 0 \tag{3.23}$$

Using the vector identity

$$\nabla \times \nabla \times F = \nabla(\nabla \cdot F) - \nabla^2 F \tag{3.24}$$

for any vector F, we obtain

$$\nabla^2 A - j\omega\sigma\mu A = -\mu J_0 \tag{3.25}$$

In the present example, in cylindrical coordinates, A and J have only z components and none of the variables depend on z or θ. This gives

$$\frac{\partial^2 A_z}{\partial r^2} + \frac{1}{r}\frac{\partial A_z}{\partial r} - \alpha^2 A_z = -\mu J_{z0} \tag{3.26}$$

where $\alpha^2 = j\omega\sigma\mu$. The solution of Equation (3.26) is given in terms of modified Bessel functions of the first and second kind of zero order.

$$A_z(r) = C_1 I_0(\alpha r) + C_2 K_0(\alpha r) + \frac{\mu J_{z0}}{\alpha^2} \tag{3.27}$$

The two constants are solved for by applying the boundary conditions. For a single homogeneous conductor in free space, we can set the vector potential at the outer radius to zero. The second condition is that the vector potential is finite at $r = 0$, giving $C_2 = 0$. Solving then, we obtain for the vector potential,

$$A_z(r) = \frac{\mu J_{z0}}{\alpha^2}\left(1.0 - \frac{I_0(\alpha r)}{I_0(\alpha a)}\right) \tag{3.28}$$

This solution can be used to analyze the problem treated in Section 3.1, in which we used the field variables instead of the vector potential. In the examples below, where we consider conductors made of composite materials to make a layered conductor, solving for the vector potential will be more convenient. Considering Equation (3.28), let us consider the example of a copper conductor of radius $a = 0.05\,\text{m}$. The conductivity is $\sigma = 5.8 \times 10^7\,\text{S}\,\text{m}^{-1}$ and the frequency is 60 HZ. In this formulation, the conductor is driven by the input current density J_{z0}. In this case, the current density is chosen such that the current in the conductor would be $1.0\,\text{A}$ without the effect of eddy currents. Therefore the current density input in Equation (3.28) is

$$J_{z0} = \frac{1.0}{\pi a^2} = 127.32\,\text{A}\,\text{m}^{-2} \tag{3.29}$$

Equation (3.28) was then solved for a number of points along a radius. The solution for the magnitude of the vector potential is shown in Figure 3.8.

Consideration of the shape of Figure 3.8 reveals some interesting aspects of the vector potential. First we note that the vector potential magnitude is changing rapidly near the surface of the conductor and then is relatively constant after around three skin depths. We expect the highest current and flux density at the surface and relatively little current or flux density near the center as this is several skin depths from the surface. To get approximately zero current at the center, the induced current must cancel the applied current. Refer back to Section 1.1 where we discussed the two different sources of electric field. The *conservative* electric field, produced by charges, is the one associated with the source current density, J_{z0}. This component of electric field is

$$E_\phi = \frac{J_{z0}}{\sigma} = \frac{127.32}{5.8 \times 10^7} = 2.195 \times 10^{-6}\,\text{V}\,\text{m}^{-1} \tag{3.30}$$

Figure 3.8 Magnitude of the magnetic vector potential.

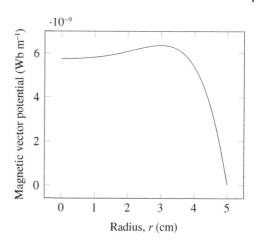

For the induced nonconservative field to cancel this applied electric field, we can deduce what value of vector potential must exist near the center. The induced field is

$$E_\psi = j\omega A \tag{3.31}$$

This tell us that the expected vector potential at the center of the conductor is

$$|A| = \frac{J_{z0}}{\omega\sigma} = 5.82 \times 10^{-9} \, \text{Wb m}^{-1} \tag{3.32}$$

This value agrees with that shown in the figure. The vector potential is, of course, a complex number and Figure 3.9 shows the real and imaginary components. Note that real and imaginary parts are referred to the input current density, which in this case has no imaginary component, $J_{z0} = (127.0 + j0.0) \, \text{A m}^{-2}$, so the real part of the vector potential is the component in phase with the input current density.

From the vector potential solution, we can easily find the current density. The eddy current density is given by

$$J_e = -j\omega\sigma A \tag{3.33}$$

For the steady-state sinusoidal case, the eddy current density is proportional to the vector potential. To find the total current density, we must add in the input current density, J_{z0}. When we evaluate this along the radius of the conductor, we obtain the results in Figure 3.10 for the magnitude of the total current density.

We can find the total current in the conductor by integrating the current density over the conductor surface. The total current is found to be $I = (0.1685 - j1.558) \, \text{A}$. We can now find the impedance per meter by dividing this current into the applied electric field.

$$Z_{ac} = \frac{E_\phi}{I} = 7.02 \times 10^{-6} + j6.43 \times 10^{-6} \, \Omega \tag{3.34}$$

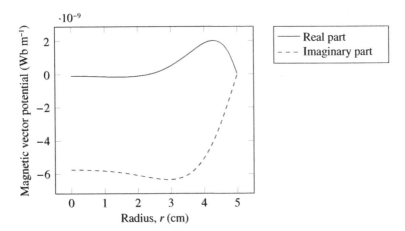

Figure 3.9 Real and imaginary parts of A_z for solid conductor example.

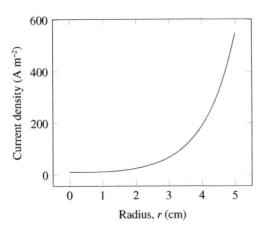

Figure 3.10 Magnitude of the current density for solid conductor example.

This agrees with the formulation of Section 3.1 which is repeated here

$$Z = \frac{\alpha\rho}{2\pi a} \frac{\mathrm{ber}(\alpha a) + j\,\mathrm{bei}(\alpha a)}{\mathrm{bei}'(\alpha a) - j\,\mathrm{bei}'(\alpha a)} \tag{3.35}$$

Evaluating this expression gives $Z_{ac} = 7.015 \times 10^{-6} + j6.40 \times 10^{-6}\ \Omega$. The vector potential can also be used to find the magnetic flux density which is given by the curl of the vector potential. The curl in cylindrical coordinates is given by

$$B_\theta = -\frac{\partial A_z}{\partial r} \tag{3.36}$$

We can evaluate this numerically by taking the values of A_z at two points, relatively close to each other, and dividing by the distance between the two points. For example, near the surface of the conductor, this process gives, $B_\theta = 6.80 \times 10^{-7} - j6.23 \times 10^{-7}$ T. Since we found the total current in the conductor, we can use Ampere's law and find the flux density at the conductor surface as

$$B_\theta = \frac{\mu_0 I}{2\pi a} = 6.80 \times 10^{-7} - j6.23 \times 10^{-7} \text{ T} \tag{3.37}$$

Now that we have applied the MVP formulation to the solid homogeneous case, we can use this formulation to consider the problem of a composite conductor with two layers as shown in Figure 3.7.

For the two region problem, we have, for the inner and outer regions respectively,

$$\frac{\partial^2 A_1}{\partial r^2} + \frac{1}{r}\frac{\partial A_1}{\partial r} - \alpha_1^2 A_1 = -\mu J_{z1} \tag{3.38}$$

and

$$\frac{\partial^2 A_2}{\partial r^2} + \frac{1}{r}\frac{\partial A_2}{\partial r} - \alpha_2^2 A_2 = -\mu J_{z2} \tag{3.39}$$

These are second-order differential equations requiring two boundary conditions each to evaluate the four constants of integration. We require

1. $A_2 = 0$ at $r = r_2$
2. $A_2 = A_1$ at $r = r_1$
3. A_1 is finite at $r = 0$
4. $\mu_1 \frac{\partial A_1}{\partial r} = \mu_2 \frac{\partial A_2}{\partial r}$ at $r = r_1$ (Tangential H is continuous).

The homogeneous solutions to Equations (3.38) and (3.39) are

$$A_{1h} = C_1 I_0(\alpha_1 r) + C_2 K_0(\alpha_1 r) \tag{3.40}$$

$$A_{2h} = C_3 I_0(\alpha_2 r) + C_4 K_0(\alpha_2 r) \tag{3.41}$$

For the particular solution, we introduce constants D_1 and D_2 such that

$$A_{1p} = D_1 = \frac{\mu_1 J_1}{\alpha_1^2} \tag{3.42}$$

$$A_{2p} = D_2 = \frac{\mu_2 J_2}{\alpha_2^2} \tag{3.43}$$

Our complete solutions are then

$$A_1 = A_{1h} + A_{1p} = C_1 I_0(\alpha_1 r) + C_2 K_0(\alpha_1 r) + \frac{\mu_1 J_1}{\alpha_1^2} \tag{3.44}$$

$$A_2 = A_{2h} + A_{2p} = C_3 I_0(\alpha_2 r) + C_4 K_0(\alpha_2 r) + \frac{\mu_2 J_2}{\alpha_2^2} \tag{3.45}$$

From the requirement that the potential remains finite at the center, we find that

$$C_2 = 0 \tag{3.46}$$

From the flux line boundary condition at r_2, we have

$$C_3 I_0(\alpha_2 r_2) + C_4 K_0(\alpha_2 r_2) = -\frac{\mu_2 J_2}{\alpha_2^2} \tag{3.47}$$

From the continuity condition at r_1, we get

$$C_1 I_0(\alpha_1 r_1) - C_3 I_0(\alpha_2 r_1) - C_4 K_0(\alpha_2 r_1) = \frac{\mu_2 J_2}{\alpha_2^2} - \frac{\mu_1 J_1}{\alpha_1^2} \tag{3.48}$$

Finally, from the continuity of tangential H, we have

$$\mu_1 C_1 \alpha_1 I_1(\alpha_1 r_1) - \mu_2 C_3 \alpha_2 I_1(\alpha_2 r_1) - C_4 \alpha_2 K_1(\alpha_2 r_1) = 0 \tag{3.49}$$

These are now solved simultaneously for the constants.

$$\begin{pmatrix} I_0(\alpha_1 r_1) & -I_0(\alpha_2 r_1) & -K_0(\alpha_2 r_1) \\ \mu_1 \alpha_1 I_1(\alpha_1 r_1) & -\mu_2 \alpha_2 I_1(\alpha_2 r_1) & -\mu_2 \alpha_2 K_1(\alpha_2 r_1) \\ 0 & I_0(\alpha_2 r_2) & K_0(\alpha_2 r_2) \end{pmatrix} \begin{pmatrix} C_1 \\ C_3 \\ C_4 \end{pmatrix} = \begin{pmatrix} \frac{\mu_2 J_2}{\alpha_2^2} - \frac{\mu_1 J_1}{\alpha_1^2} \\ 0 \\ -\frac{\mu_2 J_2}{\alpha_2^2} \end{pmatrix} \tag{3.50}$$

The determinant of the matrix is

$$\Delta = I_0 \left(-\mu_2 \alpha_2 I_1(\alpha_2 r_1) K_0(\alpha_2 r_2) + \alpha_2 K_1(\alpha_2 r_1) I_0(\alpha_2 r_2) \right)$$
$$- \mu_1 \alpha_1 I_1(\alpha_1 r_1) \left(-I_0(\alpha_1 r_1) K_0(\alpha_2 r_2) + I_0(\alpha_2 r_2) K_0(\alpha_2 r_1) \right) \tag{3.51}$$

Using Cramer's rule

$$C_1 = \frac{1}{\Delta} \begin{vmatrix} \frac{\mu_2 J_2}{\alpha_2^2} - \frac{\mu_1 J_1}{\alpha_1^2} & -I_0(\alpha_1 r_1) & -K_0(\alpha_2 r_1) \\ 0 & -\mu_2 \alpha_2 I_1(\alpha_2 r_1) & -\mu_2 \alpha_2 K_1(\alpha_2 r_1) \\ \frac{-\mu_2 J_2}{\alpha_2^2} & I_0(\alpha_2 r_2) & K_0(\alpha_2 r_2) \end{vmatrix} \tag{3.52}$$

Evaluating this, we have

$$C_1 = \left(\left(\frac{\mu_2 J_2}{\alpha_2^2} - \frac{\mu_1 J_1}{\alpha_2^2} \right) \left(-\mu_2 \alpha_2 I_1(\alpha_2 r_1) K_0(\alpha_2 r_2) + \mu_2 \alpha_2 K_1(\alpha_2 r_1) I_0(\alpha_2 r_2) \right) \right.$$
$$\left. - \frac{\mu_2 J_2}{\alpha_2^2} \left(I_0(\alpha_1 r_1) \mu_2 \alpha_2 K_1(\alpha_2 r_1) - \mu_2 \alpha_2 I_1(\alpha_2 r_1) K_0(\alpha_2 r_1) \right) \right) \frac{1}{\Delta} \tag{3.53}$$

Similarly

$$C_3 = \left(\left(\frac{\mu_2 J_2}{\alpha_2^2} - \frac{\mu_1 J_1}{\alpha_1^2} \right) \left(-\mu_1 \alpha_1 I_1(\alpha_1 r_1) K_0(\alpha_2 r_2) \right) \right.$$
$$\left. - \frac{\mu_2 J_2}{\alpha_2^2} \left(-I_0(\alpha_1 r_1) \mu_2 \alpha_2 K_1(\alpha_2 r_1) + \mu_1 \alpha_1 I_1(\alpha_1 r_1) K_0(\alpha_2 r_1) \right) \right) \frac{1}{\Delta} \tag{3.54}$$

and

$$C_4 = \left(\left(\frac{\mu_2 J_2}{\alpha_2^2} - \frac{\mu_1 J_1}{\alpha_1^2} \right) \left(\mu_1 \alpha_1 I_1(\alpha_1 r_1) \right) \right.$$

$$\left. - \frac{\mu_2 J_2}{\alpha_2^2} \left(-I_0(\alpha_1 r_1) \mu_1 \alpha_1 I_1(\alpha_1 r_1) + I_0(\alpha_1 r_1) \mu_2 \alpha_2 I_1(\alpha_2 r_1) \right) \right) \frac{1}{\Delta} \qquad (3.55)$$

With these constants now known, the solution can be evaluated. This process can be now extended to geometries with more layers. The governing equation for the regions remains the same. One need only match the vector potential and tangential fields at the interfaces in order to evaluate the constants in the solution.

We can now apply the formulation to a numerical example. First, let us check the results for a homogeneous problem. We will use the example we solved with the copper cylindrical conductor, $r = 0.05$ m, carrying 60 HZ current. In this case, we select an arbitrary boundary (half of the radius) between the two materials and use the two layer formulation to find the vector potential. The magnitude of the vector potential is plotted against radius for this case in Figure 3.11. This agrees with Figure 3.9 which was computed using the homogeneous material analysis.

For the second example, we consider the problem of an overhead transmission line ACSR conductor. The inner region, ($r_1 = 0.01$ m) is steel with $\sigma_1 = 10 \times 10^7 \, \text{S m}^{-1}$ and relative permeability of $\mu_r = 250$. The outer layer

Figure 3.11 Magnitude of vector potential vs. radius using the two layer formulation for a solid copper conductor.

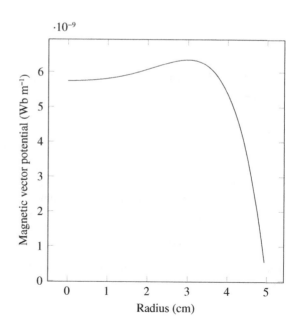

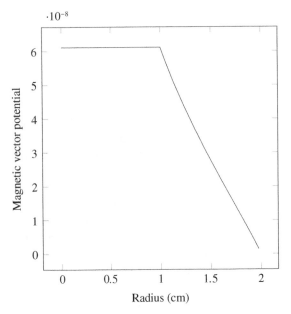

Figure 3.12 Magnitude of vector potential vs. radius using the two layer formulation for a composite conductor.

$(r_2 = 0.02\,\text{m})$ has conductivity $\sigma_2 = 3.5 \times 10^7\,\text{S}\,\text{m}^{-1}$. The frequency of the current is 60 HZ.

The magnitude of the vector potential vs. radius is plotted in Figure 3.12. The current density magnitude vs. radius is shown in Figure 3.13.

3.3 Circular Conductor with Applied Axial Flux: Resistance-Limited Case

Consider a long cylinder of radius b and resistivity ρ. We apply a uniform axial magnetic flux density of $B_z = B_0 \cos\omega t$. This is illustrated in Figure 3.14.

Based on our resistance limited assumptions and symmetry, the current density has only a θ component and all of the current is in phase. In computing the loss, we need only find the RMS value of the emf. Considering the circular path in Figure 3.14, the flux linked by the current element is

$$\psi(t) = B_0 \pi r^2 \cos(\omega t) \tag{3.56}$$

Then the induced voltage, or emf, is the negative derivative of ψ with respect to time

$$\mathcal{E}(t) = \omega B_0 \pi r^2 \sin(\omega t) \tag{3.57}$$

Figure 3.13 Magnitude of current density vs. radius using the two layer formulation for a composite conductor.

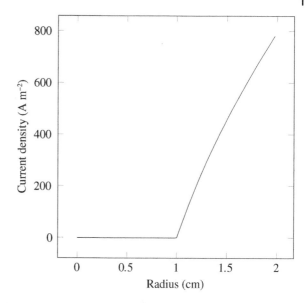

Figure 3.14 Long cylinder with axial flux.

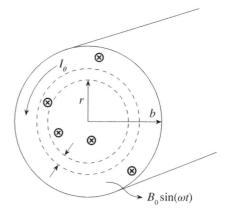

the RMS value of which is

$$\mathcal{E}_{rms} = \omega B_0 \pi r^2 / \sqrt{2} \tag{3.58}$$

The resistance of the element per unit depth is

$$dR = 2\pi r \rho / dr \tag{3.59}$$

The loss per unit depth in the thin element is then found by combining (3.58) and (3.59):

$$dW = \frac{\mathcal{E}_{rms}^2}{dR} = \frac{\pi \omega^2 B_0^2 r^3 dr}{4\rho} \tag{3.60}$$

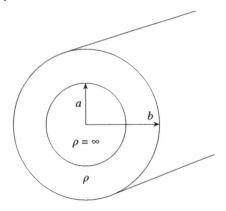

Figure 3.15 Long hollow cylinder with axial flux.

We find the total loss per unit depth by integrating over the radius of the cylinder

$$W = \int_0^b \frac{\pi\omega^2 B_0^2 r^3 dr}{4\rho} = \frac{\pi\omega^2 B_0^2 b^4}{16\rho} \tag{3.61}$$

We can easily find the current density by dividing the emf (3.58) by the path length to obtain the electric field and then dividing by the resistivity. The RMS current density is then

$$J_{rms} = \frac{\mathcal{E}_{rms}}{2\pi r} = \frac{\omega B_0 r}{2\sqrt{2}\rho} \tag{3.62}$$

The current density is zero at the center and increases linearly as we approach the edge.

We can extend this analysis to include some other cases of interest. For example, if the cylinder is hollow, as shown in Figure 3.15, then the analysis is basically the same, but in (3.61) we integrate from a to b instead of 0 to b, giving

$$W = \frac{\pi\omega^2 B_0^2 \left(b^4 - a^4\right)}{16\rho} \tag{3.63}$$

Another variation is if we have a cylinder made of different materials, such as shown in Figure 3.16. This solution is a combination of the hollow cylinder (the outer one) and the solid cylinder (the inner one).

3.4 Circular Conductor with Applied Axial Flux: Reactance-Limited Case

Now let us consider the case of a long rod with circular cross section, being excited by a solenoidal field in the axial direction using reactance-limited analysis.

Figure 3.16 Long composite cylinder
with axial flux.

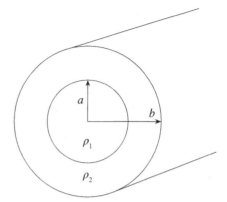

Figure 3.17 Long cylinder
with solenoidal applied field.

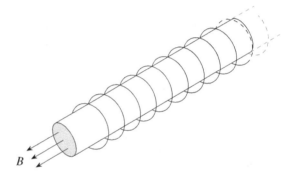

Figure 3.17 shows a slice of an infinitely long solenoid with a conducting core of circular cross section. The excitation is provided by a thin current sheet representing the winding of the solenoid. The current sheet contains NI ampere-turns per unit length. The problem is to determine the magnetic field, flux-density, and the eddy current density and power loss in the conducting circular rod.

We assume here that the material properties are linear, homogeneous, and isotropic. The permeability and the conductivity are taken as constant. The conditions are steady-state sinusoidal and the problem can be described in cylindrical coordinates.

From Ampere's law

$$\nabla \times H = J_s \tag{3.64}$$

Since the solenoid is assumed infinitely long, we have only a z component of the magnetic field, which is not a function of z or θ. Therefore

$$\nabla \times H = -\frac{\partial H_z}{\partial r} = J_s \hat{\mathbf{u}}_\theta \tag{3.65}$$

Taking the curl of both sides of Equation (3.65), we get

$$\nabla \times \nabla H = \nabla \times \left(-\frac{\partial H_z}{\partial r} \right) = \nabla \times J_s = \sigma \nabla \times E \tag{3.66}$$

We recognize that

$$\nabla \times \left(-\frac{\partial H_z}{\partial r} \right) = -\frac{1}{r} \frac{\partial}{\partial r} \left(r \frac{\partial H_z}{\partial r} \right) \tag{3.67}$$

Faraday's law states that

$$\nabla \times E = -\frac{\partial B}{\partial t} = -\mu \frac{\partial H}{\partial t} \tag{3.68}$$

For the steady-state time-harmonic case, we have

$$\nabla \times E = -j\omega\mu H \tag{3.69}$$

From Equations (3.66) through (3.69), we have

$$\nabla \times \nabla \times H = -\frac{1}{r} \frac{\partial}{\partial r} \left(r \frac{\partial H_z}{\partial r} \right) = -j\omega\sigma H_z \tag{3.70}$$

Expanding Equation (3.70) we get

$$\frac{\partial^2 H_z}{\partial r^2} + \frac{1}{r} \frac{\partial H_z}{\partial r} - j\omega\mu\sigma H_z = 0 \tag{3.71}$$

Equation (3.71) is a Bessel equation of zeroth order and the solution is in terms of Kelvin functions.

$$H_z = C\left(\text{ber}(kr) + j\,\text{bei}(kr)\right) \tag{3.72}$$

where $k^2 = \omega\mu\sigma$ and C is a constant of integration.

We now apply the boundary conditions. At the outer radius, $r = r_0$, we have

$$\oint H \cdot d\ell = NI \tag{3.73}$$

Therefore, at $r = r_0$ we have

$$H_z = \frac{NI}{\ell} = C\left(\text{ber}(kr_0) + j\,\text{bei}(kr_0)\right) \tag{3.74}$$

and

$$C = \frac{NI}{\ell} \frac{1}{\left(\text{ber}(kr_0) + j\,\text{bei}(kr_0)\right)} \tag{3.75}$$

From Equations (3.74) and (3.75), we find

$$H_z(r) = \frac{NI}{\ell} \frac{\left(\text{ber}(kr) + j\,\text{bei}(kr)\right)}{\left(\text{ber}(kr_0) + j\,\text{bei}(kr_0)\right)} \tag{3.76}$$

The eddy current density is

$$J_\theta = -\frac{\partial H_z}{\partial r} = -\frac{NI}{\ell}k\frac{\left(\text{ber}'(kr)+j\,\text{bei}'(kr)\right)}{\left(\text{ber}(kr_0)+j\,\text{bei}(kr_0)\right)} \tag{3.77}$$

The power dissipated is

$$P = \int_0^{2\pi}\int_0^{r_0}\int_0^{\ell}\frac{J_\theta J_\theta^* \rho}{2}\,dv \tag{3.78}$$

Substituting from Equation (3.77) we get

$$P = 2\pi\ell\rho\left(\frac{NI}{\ell}\right)^2\frac{k^2}{2}\int_0^{r_0}\frac{\left(\text{ber}'(kr)+j\,\text{bei}'(kr)\right)\left(\text{ber}'(kr)-j\,\text{bei}'(kr)\right)}{\left(\text{ber}(kr_0)+j\,\text{bei}(kr_0)\right)\left(\text{ber}(kr_0)-j\,\text{bei}(kr_0)\right)}r\,dr \tag{3.79}$$

or

$$P = \pi\ell\rho\left(\frac{NI}{\ell}\right)^2 k^2\int_0^{r_0}\frac{\left(\text{ber}'^2(kr)+\text{bei}'^2(kr)\right)}{\left(\text{ber}^2(kr_0)+\text{bei}^2(kr_0)\right)}r\,dr \tag{3.80}$$

Then the loss per unit length is

$$P = \pi\rho\left(\frac{NI}{\ell}\right)^2 kr_0\frac{\left(\text{ber}(kr_0)\,\text{ber}'(kr_0)+\text{bei}(kr_0)\,\text{bei}'(kr_0)\right)}{\left(\text{ber}^2(kr_0)+\text{bei}^2(kr_0)\right)} \tag{3.81}$$

Let us consider a numerical example. We have a long copper ($\sigma = 5.8 \times 10^7\,\text{S}\,\text{m}^{-1}$) rod of circular cross section of radius $r_0 = 0.03\,\text{m}$. The solenoidal applied field is $H_0 = 1.0\,\text{A}\,\text{m}^{-1}$. The frequency is 60 HZ. From these parameters $\delta = 0.0085\,\text{m}$, so that the radius is about three times the skin depth. Evaluating Equation (3.77), we see in Figure 3.18, the real part, imaginary part, and the magnitude of the current density. As shown in the figure, the current density is much greater near the outer radius and is quite small near the center. In Figure 3.19, we see the real and imaginary parts and the magnitude of the magnetic field. Note that the field at the outer radius is $H_z = 1.0\,\text{A}\,\text{m}^{-1}$ as we expect. Also we see that the field decays as we mover closer to the center of the conductor.

This reduction in the field is due to the field produced by the eddy currents opposing the uniform axial field produced by the external coil. This is clear evidence that the reactance-limited assumptions are valid here. To further illustrate this point, let us apply the equations for the same conductor and solenoid, but with the frequency of 1.0 HZ. Now the skin depth is $\delta = 0.066\,\text{m}$. This is larger than the radius and we would expect the resistance limited assumptions to be valid. In Figure 3.20, we see the magnitude of the current density vs. radius. Note that the curve is almost a straight line as predicted by the resistance-limited analysis in

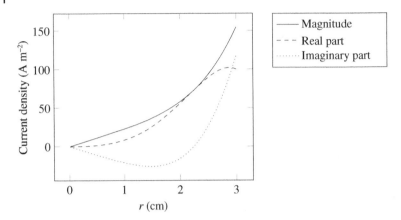

Figure 3.18 Magnitude, real part, and imaginary part of the current density vs. radius at 60 Hz.

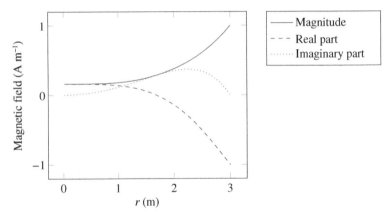

Figure 3.19 Magnitude, real part, and imaginary part of the magnetic field vs. radius at 60 Hz.

Section 3.3. The peak current density in the resistance-limited case is

$$J(r_0) = \frac{\omega \mu \sigma r_0 H_0}{2} = 6.86 \, \text{A m}^{-2} \tag{3.82}$$

From the figure, we have for the peak current density, $J = 6.85 \, \text{A m}^{-2}$. This agrees extremely well with the resistance-limited theory. If we look at the magnetic field magnitude as a function of the radius in Figure 3.21, we see that the field is practically constant and is equal to the applied field $H_0 = 1.0 \, \text{A m}^{-1}$. This tells us that the assumption of the eddy current field having no effect on the applied field is valid. This confirms our previous argument that if the skin depth

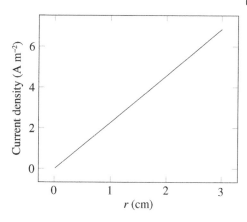

Figure 3.20 Magnitude of the current density vs. radius for 1.0 Hz.

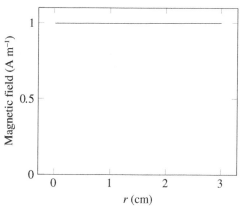

Figure 3.21 Magnitude of the magnetic field vs. radius for 1.0 Hz.

is smaller than the radius, we have a reactance-limited case and if the skin depth is larger than the radius, we can use the much simpler resistance-limited analysis.

3.5 Shielding with a Conducting Tube in an Axial Field

We can investigate an example of eddy current shielding by considering a conducting tube oriented along the axis of a solenoid as shown in Figure 3.22. The eddy currents in the tube will oppose the field inside the tube and there will be some cancellation. The amount of shielding will depend on the frequency, conductivity, permeability, and thickness of the tube. From Ampere's law, we can deduce that the field inside the tube will be constant since there are no sources inside. The currents induced in the tube, in the peripheral direction, will produce a field only inside the tube. The ratio of the magnitude of the field inside and outside the tube is the *shielding factor*.

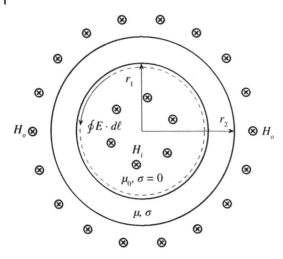

Figure 3.22 Shielding cylinder in axial field.

In this example, it is not necessary for us to evaluate the field inside the conducting region. We will only find the field in the inner nonconducting region and compare this with the applied field. Considering the problem illustrated in Figure 3.22, we will assume steady-state sinusoidal behavior. The solenoidal source field, H_o, is constant and in the z direction. The field in the hollow of the tube, H_i, is also constant over the cross section.

The problem we consider here is axisymmetric. There is no variation of the field or current in the θ direction. There is also no variation of any variable in the axial or z direction. The magnetic field has only a z component and is only a function of r. The current and electric field have only θ components.

The magnetic field in the conductor is described in cylindrical coordinates by

$$\frac{\partial^2 H}{\partial r^2} + \frac{1}{r}\frac{\partial H}{\partial r} = \alpha^2 H \tag{3.83}$$

where

$$\alpha = \sqrt{j\omega\mu\sigma} \tag{3.84}$$

The solution to Equation (3.83) is well know in terms of Bessel functions and given in Equation (3.85).

$$H(r) = CI_0(\alpha r) + DK_0(\alpha r) \tag{3.85}$$

The constants C and D are found by applying the interface conditions at the boundaries of the conducting tube. The tangential magnetic field must be continuous at the boundary. At $r = r_2$, the outer radius of the tube, we have

$$H_o = H_s(r_2) = CI_0(\alpha r_2) + DK_0(\alpha r_2) \tag{3.86}$$

At the inner radius, r_1 we have

$$H_i = H_s(r_1) = CI_0(\alpha r_1) + DK_0(\alpha r_1) \tag{3.87}$$

We must now find the unknown coefficients C and D. We can use Faraday's law to find a relationship between the field in the shielded region, H_i, and the current density on the inner boundary, r_1. From Faraday's law, we can write the integral of the electric field along the inner boundary as

$$\oint E \cdot d\ell = -\frac{d\psi}{dt} \tag{3.88}$$

where ψ is the flux linking the path of integration. Since, by symmetry, the electric field is constant along the path, this gives

$$2\pi r_1 E = -j\omega\mu_0 H_1 \pi r_1^2 \tag{3.89}$$

Using

$$J = \sigma E \tag{3.90}$$

we have

$$J(r_1) = -\frac{\partial H}{\partial r}\Big|_{r=r_1} = -\frac{\alpha^2 r_1}{2}H_i \tag{3.91}$$

Substituting

$$\frac{\partial H_i}{\partial r} = C\alpha I_0'(\alpha r_1) + D\alpha K_0'(\alpha r_1) = -\frac{\alpha^2 r_1}{2}H_i \tag{3.92}$$

Combining these equations, we obtain

$$H_i = \frac{2H_0}{\alpha^2 r_1^2} \frac{1}{I_0(\alpha r_1)K_2(\alpha r_2) - K_0(\alpha r_1)I_2(\alpha r_2)} \tag{3.93}$$

We can simplify Equation (3.93) if r_1 and r_2 are much greater than δ and $r_1 \approx r_2$. In other words, the radius of the tube is large compared to a skin depth, and the wall thickness is smaller than the skin depth [26, 43]. In this case

$$H_i \approx \sqrt{\frac{r_1}{r_2}}H_0\left(\cosh(\alpha(r_2 - r_1)) + \frac{1}{2}\alpha 1 r_2 \sinh(\alpha(r_2 - r_1))\right) \tag{3.94}$$

The ratio of $|H_i|/|H_o|$ is the shielding factor. If the ratio is close to 1, then there is little effect of the shield. If the ratio is close to 0, then the shielding is very effective. We expect that the effectiveness of the shield would improve as the frequency increases, the thickness of the shield increases and the conductivity of the tube increases.

As a numerical example, let us consider a copper tube with inner diameter $r_1 = 0.2$ m. In this case, the tube thickness is 0.005 m. In Figure 3.23, we see the shielding factor plotted vs. frequency using the exact and approximate formulas.

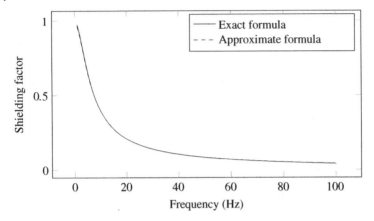

Figure 3.23 Exact and approximate formulas for $r_1 = 0.2$ m and $r_2 = 0.205$ m.

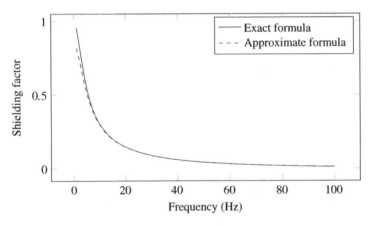

Figure 3.24 Exact and approximate formulas for $r_1 = 0.05$ m and $r_2 = 0.07$ m.

As we expect, at low frequency, the shielding factor approaches 1.0 (no shielding), while at high frequency, the shielding factor goes to zero (perfect shielding). Whether we use Equation (3.93) or (3.94), we get the same results down to 1.0 HZ. This example does meet the criterion of the asymptotic expansion stated above since the radii are large and the wall is thin. If we consider a copper tube of inner radius $r_1 = 0.05$ m and $r_2 = 0.07$ m, then the assumptions are not as good for the approximation to be valid. We investigate this in Figure 3.24, by plotting the shielding factor using both Equations (3.93) and (3.94). We can see that the approximate formula is still quite good until we get to very low frequencies.

Figure 3.25 shows the shielding factor for different tube thicknesses plotted against the logarithm of the frequency. As expected, as the shield thickness

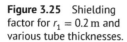

Figure 3.25 Shielding factor for $r_1 = 0.2$ m and various tube thicknesses.

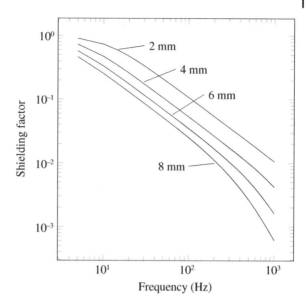

gets greater, the shielding factor is lower, meaning more flux is canceled by the shield.

3.6 Circular Conductors with Transverse Applied Field: Resistance-Limited Case

Consider the case of Figure 3.26 in which we have a long conducting cylinder of circular cross section with a uniform sinusoidally time varying flux density, $B_0 \cos(\omega t)$, applied in the transverse direction. The conductor has radius b, and resistivity, ρ. In this case, the eddy currents are axial and in opposite directions on the top and bottom of Figure 3.26.

The current loop is of rectangular cross section with sides of $2r \sin \theta$. For the resistance limited assumption, the eddy currents do not affect the applied field. The flux linked by the current loop, per unit axial depth, is then

$$\psi = 2B_0 r \sin \theta \cos(\omega t) \tag{3.95}$$

The induced emf per unit axial length is therefore

$$\mathcal{E}(t) = -\frac{d\psi}{dt} = 2\omega B_0 r \sin(\omega t) \tag{3.96}$$

and the RMS value of the induced voltage per unit axial length is

$$\mathcal{E}_{rms} = \sqrt{2}\omega B_0 r \sin \theta \tag{3.97}$$

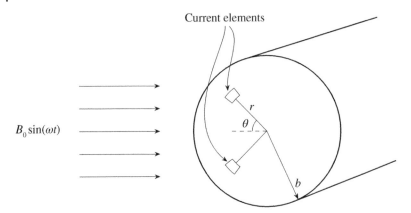

Figure 3.26 Long cylinder with transverse flux.

The resistance of the infinitesimal element per unit axial length is

$$dR = \frac{2\rho}{r\,dr\,d\theta} \tag{3.98}$$

where the factor of 2 accounts for the top and bottom sides of the loop.

The loss in the differential element per unit length is then found by combining Equations (3.97) and (3.98)

$$dW = \frac{\mathcal{E}_{rms}^2}{dR} = \frac{\omega^2 B_0^2 \sin^2\theta}{\rho} r^3\,dr\,d\theta \tag{3.99}$$

Then the total loss per unit length is

$$W = \frac{2\omega^2 B_0^2}{\rho} \int_0^{\pi/2} \sin^2\theta\,d\theta \int_0^b r^3\,dr \tag{3.100}$$

which gives[1]

$$W = \frac{\pi\omega^2 B_0^2 b^4}{8\rho} \tag{3.101}$$

The current density is found by dividing the emf by the path length to obtain the electric field and then dividing by the resistivity. The emf of (3.97) is already in the form of voltage per unit axial length. Therefore, the electric field is found simply by dividing (3.97) by 2 to obtain the value for one side of the path. The RMS current density, which is now a function of r and θ, is then

$$J_{rms} = \frac{\omega B_0 r \sin\theta}{\sqrt{2\rho}} \tag{3.102}$$

---

1 Convert $\sin^2\theta = \frac{1}{2}(1 - \cos(2\theta))$.

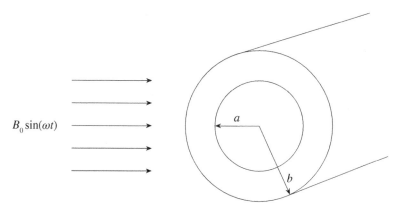

Figure 3.27 Long composite cylinder with transverse flux.

Equation (3.101) can be adapted to the problem of a hollow conducting cylinder with inner radius a and outer radius b as shown in Figure 3.27. In this case, the integral over r is from a to b instead of 0 to b. We obtain for the loss per unit length

$$W = \frac{\pi \omega^2 B_0^2 (b^4 - a^4)}{8\rho} \tag{3.103}$$

The solution for a composite cylinder made of two (or more) layers can be found from superposition of the solid and hollow cylinders (see Figure 3.27). Recall that in the resistance-limited analysis, the eddy currents in one section have no effect on the field or losses in another section.

3.7 Cylindrical Conductor with Applied Transverse Field: Reactance-Limited Case

In Section 3.6 we found the solution of a long conductor with circular cross-section in a transverse sinusoidally time-varying field for the resistance-limited case. Due to the assumption that the eddy current field does not affect the source field, the field remained one-dimensional viewed in Cartesian coordinates. The field was only in the x direction. The eddy currents were only in the z direction and the eddy currents were only a function of y. If we now remove that restriction, and include the effects of the eddy current produced field, we will get a net field in both the x and y directions, or in cylindrical coordinates, we will have a field in the r and θ directions. This makes the solution more complicated. We can simplify the analysis by solving for the magnetic vector potential. The vector potential has only a z component, but by taking the curl, we obtain the radial and peripheral components of the flux density. The time-harmonic form of the diffusion equation for the vector potential in cylindrical coordinates is

$$\frac{\partial^2 A_1}{\partial r^2} + \frac{1}{r}\frac{\partial A_1}{\partial r} + \frac{1}{r^2}\frac{\partial^2 A_1}{\partial \theta^2} = j\omega\mu\sigma A_1 \tag{3.104}$$

In the free-space region surrounding, the conductor we have $\sigma = 0$, and therefore the relevant equation is

$$\frac{\partial^2 A_2}{\partial r^2} + \frac{1}{r}\frac{\partial A_2}{\partial r} + \frac{1}{r^2}\frac{\partial^2 A_2}{\partial \theta^2} = 0 \tag{3.105}$$

In region 1, the conductor, we will assume a solution of the form

$$A_1 = R(r)\sin(n\theta) \tag{3.106}$$

Substituting this into the diffusion equation gives the Bessel equation

$$\frac{\partial^2 R}{\partial r^2} + \frac{1}{r}\frac{\partial R}{\partial r} - \left(j\omega\mu\sigma + \frac{n^2}{r^2} \right)R = 0 \tag{3.107}$$

Let $\kappa = \sqrt{\omega\mu\sigma}$. The general solution for R_1 is

$$R_1 = C_n J_n\left(j^{\frac{3}{2}}\kappa r \right) \quad n = 1, 2, 3, \dots \tag{3.108}$$

Since each one of these terms is a solution to the equation, the sum of all the terms is the general solution.

$$A_1 = \sum_{n=1}^{\infty} C_n J_n\left(j^{\frac{3}{2}}\kappa r \right)\sin(n\theta) \tag{3.109}$$

In the surrounding region, region 2, we use the same process. With $\sigma = 0$ in this region, we assume a solution of the form

$$A_2(r, \theta) = D_n r^{-n} + E_n r^n \quad n = 1, 2, 3, \dots \tag{3.110}$$

The flux density far from the conductor must be the applied field, B_0. Therefore the vector potential in the far field must have the form

$$A(r, \theta) = B_0 r \sin\theta \tag{3.111}$$

This tells us that the only possible positive n term in Equation (3.110) must be the $n = 1$ term. Therefore

$$A_2(r, \theta) = B_0 r \sin\theta + \sum_{n=1}^{\infty} D_n r^{-n}\sin(n\theta) \tag{3.112}$$

To evaluate the constants, we use the curl operator to find the normal, r, and tangential, θ, components of the flux density and magnetic field and match these at the interface.

$$B_r = \frac{1}{r}\frac{\partial A}{\partial \theta}$$

$$B_\theta = -\frac{\partial A}{\partial r} \tag{3.113}$$

After many algebraic manipulations, (see [18]) we obtain, for the vector potential in the conductor

$$A_1(r,\theta) = \frac{4B_0}{j^{\frac{3}{2}}\kappa\left(\mu_1 + \mu_2\right)J_0\left(j^{\frac{3}{2}}\kappa a\right) + (\mu_1 - \mu_2)J_2\left(j^{\frac{3}{2}}\kappa a\right)} \sin\theta \tag{3.114}$$

With Equation (3.114), we can find the eddy current density as

$$J(r,\theta) = j\omega\sigma A(r,\theta)$$

$$= 4B_0 j^{\frac{3}{2}}\kappa \frac{J_1\left(j^{\frac{3}{2}}\kappa r\right)}{(\mu_1 + \mu_2)J_0\left(j^{\frac{3}{2}}\kappa a\right) + (\mu_1 - \mu_2)J_2\left(j^{\frac{3}{2}}\kappa a\right)}\sin\theta \tag{3.115}$$

We can now evaluate Equation (3.115) for a particular example. We will use a copper conductor with radius $a = 0.05\,\text{m}$, conductivity $\sigma = 5.8 \times 10^7\,\text{S m}^{-1}$, and a frequency of 60 HZ. The conductor is excited by a transverse field of $H_0 = 1.0\,\text{A m}^{-1}$. The magnitude of the eddy current density along a radial line from the center in the positive y direction is shown in Figure 3.28. As we expect, the current density is highest at the surface and drops to zero at the center. Since the radius is much larger than the skin depth, we have the approximate exponential decay. Comparing the current density at one skin depth from the surface ($r = 0.0415\,\text{m}$) and the current density at the surface, the ratio is 0.379 or

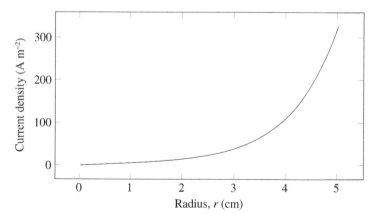

Figure 3.28 Magnitude of current density vs. r, for 60 HZ in copper conductor.

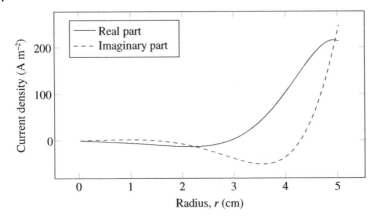

Figure 3.29 Real and imaginary current density vs. r, for 60 HZ in copper conductor.

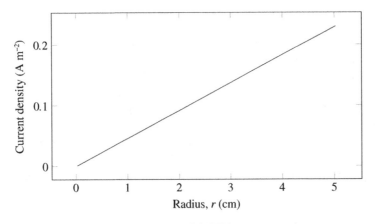

Figure 3.30 Current density vs. r, for 0.01 HZ in copper conductor.

almost exactly $1/e$. Figure 3.29 shows the real and imaginary parts of the current density along the radius.

If we lower the frequency to 0.01 HZ, we obtain the result shown in Figure 3.30. As we found in Section 3.6, the current density distribution is linear, starting at zero at the center and reaching a maximum at the surface. We can easily use the resistance-limited analysis to check the result. If we apply a flux density of 1.0 T, and the area of the loop (from top to bottom) is $2a = 0.1$ m, then the flux linkage is $\psi = B_0 S = 0.01 \, \text{Wb m}^{-1}$. We find the induced voltage magnitude as emf $= \omega \psi = 0.00628$ V. The resistance of the path is $dR = 2/\sigma = 3.448 \times 10^{-8} \, \Omega$. The current density is $J = \text{emf}/dR = 1.822 \times 10^{5} \, \text{A m}^{-2}$ which agrees with the result on Figure 3.30.

Using this expression for the current density, the losses have been computed [18] in terms of the Kelvin functions, ber and bei. The result is

$$P = \frac{4\pi}{\sigma}(\kappa a)^2 B_0^2 \left(\frac{\text{ber}(\kappa a)\,\text{bei}'(\kappa a) - \text{ber}'(\kappa a)\,\text{bei}(\kappa a)}{D} \right) \tag{3.116}$$

where the denominator, D, is

$$D = \left((\mu_1 + \mu_2)\,\text{ber}(\kappa a) + (\mu_1 - \mu_2)\text{ber}_2(\kappa a) \right)^2$$
$$+ \left((\mu_1 + \mu_2)\,\text{bei}(\kappa a) + (\mu_1 - \mu_2)\text{bei}_2(\kappa a) \right)^2 \tag{3.117}$$

As this expression is rather complicated, reference [18] offer asymptotic expansions for the cases in which the $a < \delta$ and $a > \delta$. For the case $a < \delta$, the loss becomes,

$$P = \frac{\pi}{2\sigma}B_0^2 \frac{(\kappa a)^4}{\left(1 + \frac{\mu_1}{\mu_2}\right)^2} \tag{3.118}$$

where μ_1 is the permeability of region 1 and μ_2 is the permeability of the surrounding region. For many of our applications, we will have $\mu_1 = \mu_2$, and in this case, the formula simplifies to

$$P = \frac{\pi}{8\sigma}B_0^2 \left(\frac{a}{\delta}\right)^4 \tag{3.119}$$

Substituting the expression

$$\delta = \sqrt{\frac{2}{\omega\mu\sigma}} \tag{3.120}$$

we obtain exactly the expression found in Section 3.6 where we used the resistance-limited analysis. For the case $a > \delta$, the loss becomes

$$P = \frac{2\pi}{\sigma}B_0^2 \left(\frac{a}{\delta} - \frac{1}{2}\right) \tag{3.121}$$

We will now apply this analysis, and use Equation (3.119) to the problem of a composite circular conductor of radius a with a large number of small insulated conductors, each of radius b which is smaller than the skin depth (see Figure 3.31). In practice, these conductors are twisted so that all of the conductors occupy the same average position in the coil. This means that all of the fine wires have the same impedance and we can safely assume that each strand carries the same current. The self-field from these conductors is peripheral and, by Ampere's law, increases linearly from zero at the center to

$$H = \frac{I}{2\pi a} \tag{3.122}$$

at the outer radius. Considering the individual small wires in the bundle, we have circular conductors in a transverse field and can use the present analysis

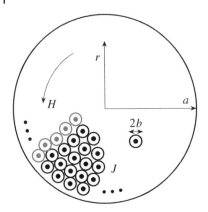

Figure 3.31 Circular composite conductor with fine wires.

to find how we can correct the resistance to account for the eddy current losses. A good analysis of losses in Litz wire is given in Perry [44] and Lammeraner and Štafl [18].

Assume that the number of small conductors in the coil is N. The process will be to find the number of small conductors in a layer, find the flux density in those layers, and use our formulas to find the local loss density. Then we can integrate this quantity over the entire conductor cross section. If the total current in the coil is I, then the current density is

$$J = \frac{I}{N} \frac{p}{\pi b^2} \tag{3.123}$$

where p is the packing factor or fill factor. The packing factor is the ratio of the total area to the area of the conductor. A typical packing factor might be in the range of 0.5–0.6. From Ampere's law, the field at any radius will be

$$H(r) = \frac{Jr}{2} = \frac{I}{2\pi b^2} \frac{p}{N} r \tag{3.124}$$

We also have

$$H = \frac{I}{2\pi a^2} r \tag{3.125}$$

We have found that the eddy current loss in each wire is

$$P = \frac{2\pi}{\sigma} H_0^2 \left(\frac{b}{\delta} \right)^4 \tag{3.126}$$

The number of wires per unit area is $\frac{N}{\pi a^2}$, but this is the same as $\frac{p}{\pi b^2}$. We can then write Equation (3.126) for the loss density as

$$P' = \frac{2\pi}{\sigma} \frac{p}{\pi b^2} H(r)^2 \left(\frac{b}{\delta} \right)^4 \tag{3.127}$$

The loss in a thin annulus as shown in Figure 3.31 is

$$dP = \frac{4\pi}{\sigma} \frac{p}{b^2} H(r)^2 \left(\frac{b}{\delta}\right)^4 r \, dr \tag{3.128}$$

We can now integrate this expression over the cross section of the coil.

$$P = \frac{4\pi}{\sigma} \frac{p}{b^2} \frac{b^4}{\delta^4} \frac{I^2}{4\pi a^4} \int_0^a r^3 \, dr = \frac{I^2}{4\pi \sigma} \frac{f}{b^2} \frac{4b^4}{\delta^4} \tag{3.129}$$

Evaluating the resistance as

$$R = \frac{2P}{I^2} \tag{3.130}$$

We find

$$R = \frac{1}{8\pi\sigma} \frac{p}{b^2} \frac{b^4}{\delta^4} = \frac{1}{8\pi\sigma} \frac{b^2}{\delta^4} \tag{3.131}$$

The losses we have just found are the circulating eddy losses due to the fields produced by the wires in the coil. In finding the resistance, we must also include the losses produced by the load current. We have found this in Section 3.1. The result for the resistance of a circular wire of radius less than the skin depth (low frequency approximation) is repeated here.

$$R'_{ac} = R_0 \left(1 + \frac{1}{48}\left(\frac{b}{\delta}\right)^4\right) \tag{3.132}$$

where

$$R_0 = \frac{1}{\pi\sigma b^2} \tag{3.133}$$

is the dc resistance. Since there are N conductors in parallel, the final contribution must be divided by N. Therefore

$$R_{ac} = \frac{R'_{ac}}{N} = \frac{1}{N\pi\sigma b^2}\left(1 + \frac{1}{48}\left(\frac{b}{\delta}\right)^4\right) \tag{3.134}$$

To find the total effective resistance, we must add Equations (3.131) and (3.134) to obtain

$$R = \frac{1}{8\pi\sigma} \frac{b^2}{\delta^4} + \frac{1}{N\pi\sigma b^2}\left(1 + \frac{1}{48}\left(\frac{b}{\delta}\right)^4\right) \tag{3.135}$$

To illustrate this point, let us consider a numerical example. We will take a copper conductor of radius $a = 0.01\,\text{m}$, composed of many smaller circular conductors. For the conductivity, we use $\sigma = 5.8 \times 10^7\,\text{S}\,\text{m}^{-1}$. Assume the packing factor is $p = 0.5$. This is the ratio of the cross section of the entire conductor, $\pi a^2 = 314.16\,\text{mm}^2$, to the cross section of all of the smaller strands combined. If the wire is made of $N = 400$ strands, then the radius of each strand is $b = a\sqrt{\frac{p}{N}} = 0.00035$ m. The total area of copper is then $N\pi b^2 = 157.08\,\text{mm}^2$. The frequency is $f = 1000\,\text{HZ}$. For copper, the skin depth is then $\delta = 0.0021$ m. We will compare the effective resistance of the stranded wire to the resistance of an equivalent solid wire of radius a.

The exact solution for the ac resistance for the solid circular conductor is

$$R_{\text{ac}_{\text{solid}}} = \Re\left(\frac{j^{3/2}\kappa \; J_0(j^{3/2}\kappa a)}{2\pi a\sigma \; J_1(j^{3/2}\kappa a)}\right) = 0.146 \times 10^{-3}\,\Omega \tag{3.136}$$

where

$$\kappa = \sqrt{\omega\mu\sigma} = 676.72\,\text{m}^{-1} \tag{3.137}$$

We have that

$$\frac{R_{\text{ac}_{\text{solid}}}}{R_{\text{dc}_{\text{solid}}}} = 2.662 \tag{3.138}$$

For the individual strand of radius b, the exact solution is

$$R_{\text{ac}_{\text{strand}}} = \frac{1}{N}\Re\left(\frac{j^{3/2}\kappa \; J_0(j^{3/2}\kappa b)}{2\pi b\sigma \; J_1(j^{3/2}\kappa b)}\right) = 0.1098 \times 10^{-3}\,\Omega \tag{3.139}$$

We find that the ratio of the ac and dc resistance for the individual strand is essentially 1.0 due to the small radius of the strand. This is reasonable since we are in the resistance-limited regime. The ratio of the ac resistance of the stranded conductor to the ac resistance of the solid conductor is then 0.0751 considering only the load current.

We have found previously that the high frequency $(a > \delta)$ approximation for the ratio of ac resistance to dc resistance for the wire is a factor of

$$A = \frac{R_{ac}}{R_{dc}} = \frac{1}{4} + \frac{a}{2\delta} + \frac{3\delta}{32a} = 2.662 \tag{3.140}$$

For the case in which the skin depth is larger that the wire radius, which applies to the small strands, we have the factor

$$B = \left(1 + \frac{1}{48}\frac{b}{\delta}\right)^4 \approx 1.0 \tag{3.141}$$

which agrees with the exact formula.

For the problem of transverse flux, for the case of the skin depth greater than the radius, we have the factor

$$C = \frac{1}{4}\left(\frac{b}{\delta}\right)^4 = 0.000021 \tag{3.142}$$

which also agrees with the exact formula. Therefore

$$\frac{R_{\text{stranded}}}{R_{\text{solid}}} = \frac{B}{p \times A}\left(1 + \frac{p^2}{2}\left(\frac{a}{b}\right)^2\frac{C}{B}\right) = 0.7686 \tag{3.143}$$

This means that, even though we are using only one half the amount of copper as a solid wire, the effective resistance is lower, by a factor of 0.7686, due to the stranding. We note that the assumptions used here are that the large wire radius is greater than the skin depth and the strand radius is smaller than the skin depth. If this is not the case then the full Bessel function solutions should be used. Comparing this result to the ratio of the ac-stranded conductor to the ac solid conductor which was 0.0751 we see the difference is only a factor of 1.02. This means that most of the reduction in resistance is due to the lower losses in the individual strands from the load current and only around 2% is due to the extra loss produced by the peripheral flux in the coil.

3.8 Shielding with a Conducting Tube in a Transverse Field

In Section 3.5, we considered the case of a conducting tube in an axial field. This problem was one-dimensional, with only one component of field (axial) and one component of current (peripheral) [44]. If we now consider a long conducting tube in a transverse field, the fields are two-dimensional. We have fields both in the radial direction (B_r) and peripheral direction, (B_θ).

For the case of a conducting cylinder or tube in a transverse field, we have two components of the magnetic flux density, B_r and B_θ. Consider the problem illustrated in Figure 3.32.

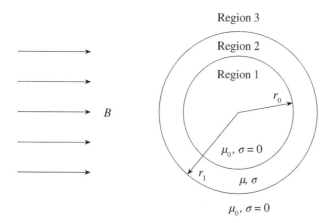

Figure 3.32 Conducting tube in transverse field.

The equation for the radial component in cylindrical coordinates is

$$r^2 \frac{\partial^2 B_r}{\partial r^2} + 3r \frac{\partial B_r}{\partial r} + B_r + \frac{\partial^2 B_r}{\partial \theta^2} + j\omega\mu\sigma r^2 B_r = 0 \tag{3.144}$$

Using $\nabla \cdot B = 0$, we can relate B_r to B_θ as

$$\frac{\partial B_r}{\partial r} + \frac{1}{r}\left(\frac{\partial B_\theta}{\partial \theta} + B_r\right) = 0 \tag{3.145}$$

In regions 1 and 3, we have no conducting material so the last term in Equation (3.144) vanishes, leaving us with

$$r^2 \frac{\partial^2 B_r}{\partial r^2} + 3r \frac{\partial B_r}{\partial r} + B_r + \frac{\partial^2 B_r}{\partial \theta^2} = 0 \tag{3.146}$$

The interface conditions require that the normal flux density (r component) and tangential magnetic field (θ component) are continuous at the inner and outer boundaries of the cylinder. The solution of Equation (3.144) in the conducting region [44] is

$$B_r = B_0 \left(C_1 I_1(\alpha r) + C_2 K_1(\alpha r)\right) \frac{r_1 \cos\theta}{r} \tag{3.147}$$

$$B_\theta = -B_0 \left(C_1 \left(\alpha r I_0(\alpha r) - I_1(\alpha r)\right) - C_2 \left(\alpha r K_0(\alpha r) + K_1(\alpha r)\right)\right) \frac{r_1 \sin\theta}{r} \tag{3.148}$$

Applying the interface conditions, we have the complete solution in the inner region as

$$B_r = -2B_0 \mu_r(\alpha r_1) \left(I_0(\alpha r_0)K_1(\alpha r_0) + K_0(\alpha r_0)I_1(\alpha r_0)\right) \frac{\cos\theta}{D} \tag{3.149}$$

$$B_\theta = 2B_0 \mu_r(\alpha r_1) \left(I_0(\alpha r_0)K_1(\alpha r_0) + K_0(\alpha r_0)I_1(\alpha r_0)\right) \frac{\sin\theta}{D} \tag{3.150}$$

For the outer free space region, we get

$$B_r = B_0 \left(1 - \left(1 - C_1 I_1(\alpha r_1) - C_2 K_1(\alpha r_1)\right) \frac{r_1^2}{r^2}\right) \cos\theta \tag{3.151}$$

$$B_\theta = -B_0 \left(1 + \left(1 - C_1 I_1(\alpha r_1) - C_2 K_1(\alpha r_1)\right) \frac{r_1^2}{r^2}\right) \sin\theta \tag{3.152}$$

The constants are given by

$$C_1 = -2\mu_r \left((1 + \mu_r)K_1(\alpha r_0) + \alpha r_0 K_0(\alpha r_0)\right) / D$$
$$C_2 = 2\mu_r \left((1 + \mu_r)I_1(\alpha r_0) - \alpha r_0 I_0(\alpha r_0)\right) / D$$
$$D = \left((\mu_r - 1)K_1(\alpha r_1) - \alpha r_1 K_0(\alpha r_1)\right)\left((\mu_r + 1)I_1(\alpha r_0) - \alpha r_0 I_0(\alpha r_0)\right)$$
$$- \left((\mu_r + 1)K_1(\alpha r_0) + \alpha r_0 K_0(\alpha r_0)\right)\left((\mu_r - 1)I_1(\alpha r_1) + \alpha r_1 I_0(\alpha r_1)\right) \tag{3.153}$$

Figure 3.33 Flux density in outer region.

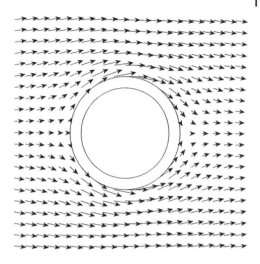

As an example, consider the case of a hollow copper conductor of inner radius $r_0 = 0.08$ m and outer radius $r_1 = 0.1$ m. The conductivity of the copper is $\sigma = 5.8 \times 10^7$ S m^{-1}. There is a transverse field applied in the x direction of magnitude 1.0 T at 60 HZ. We see in Figure 3.33 the flux density in the outer region computed using Equations (3.151) and (3.152). The arrows indicate the direction of the field and the length of the arrows is proportional to the field strength. We see that the eddy currents in the conductor are shielding the interior of the conductor.

We now consider the same geometry and look at the shielding factor as a function of frequency. Since we are applying an exterior field of 1.0 T, we use Equations (3.149) and (3.150) to evaluate the field in the interior region, which is constant. This interior field is then the shielding ratio. A ratio of 1.0 means there is no shielding and a ratio of 0.0 means perfect shielding. We see in Figure 3.34 the shielding ratio vs. frequency. As we expect, for low frequency, we have very little shielding, and as the frequency increases the shielding increases.

3.9 Spherical Conductor in a Uniform Sinusoidally Time-Varying Field: Resistance-Limited Case

Consider now the case of a conducting sphere in a uniform time-varying field. We will consider the field applied in the z direction, but of course, with a sphere the direction is arbitrary. Using the resistance-limited assumptions, the eddy currents in the sphere will not affect the applied field. The geometry is shown in Figure 3.35.

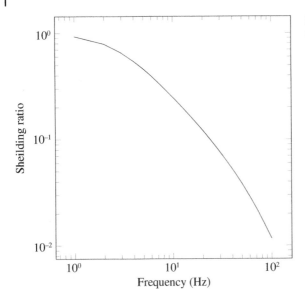

Figure 3.34 Shielding ratio vs. frequency.

To find the emf in the infinitesimal loop in Figure 3.35, we see that the area circumscribed by the loop is

$$S = \pi (r \sin \theta)^2 \tag{3.154}$$

The flux linked by the loop is then

$$\psi(t) = B_0 \pi (r \sin \theta)^2 \cos(\omega t) \tag{3.155}$$

The emf induced in the loop is therefore

$$\mathcal{E}(t) = -\frac{d\psi}{dt} = B_0 \omega \pi (r \sin \theta)^2 \sin(\omega t) \tag{3.156}$$

the RMS value of which is

$$\mathcal{E}_{\text{rms}} = \frac{B_0}{\sqrt{2}} \omega \pi r^2 \sin^2 \theta \tag{3.157}$$

The elemental resistance of the filament is

$$dR = \frac{\rho 2\pi r \sin \theta}{r \, dr \, d\theta} = \frac{\rho 2\pi \sin \theta}{dr \, d\theta} \tag{3.158}$$

The loss in the differential loop is

$$dW = \frac{\mathcal{E}_{\text{rms}}^2}{dR} = \frac{B_0^2 \omega^2 \pi r^4 \sin^3 \theta}{4\rho} \, dr \, d\theta \tag{3.159}$$

Figure 3.35   Conducting sphere with applied field.

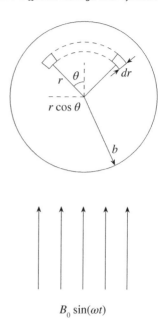

$B_0 \sin(\omega t)$

and the total loss in the sphere is then[2]

$$W = \int_0^\pi \int_0^b \frac{B_0^2 \omega^2 \pi r^4 \sin^3\theta}{4\rho} \, dr \, d\theta = \frac{\pi B_0^2 \omega^2 b^5}{15\rho} \tag{3.160}$$

If we have a hollow sphere as shown in Figure 3.36, the process is the same but we would integrate over r from a to b instead of 0 to b. The resulting loss is

$$W = \frac{\pi B_0^2 \omega^2 (b^5 - a^5)}{15\rho} \tag{3.161}$$

As in the case of the cylinder, we can now make a composite sphere of two or more materials by combining the solutions for the hollow and solid spheres.

3.10   Diffusion Through Thin Cylinders

We have considered the eddy currents produced by alternating flux impinging on conducting hollow cylinders. These previously examples considered only steady-state ac excitation. In this section, we will analyze the eddy currents in

---

2 For the integral of $\sin^3\theta$ we use $\sin^3\theta = \sin^2\theta \sin\theta = (1 - \cos^2\theta)\sin\theta$. Then, let $x = \cos\theta$, $dx = -\sin\theta d\theta$. We are then integrating $-\int(1 - x^2)dx = -x + x^3/3$. We now evaluate $-\cos\theta + \frac{\cos^3\theta}{3}$ from 0 to π.

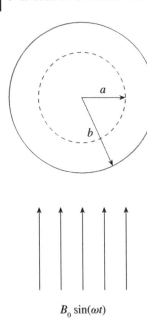

Figure 3.36 Conducting hollow sphere with applied uniform field.

$B_0 \sin(\omega t)$

thin conducting tubes with a step function of field applied in the axial direction and in the transverse direction. One feature of this analysis is that since the conductor is assumed to be thin, we can assume that the current density in the tube is constant in the radial direction. This also allows us to use scalar potential analysis to find the field in the nonconducting regions. The discontinuity in the magnetic potential will give us the current in the cylinder.

3.10.1 Diffusion Through Hollow Cylinder with Applied Axial Magnetic Field

First, let us consider the long cylinder with a thin wall shown in Figure 3.37. If we apply an axial field, we will induce currents in the peripheral direction. In this case, by symmetry, the field inside the tube is constant. We will assume that the cylinder is made of conducting material with homogeneous conductivity, σ. The relative permeability of the cylinder is $\mu_r = 1.0$. The cylinder has radius a and thickness b and is thin enough that we may consider the current density in the conductor uniform through the thickness. We apply a magnetic field, H_o, in the y direction, parallel to the axis of the cylinder. Applying Ampere's law to the contour shown in the figure, we obtain an expression relating the field inside the cylinder H_i to the field outside H_o and the surface current density, J_s. Using

$$\oint H \cdot d\ell = I_{enclosed} \tag{3.162}$$

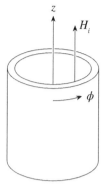

Figure 3.37 Thin conducting cylinder in applied axial field.

we find

$$-H_o + H_i = J_s \tag{3.163}$$

From Faraday's law, we have

$$\oint E \cdot d\ell = -\frac{d}{dt} \oint B \cdot dS \tag{3.164}$$

The current density in the conductor can be written in terms of the surface current as

$$J_s = Jb = \sigma Eb \tag{3.165}$$

So that

$$E = \frac{J_s}{\sigma b} \tag{3.166}$$

Applying Faraday's law

$$\frac{2\pi a J_s}{\sigma b} = -\frac{d}{dt}\left(\mu_0 2\pi a^2 H_i^2\right) \tag{3.167}$$

giving

$$-\frac{d}{dt}H_i = \frac{2J_s}{\mu_0 \sigma ba} = \frac{2\left(-H_o + H_i\right)}{\mu_0 \sigma ba} \tag{3.168}$$

Defining the time constant as

$$\tau = \frac{1}{2}\mu_0 \sigma ab \tag{3.169}$$

We have

$$\frac{dH_i}{dt} + \frac{H_i}{\tau} = \frac{H_o}{\tau} \tag{3.170}$$

This first-order differential equation has a particular solution given by the applied constant field, $H_i = H_o$. The homogeneous solution is $H_o e^{-\frac{t}{\tau}}$. Resulting in the complete solution

$$H_i = H_o \left(1 - e^{-\frac{t}{\tau}}\right) \tag{3.171}$$

We also have an expression for the current sheet

$$J_s = -H_o e^{-\frac{t}{\tau}} \tag{3.172}$$

By looking at the form of Equation (3.171), we see that at the instant of the application of the external field, the conductor *perfectly* shields the interior of the cylinder. As time progresses, we reach a steady state, in which the field completely penetrates the cylinder and the field inside the cylinder is equal to the external applied field.

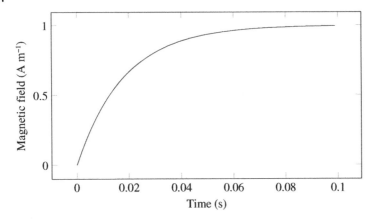

Figure 3.38 Magnetic field inside copper cylinder vs. time.

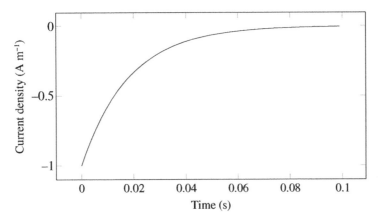

Figure 3.39 Surface current density vs. time.

Let us consider a numerical example. We will apply a step function of magnetic field along the axis of a conducting tube at $t = 0^+$. The cylinder is made of copper. The radius is 0.1 m and the thickness is 0.005 m. The applied magnetic field is $1.0\,\mathrm{A\,m^{-1}}$. The time constant is found from Equation (3.169) and is equal to $\tau = 0.01822\,\mathrm{s}$. The solution for the magnetic field is shown in Figure 3.38. As we expect, the field starts out at zero and then approaches the applied value of $H = 1.0\,\mathrm{A\,m^{-1}}$ exponentially.

In Figure 3.39, we see the results for the current sheet in the conductor vs. time. As the figure shows, the initial current sheet density is $-1.0\,\mathrm{A\,m^{-1}}$ which is exactly the magnitude of the current sheet necessary to completely cancel the applied field. As the current decays to zero, the field inside the tube approaches

Figure 3.40 Thin conducting cylinder with magnetic core.

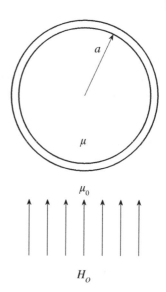

the applied value of field. The difference between the outside and inside tangential field is equal to the value of the surface current sheet.

3.10.2 Diffusion Through Hollow Cylinder with Applied Transverse Magnetic Field

Previously, we have looked at the problem of a hollow conducting cylinder with transverse applied field. We have done this for steady-state sinusoidal excitation. We will now look at this problem with the application of a step function of transverse field. Haus and Melcher [45] describes the problem of a thin conducting cylinder with a permeable, non-conducting core, excited by the sudden application of a dc magnetic field. The geometry is shown in Figure 3.40.

In this case, since the conducting region is very thin and we will not be finding the field inside this region, we can represent the exterior magnetic field as the gradient of a scalar potential. We recall that the magnetic potential and field are related by

$$H = -\nabla \psi \tag{3.173}$$

If we assume that far from the cylinder, the field is uniform with a value of H_0, we can represent the potential as [3]

$$\psi = -H_0 r \cos \phi \tag{3.174}$$

---

[3] To verify this, we use $r^2 = x^2 + y^2$, $\cos(\phi) = x/r$ and $\sin(\phi) = y/r$ which gives for the gradient in rectangular coordinates, $\frac{\partial f}{\partial x} = \cos(\phi)\hat{\mathbf{u}}_r - \frac{1}{r}\sin(\phi)\hat{\mathbf{u}}_\phi$ and $\frac{\partial f}{\partial y} = \sin(\phi)\hat{\mathbf{u}}_r + \frac{1}{r}\cos(\phi)\hat{\mathbf{u}}_\phi$. Substituting Equation (3.174) we obtain $H_x = H_0$ and $H_y = 0$.

We will assume a solution of the form

$$\psi_o = -H_0 r \cos\phi + C_1 \frac{\cos(\phi)}{r} \tag{3.175}$$

Note that the first term is the potential far from the cylinder. Inside the cylinder, we assume

$$\psi = C_2 r \cos\phi \tag{3.176}$$

The constants C_1 and C_2 are found from matching interface conditions.

Substituting Equations (3.175) and (3.176) into Equation (3.173) gives

$$B_o = \mu_0 \left(H_0 + \frac{C_1}{r^2}\right)\cos\phi\,\hat{\mathbf{u}}_r - \mu_0 \left(H_0 - \frac{C_1}{r^2}\right)\sin\phi\,\hat{\mathbf{u}}_\phi \tag{3.177}$$

and

$$B_i = -\mu C_2 \left(\cos\phi\,\hat{\mathbf{u}}_r - \sin\phi\,\hat{\mathbf{u}}_\phi\right) \tag{3.178}$$

We require that the normal component of flux density is continuous and the tangential components of field are related by the surface current density. In matching the boundary conditions, we note that both the inside and outside flux densities vary as $\sin\phi$.

$$\mu_0 \left(H_0 + \frac{C_1}{a^2}\right) = -\mu C_2 \tag{3.179}$$

$$-\frac{1}{ab\sigma}\left(H_0 - \frac{C_1}{a^2}\right) - \frac{\mu_0}{\mu ab\sigma}\left(H_0 - \frac{C_1}{a^2}\right) = \mu_0 \left(\frac{dH_0}{dt} + \frac{1}{a^2}\frac{dC_1}{dt}\right) \tag{3.180}$$

Rearranging we have

$$\frac{dC_1}{dt} + \frac{C_1}{\tau} = -a^2\frac{dH_0}{dt} - \frac{H_0 a}{\mu_0 b\sigma}\left(\frac{\mu - \mu_0}{\mu_0}\right) \tag{3.181}$$

where

$$\tau = \frac{\mu}{\mu_0 + \mu}\mu_0\sigma ba \tag{3.182}$$

As an example, let us consider the application of a step function of field produced by an ideal source. The time-domain solution is made of the steady-state solution plus the transient solution. We have a solution of the form

$$C_1 = H_0 a^2 \left(\frac{r}{a} - \frac{a}{r}\left(1 - e^{-t/\tau}\right) - e^{-t/\tau}\right) \tag{3.183}$$

We now evaluate C_2 from (3.179) which results in

$$\psi_o = -H_0 a \left(\frac{r}{a} - \frac{a}{r}\left(\frac{(\mu - \mu_0)}{(\mu + \mu_0)}\left(1 - e^{-t/\tau}\right) - e^{-t/\tau}\right)\right)\cos\phi \tag{3.184}$$

$$\psi_i = -H_0 a \frac{r}{a}\frac{2\mu_0}{a(\mu + \mu_0)}\left(1 - e^{-t/\tau}\right)\cos\phi \tag{3.185}$$

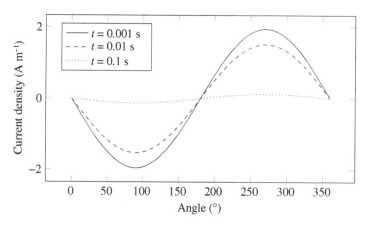

Figure 3.41 Current density around the cylinder at different times.

Recognizing that the current sheet at the cylinder surface is equal to the difference between the outside field and the inside field. This gives

$$J(\theta, t) = -2H_0 e^{-t/\tau} \sin \theta \tag{3.186}$$

Consider the case of a conducting cylinder with a permeable core with $\mu_r = 100$. The conducting cylinder is made of copper and has a radius of 0.1 m and a thickness of 0.005 m. We apply a step function of field equal to $H = 1.0\,A\,m^{-1}$. From Equation (3.182), we find $\tau = 0.0361$ s. In Figure 3.41, we see the current density around the cylinder at different times. As expected, initially the current is high and will cancel the interior field. At later times, we see that the current decays and after a few times constants, the current is quite small.

3.11 Surface Impedance Formulation for Electric Machines

Using the methods we have introduced in this chapter, we can treat a problem that has been useful in the analysis of electric machines. The geometry will be cylindrical, but the fields will vary sinusoidally in the peripheral direction. This has applications in solid rotor induction machines as well as the study of shielding in superconducting machines. Consider a circular cross section of an electrical machine comprising three regions as shown in Figure 3.42.

The inner-most radius, r_1, is the radius of the rotor indicating region 1, the middle radius, r_2, is the outer radius of the air-gap region, and r_3 is the outer radius of the nonconducting stator region. The inner circular cross section (region 1) represents the eddy current region, such as a round rotor of an electrical machine.

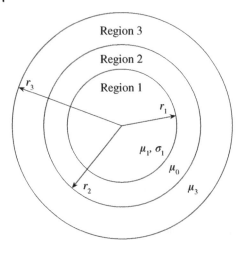

Figure 3.42 Cylindrical electric machine geometry.

The middle region is the air-gap annulus and the outer region is the non-eddy current stator region consisting of thin laminations stacked in the axial direction. We shall now investigate the magnetic field solution in the cross section.

The assumptions made in this analysis are that the problem is two-dimensional. There is no variation of any variable in the z direction. We also assume that the material properties are linear and homogeneous. The analysis assumes only time-harmonic or phasor quantities. The appropriate field equations which apply are

$$\nabla \times A = B \tag{3.187}$$

$$\nabla \times H = J_e \tag{3.188}$$

$$\nabla \times E = -\frac{\partial B}{\partial t} = -\frac{\partial}{\partial t}\nabla \times A \tag{3.189}$$

$$B = \mu_r \mu_0 H \tag{3.190}$$

$$J_e = \sigma E \tag{3.191}$$

Combining these gives

$$\nabla \times H = \frac{1}{\mu_r \mu_0}\nabla \times B = \frac{1}{\mu_r \mu_0}\nabla \times \nabla \times A = J_e \tag{3.192}$$

$$E = -\frac{\partial A}{\partial t} \tag{3.193}$$

In cylindrical coordinates

$$B = \nabla \times A = \frac{1}{r}\frac{\partial A}{\partial \theta}\hat{u}_r - \frac{\partial A}{\partial r}\hat{u}_\theta \tag{3.194}$$

Manipulating Equations (3.191), (3.193), and (3.194), and using vector identities, we obtain the partial differential equation in cylindrical polar coordinates as

$$r^2\frac{\partial^2 A_z}{\partial r^2} + r\frac{\partial A_z}{\partial r} + \frac{\partial^2 A_z}{\partial \theta^2} = j\omega\mu\sigma r^2 A_z \tag{3.195}$$

Using separation of variables, we assume the solution is of the form

$$A_z = R(r)\Theta(\theta) \tag{3.196}$$

The diffusion equation can now be written in the form

$$r^2\frac{\partial^2 R(r)}{\partial r^2}\Theta(\theta) + r\frac{\partial R(r)}{\partial r}\Theta(\theta) + \frac{\partial^2 \Theta}{\partial \theta^2}R(r) = j\omega\sigma\mu r^2 R(r)\Theta(\theta) \tag{3.197}$$

Dividing Equation (3.197) by $R(r)\Theta(\theta)$, we get

$$\frac{r^2}{R(r)}\frac{\partial^2 R(r)}{\partial r^2} + \frac{r}{R(r)}\frac{\partial R(r)}{\partial r} + \frac{1}{\Theta(\theta)}\frac{\partial^2 \Theta}{\partial \theta^2} = j\omega\sigma\mu r^2 R \tag{3.198}$$

We now separate the terms with $R(r)$ and $\Theta(\theta)$ so that

$$r^2\frac{\partial^2 R(r)}{\partial r^2} + r\frac{\partial R(r)}{\partial r} = \left(j\omega\mu\sigma r^2 + n^2\right)R(r) \tag{3.199}$$

$$\frac{\partial^2 \Theta(\theta)}{\partial \theta^2} = -\left(n^2\right)\Theta(\theta) \tag{3.200}$$

Equation (3.199) is the modified Bessel equation of order n. The solution is

$$R(r) = \sum_{n=1}^{\infty} C\left(\mathrm{ber}_n(kr) + j\,\mathrm{bei}_n(kr)\right) \tag{3.201}$$

where $k^2 = \omega\sigma\mu$. The solution to Equation (3.200) is

$$\Theta(\theta) = F\cos(n\theta) + G\sin(n\theta) \tag{3.202}$$

In the current problem, we have even symmetry, so only the cosine terms are considered. We conclude that the magnetic vector potential is described by

$$A_z = \sum_{n=1}^{\infty} C\left(\mathrm{ber}_n(kr) + j\,\mathrm{bei}_n(kr)\right)\cos(n\theta) \tag{3.203}$$

For the two-dimensional problem, under consideration

$$B_r = \frac{1}{r}\frac{\partial A_z}{\partial \theta} \tag{3.204}$$

$$B_\theta = -\frac{\partial A_z}{\partial r} \tag{3.205}$$

so

$$B_r = \sum_{n=1}^{\infty} -\frac{n}{r} \left(\text{ber}_n(kr) + j\,\text{bei}_n(kr) \right) \sin(n\theta) \tag{3.206}$$

$$B_\theta = \sum_{n=1}^{\infty} -k \left(\text{ber}'_n(kr) + j\,\text{bei}'_n(kr) \right) \cos(n\theta) \tag{3.207}$$

In regions 2 and 3, we have $k = 0$, and therefore

$$A_z = \sum_{n=1}^{\infty} C \left(Cr^n + Dr^{-n} \right) \cos(n\theta) \tag{3.208}$$

To summarize, we find the following expressions for the vector potential A_z and the flux density components B_r and B_θ.

For region 1:

$$A_{z1} = \sum_{n=1}^{\infty} C_1 \left(\text{ber}_n(k_1 r) + j\,\text{bei}_n(k_1 r) \right) \cos(n\theta) \tag{3.209}$$

$$B_{r_1} = \sum_{n=1}^{\infty} -\frac{n}{r} C_1 \left(\text{ber}_n(k_1 r) + j\,\text{bei}_n(k_1 r) \right) \sin(n\theta) \tag{3.210}$$

$$B_{\theta_1} = \sum_{n=1}^{\infty} -k_1 C_1 \left(\text{ber}'_n(k_1 r) + j\,\text{bei}'_n(k_1 r) \right) \cos(n\theta) \tag{3.211}$$

In region 2:

$$A_{z2} = \sum_{n=1}^{\infty} \left(C_2 r^n + D_2 r^{-n} \right) \cos(n\theta) \tag{3.212}$$

$$B_{r_2} = \sum_{n=1}^{\infty} -n \left(C_2 r^{n-1} + D_2 r^{-n-1} \right) \sin(n\theta) \tag{3.213}$$

$$B_{\theta_2} = \sum_{n=1}^{\infty} -n \left(C_2 r^{n-1} - D_2 r^{-n-1} \right) \cos(n\theta) \tag{3.214}$$

In region 3:

$$A_{z3} = \sum_{n=1}^{\infty} \left(C_3 r^n + D_3 r^{-n} \right) \cos(n\theta) \tag{3.215}$$

$$B_{r_3} = \sum_{n=1}^{\infty} -n \left(C_3 r^{n-1} + D_3 r^{-n-1} \right) \sin(n\theta) \tag{3.216}$$

$$B_{\theta_3} = \sum_{n=1}^{\infty} -n \left(C_3 r^{n-1} - D_3 r^{-n-1} \right) \cos(n\theta) \tag{3.217}$$

We can evaluate the constants by applying the appropriate interface conditions at the boundaries. At $r = r_1$, the normal flux density is continuous and the tangential magnetic field is also continuous.

Therefore at $r = r_1$

$$B_{r_1} = B_{r_2} \tag{3.218}$$

and

$$B_{\theta_1} = B_{\theta_2}\mu_{r1} \tag{3.219}$$

And at $r = r_2$

$$B_{r_2} = B_{r_3} \tag{3.220}$$

and

$$B_{\theta_3} = B_{\theta_2}\mu_{r3} \tag{3.221}$$

First, considering the $r = r_1$ interface

$$-\frac{n}{r_1}C_1\left(\mathrm{ber}_n(k_1 r_1) + j\,\mathrm{bei}_n(k_1 r_1)\right) = -n\left(C_2 r_1^{n-1} + D_2 r_1^{-n-1}\right) \tag{3.222}$$

$$-\frac{k_1}{\mu_{r1}}C_1\left(\mathrm{ber}_n'(k_1 r_1) + j\,\mathrm{bei}_n'(k_1 r_1)\right) = -n\left(C_2 r_1^{n-1} - D_2 r_1^{-n-1}\right) \tag{3.223}$$

With these two equations we find

$$C_2 = \frac{C_1}{2}r^{-n+1}\left(\frac{n}{r_1}C_1\left(\mathrm{ber}_n(k_1 r_1) + j\,\mathrm{bei}_n(k_1 r_1)\right) + \frac{k_1}{\mu_{r1}}\left(\mathrm{ber}_n'(k_1 r_1) + j\,\mathrm{bei}'(k_1 r_1)\right)\right) \tag{3.224}$$

$$D_2 = \frac{C_1}{2}r^{n+1}\left(\frac{n}{r_1}C_1\left(\mathrm{ber}_n(k_1 r_1) + j\,\mathrm{bei}_n(k_1 r_1)\right) - \frac{k_1}{\mu_{r1}}\left(\mathrm{ber}_n'(k_1 r_1) + j\,\mathrm{bei}'(k_1 r_1)\right)\right) \tag{3.225}$$

Now, for the $r = r_2$ boundary

$$C_3 r_2^{n-1} + D_3 r_2^{-n-1} - C_2 r_2^{n-1} - D_2 r_2^{-n-1} = 0 \tag{3.226}$$

$$\frac{C_3 r_2^{n-1} - D_3 r_2^{-n-1}}{\mu_{r3}} - C_2 r_2^{n-1} + D_2 r_2^{-n-1} = \mu_0 J_s \tag{3.227}$$

These two simultaneous equations give

$$C_3 r_2^{n-1}\left(1 + \frac{1}{\mu_{r3}}\right) + D_3 r_2^{-n-1}\left(1 - \frac{1}{\mu_{r3}}\right) - 2C_2 r_2^{n-1} = \mu_0 J_s \tag{3.228}$$

$$C_3 r_2^{n-1}\left(1 - \frac{1}{\mu_{r3}}\right) + D_3 r_2^{-n-1}\left(1 + \frac{1}{\mu_{r3}}\right) - 2D_2 r_2^{-n-1} = -\mu_0 J_s \tag{3.229}$$

We notice that at $r = r_3$, we have $A_{z3} = 0$, so that

$$D_3 = -C_3 r_3^{2n} \tag{3.230}$$

Defining

$$P = \left(1 + \frac{1}{\mu_{r3}}\right)$$

$$Q = \left(1 - \frac{1}{\mu_{r3}}\right) \tag{3.231}$$

After some simplification, we obtain

$$C_3 \left[P - Q\left(\frac{r_3}{r_2}\right)^{2n}\right] - 2C_2 = \mu_0 J_s r_2^{-n+1} \tag{3.232}$$

$$C_3 \left[Q - P\left(\frac{r_3}{r_2}\right)^{2n}\right] - 2D_2 r_2^{-2n} = -\mu_0 J_s r_2^{-n+1} \tag{3.233}$$

Eliminating C_3 from these equations, and after simplification, we get

$$C_2 - D_2 \frac{\left[P - Q\left(\frac{r_3}{r_2}\right)^{2n}\right] r_2^{-2n}}{\left[Q - P\left(\frac{r_3}{r_2}\right)^{2n}\right]} = -\frac{\mu_0 J_s}{2} r_2^{-n+1}(P + Q)\frac{\left[1 - \left(\frac{r_3}{r_2}\right)^{2n}\right]}{\left[Q - P\left(\frac{r_3}{r_2}\right)^{2n}\right]} \tag{3.234}$$

Defining

$$S = \frac{n}{r_1}\left(\mathrm{ber}_n(k_1 r_1) + j\,\mathrm{bei}_n(k_1 r_1)\right) + \frac{k_1}{\mu_{r1}}\left(\mathrm{ber}'_n(k_1 r_1) + j\,\mathrm{bei}'_n(k_1 r_1)\right) \tag{3.235}$$

and

$$T = \frac{n}{r_1}\left(\mathrm{ber}_n(k_1 r_1) + j\,\mathrm{bei}_n(k_1 r_1)\right) + -\frac{k_1}{\mu_{r1}}\left(\mathrm{ber}'_n(k_1 r_1) + j\,\mathrm{bei}'_n(k_1 r_1)\right) \tag{3.236}$$

We find

$$C_2 = \frac{C_1}{2n} r_1^{-n+1}(S + T) \tag{3.237}$$

and

$$D_2 = \frac{C_1}{2n} r_1^{n+1}(S - T) \tag{3.238}$$

Solving for C_1

$$C_1 = -\frac{\mu_0 J_s n \left(\frac{r_1^{n-1}}{r_2^{n+1}}\right)(P + Q)\left[1 - \left(\frac{r_3}{r_2}\right)^{2n}\right]}{\left\{\left[(S + T)\left[Q - P\left(\frac{r_3}{r_2}\right)^{2n}\right]\right] - (S - T)\right\}} \tag{3.239}$$

So that

$$A_{z1} = \sum_{n=1}^{\infty} C_1 \left(\mathrm{ber}_n(k_1 r) + j\,\mathrm{bei}_n(k_1 r)\right) \cos(n\theta) \tag{3.240}$$

At $r = r_1$

$$A_{z1} = \sum_{n=1}^{\infty} C_1 \left(\mathrm{ber}_n(k_1 r_1) + j\,\mathrm{bei}_n(k_1 r_1) \right) \cos(n\theta) \tag{3.241}$$

The electric field is found as $E = -j\omega A_{z1}$ and the total current is then

$$I = \int_0^{\frac{\pi}{2}} J_s \cos(n\theta)\, d\theta = \frac{J_s}{n} \sin\left(\frac{n\pi}{2} \right) \tag{3.242}$$

The surface impedance per unit length is then defined as

$$Z_s = -\frac{j\omega A_{z1}}{I} \tag{3.243}$$

Substituting we get

$$Z_s = \sum_{n=1}^{\infty} \frac{-j\omega \mu_0 n^2 \left(\frac{r_1^{n-1}}{r_2^{n+1}} \right)(P+Q)\left[1 - \left(\frac{r_3}{r_2} \right)^{2n} \right] \left(\mathrm{ber}_n(k_1 r_1) + j\,\mathrm{bei}_n(k_1 r_1) \right)}{\sin\left(\frac{n\pi}{2} \right)\left[(S+T)\left[Q - P\left(\frac{r_3}{r_2} \right)^{2n} \right] - (S-T) \right]} \cos(n\theta) \tag{3.244}$$

This formulation can now be used to find the field and eddy current distribution in the machine and the field in the air gap. From these, other quantities such as energy and torque can also be evaluated.

3.12 Summary

In this chapter, we have applied our techniques to analyze conductor geometry with circular cross sections and cylindrical symmetry. We have analyzed these using field variables as well as vector and scalar potentials. The problems include conductors with axial current, applied axial fields, and applied transverse fields. These were done in the steady-state ac domain using complex phasor analysis and in the time domain in which the differential equations were solved directly. We discussed the shielding effects of the eddy currents and the diffusion time constants involved. Many of these solutions involved Bessel functions or modified Bessel functions. In some cases, asymptotic expansions were introduced to replace the Bessel functions and these considerably simplify the analysis. We also considered multi-layered circular conductors and stranded conductors. Finally, an application involving sinusoidally distributed fields that is applicable to electric machines was introduced.

Part II

Modeling

4

Formulations

In Chapters 2 and 3 we have presented a number of examples of practical importance but with simple regular geometries. These also had rather simple boundary conditions and with some exceptions, had linear homogeneous material properties. Many problems however have irregular boundaries, non-homogeneous and non-isotropic materials, and perhaps complex boundary conditions. These problems have no closed-form solutions. These problems require numerical methods which will typically involve large systems of equations and often iterative techniques. Numerical methods give approximate solutions to the field and eddy current problems, but ones which have proven very useful in the analysis of important practical applications. Surprisingly, the methods that we will introduce in Chapters 5-7, the finite difference method, finite element method, and integral equation formulations, often result in physical models that can be used to understand the eddy current solutions. We shall see that all of these methods result in equivalent circuit interpretations that help in the understanding of the eddy current or magnetic diffusion results.

We have seen in Chapters 2 and 3 that eddy current problems can be described by the diffusion equation for the magnetic field or the current density. For numerical method applications, we often rewrite the equations in terms of potentials, either scalar, vector, or a combination of the two. These are often easier to formulate and allow more freedom in incorporating the boundary conditions. In this chapter, we will introduce some of the most common formulations for eddy currents.

4.1 Mathematical Formulations for Eddy Current Modeling

Many problems have both conducting and non-conducting regions. In the conducting regions, where we may have eddy currents, we cannot use a scalar potential formulation. A vector formulation results in three unknowns at each

Eddy Currents: Theory, Modeling, and Applications, First Edition.
Sheppard J. Salon, M. V. K. Chari, Lale T. Ergene, David Burow, and Mark DeBortoli.
© 2024 The Institute of Electrical and Electronics Engineers, Inc. Published 2024 by John Wiley & Sons, Inc.

nodal point, while in scalar potential formulations, we need a single unknown at the node. It is often possible to use vector unknowns in the current-carrying regions and scalar unknowns in the nonconducting regions. We have seen that in two-dimensional eddy current problems we often use the diffusion equation for the magnetic vector potential. This has the advantage that we only need to compute one component of the vector potential so that the mathematics is essentially the same as in a scalar solution. We have already made the argument for ignoring the displacement currents. For the steady-state time-harmonic case, we have, in Chapters 2 and 3, used the single component MVP formulation.

$$\frac{\partial^2 A_z}{\partial x^2} + \frac{\partial^2 A_z}{\partial y^2} = -\mu J_0 + j\omega\mu\sigma A_z \tag{4.1}$$

This is not the only choice available to us, and we will consider some of the other options that are currently in use for eddy current analysis. First, let us look at the full set of Maxwell's equations and formulate the problem in terms of the field variables directly.

$$\nabla \times \mathbf{E} = -\frac{\partial \mathbf{B}}{\partial t} \tag{4.2}$$

$$\nabla \times \mathbf{H} = \mathbf{J} \tag{4.3}$$

$$\nabla \cdot \mathbf{D} = \rho \tag{4.4}$$

$$\nabla \cdot \mathbf{B} = 0 \tag{4.5}$$

We have five variables and four equations. For the general three-dimensional analysis, the unknowns will all be vectors. This means that there are 15 variables. We can use the constitutive relationships to eliminate some of these unknowns.

$$\mathbf{D} = \varepsilon\mathbf{E} \tag{4.6}$$

$$\mathbf{B} = \mu\mathbf{H} \tag{4.7}$$

$$\mathbf{J} = \sigma\mathbf{E} \tag{4.8}$$

For example, we can eliminate flux densities, $\mathbf{D}$ and $\mathbf{B}$, and solve for the six components of fields, $\mathbf{E}$ and $\mathbf{H}$.

To reduce the number of unknowns, we transform these first-order equations into second-order equations. Taking the curl of Equation (4.3)

$$\nabla \times \nabla \times \mathbf{H} = \nabla \times \mathbf{J} = \nabla \times \sigma\mathbf{E} = -\sigma\frac{\partial \mathbf{B}}{\partial t} \tag{4.9}$$

We use the vector identity

$$\nabla \times \nabla \times \mathbf{F} = \nabla\nabla \cdot \mathbf{F} - \nabla^2\mathbf{F}$$

to get

$$\nabla^2\mathbf{H} - \sigma\mu\frac{\partial \mathbf{H}}{\partial t} = \nabla\nabla \cdot \mathbf{H} = 0 \tag{4.10}$$

If we have linear isotropic materials, then $\nabla \cdot \mathbf{H} = 0$. Then Equation (4.10) is a second-order equation with only one vector unknown.

4.1.1 Uniqueness of the Solution

In modeling electromagnetic phenomena, many of our choices depend on ensuring that the boundary conditions are satisfied. Any vector field that satisfies Maxwell's equations and also satisfies tangential components of the field on the boundaries is the only possible solution. For example, assume that $\mathbf{E}_1$ and $\mathbf{H}_1$ are solutions that meet these criteria. Then

$$\nabla \cdot \varepsilon \mathbf{E}_1 = \rho \tag{4.11}$$

$$\nabla \cdot \mu \mathbf{H}_1 = 0 \tag{4.12}$$

$$\nabla \times \mathbf{E}_1 = -\mu \frac{\partial \mathbf{H}_1}{\partial t} \tag{4.13}$$

$$\nabla \times \mathbf{H}_1 = \mathbf{J}_0 + \sigma \mathbf{E}_1 \tag{4.14}$$

If there is another solution, say $\mathbf{E}_2$ and $\mathbf{H}_2$, then

$$\nabla \cdot \varepsilon \mathbf{E}_2 = \rho \tag{4.15}$$

$$\nabla \cdot \mu \mathbf{H}_2 = 0 \tag{4.16}$$

$$\nabla \times \mathbf{E}_2 = -\mu \frac{\partial \mathbf{H}_2}{\partial t} \tag{4.17}$$

$$\nabla \times \mathbf{H}_2 = \mathbf{J}_0 + \sigma \mathbf{E}_2 \tag{4.18}$$

Now take the differences between the two solutions.

$$\delta \mathbf{E} = \mathbf{E}_1 - \mathbf{E}_2 \tag{4.19}$$

$$\delta \mathbf{H} = \mathbf{H}_1 - \mathbf{H}_2 \tag{4.20}$$

Since the system is linear, $\delta \mathbf{E}$ and $\delta \mathbf{H}$ must also be solutions.

$$\nabla \cdot \varepsilon \delta \mathbf{E} = \rho \tag{4.21}$$

$$\nabla \cdot \mu \delta \mathbf{H} = 0 \tag{4.22}$$

$$\nabla \times \delta \mathbf{E} = -\mu \frac{\partial \delta \mathbf{H}}{\partial t} \tag{4.23}$$

$$\nabla \times \delta \mathbf{H} = \mathbf{J}_0 + \sigma \delta \mathbf{E} \tag{4.24}$$

If we take the dot product of the last equations with $\delta \mathbf{E}$

$$\delta \mathbf{E} \cdot \nabla \times \delta \mathbf{H} = \sigma \delta \mathbf{E}^2 \tag{4.25}$$

and use the identity

$$\mathbf{F} \cdot \nabla \times \mathbf{G} = \mathbf{G} \cdot \nabla \times \mathbf{F} - \nabla \cdot (\mathbf{F} \times \mathbf{G}) \tag{4.26}$$

we get

$$\delta\mathbf{H} \cdot \nabla \times \delta\mathbf{E} = \nabla \cdot (\delta\mathbf{E} \times \delta\mathbf{H}) + \sigma\delta\mathbf{E}^2 \tag{4.27}$$

The left-hand side becomes

$$\delta\mathbf{H} \cdot \nabla \times \delta\mathbf{E} = -\frac{1}{2}\mu\frac{d\delta\mathbf{H}^2}{dt} \tag{4.28}$$

Integrating this over the volume, Ω, and applying the divergence theorem we find

$$\int_S (\delta\mathbf{E} \times \delta\mathbf{H}) \cdot dS = -\frac{1}{2}\mu\int_\Omega \left(\frac{d\delta\mathbf{H}^2}{dt}\right) d\Omega + \int_\Omega \left(\sigma\delta\mathbf{E}^2\right) d\Omega \tag{4.29}$$

If the tangential components of **E** and **H** on the boundary are equal in both solutions, then the left-hand side of Equation (4.29) vanishes. We notice that each individual term on the right-hand side must also vanish. This demonstrates that if we specify the tangential fields of the boundaries, we have completely defined the problem and therefore the solution is unique.

4.1.2 Total Magnetic Potential and Reduced Magnetic Potential

We cannot generally use a scalar potential to solve eddy current problems. This can easily be seen since, for any scalar

$$\nabla \times \nabla f = 0 \tag{4.30}$$

and from Ampere's law, ignoring displacement current,

$$\nabla \times \mathbf{H} = \mathbf{J} \tag{4.31}$$

This means that the magnetic field in the current carrying region cannot be described as the gradient of a scalar. However, we can often solve problems in regions with current using mixed formulations. In regions with no current, it is valid to use a scalar potential formulation. One needs to be careful in these cases as the numerical solution may not be unique. We, therefore, include the following discussion.

If we consider a current free region, we can write

$$\nabla \times \mathbf{H} = 0 \tag{4.32}$$

We can, in this case, describe the field as the gradient of a scalar potential.

$$\mathbf{H} = -\nabla\Omega \tag{4.33}$$

where Ω, *the magnetic scalar potential*, has units of amperes. Because

$$\nabla \times \mathbf{H} = 0 \tag{4.34}$$

and

$$\nabla \cdot \mathbf{B} = 0 \tag{4.35}$$

and

$$\mathbf{B} = \mu \mathbf{H} \tag{4.36}$$

we obtain

$$\nabla \cdot \mu \nabla \Omega = 0 \tag{4.37}$$

The scalar potential described in Equation (4.37) is not unique in current-carrying regions. This is a serious issue for numerical computation and often results in singular matrices. One way to avoid this difficulty is to define a new variable. Therefore, we now define a *reduced scalar potential* so that the gradient gives only part of the magnetic field. We do this in the following manner. We consider the component of **H** which is produced by currents and which we will call $\mathbf{H_c}$. Then

$$\nabla \times \mathbf{H_c} = \mathbf{J} \tag{4.38}$$

so

$$\nabla \times (\mathbf{H} - \mathbf{H_c}) = 0 \tag{4.39}$$

and

$$\mathbf{H} - \mathbf{H_c} = -\nabla \Omega_r \tag{4.40}$$

The reduced magnetic potential, Ω_r, exists and is unique in current-free regions. So

$$\nabla \cdot \mu (\mathbf{H} - \mathbf{H_c}) = -\nabla \cdot \mu \nabla \Omega_r \tag{4.41}$$

Since $\nabla \cdot \mu \mathbf{H} = 0$, we have

$$\nabla \cdot \mu \nabla \Omega_r = \nabla \cdot \mu \mathbf{H_c} \tag{4.42}$$

One common way of evaluating $\mathbf{H_c}$ [46] is to find the field produced by known current sources in an open boundary region containing uniform permeability. In this case, we can use the Biot–Savart law to compute $\mathbf{H_c}$ as

$$\mathbf{H_c} = \frac{1}{4\pi} \int_V \frac{\mathbf{J} \times \mathbf{r}}{r^3} dV \tag{4.43}$$

This field has no divergence, so

$$\nabla \cdot \mu \nabla \Omega_r = \nabla \cdot \mu \mathbf{H_c} = \mu \nabla \cdot \mathbf{H_c} + \mathbf{H_c} \cdot \nabla \mu \tag{4.44}$$

and then

$$\nabla \cdot \mu \nabla \Omega_r = \mathbf{H_c} \cdot \nabla \mu \tag{4.45}$$

4.1.3 Using a Mixed Scalar Potential Formulation

We have just seen that the magnetic scalar potential is not unique in regions with currents, but that we could use a reduced scalar potential in these regions. This would theoretically solve our problem. However, the magnetic scalar potential is prone to error in regions with magnetic material. In these regions $\mu \gg \mu_0$, and the magnetic field can be very small. Using the reduced scalar potential, the field is found as the sum of the source field (produced by currents) and the negative gradient of the potential. Since $\mathbf{H} \approx 0$, these two fields must almost exactly cancel in regions of high permeability, so that small errors in either one can result in large errors in $\mathbf{H}$. We can overcome this by using a mixed formulation. The reduced magnetic scalar potential is used in the current-carrying regions, and the total potential is used in the high-permeability regions. For more information on the topic, the reader is directed to [46]. We illustrate this idea with some examples in Figures 4.1–4.3.

These figures illustrate the problem of connectivity. The total scalar potential can be used only in curl-free regions ($\nabla \times H = 0$). This means that the formulation is not valid if there exists a path that links a current. For example, in Figure 4.1, we can use a mixed formulation, since there is no path in the total scalar region that encloses a current. In the total scalar potential domain

$$\oint_l \mathbf{H} \cdot dl = 0 \tag{4.46}$$

for any closed path.

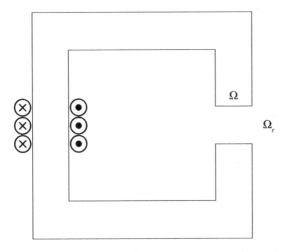

Figure 4.1 Correct use of mixed scalar potential formulation.

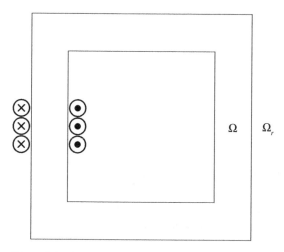

Figure 4.2 Incorrect use of scalar potential formulation.

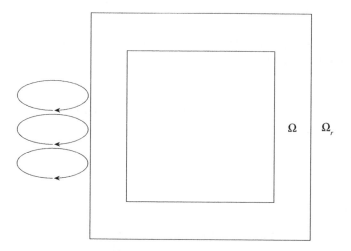

Figure 4.3 Closed magnetic circuit linking no current.

This is valid since we have a break in the magnetic circuit, in which we are using the reduced potential formulation. In Figure 4.2, there is a continuous magnetic circuit, but because this circuit is linking the current source, we cannot apply the total scalar potential. However, in Figure 4.3, we can apply the mixed formulation because the path in the scalar potential region does not link any current. There is sometimes a trick that may be used. In Figure 4.4, we see the same geometry as in Figure 4.2. We have used symmetry here to solve only one-half of the problem.

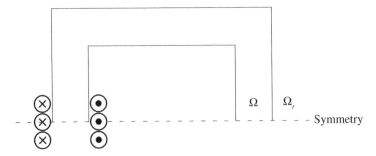

Figure 4.4 Scalar potential is valid with symmetry condition.

There is now no path in the magnetic region that links the current in the problem domain. In this case, the mixed scalar potential can be applied.

4.1.4 Combined Vector and Scalar Potential, the A − V Formulation

We often use a formulation involving both vector and scalar potentials. One very popular example is the $A - V$ formulation. This uses the three-component magnetic vector potential and the electric scalar potential. We will develop the curl–curl equation by starting with

$$\nabla \cdot \mathbf{B} = 0 \tag{4.47}$$

As we have seen, we can write the magnetic flux density as the curl of a vector. This is because for any vector $\mathbf{F}$

$$\nabla \cdot \nabla \times \mathbf{F} = 0 \tag{4.48}$$

This gives

$$\nabla \times \mathbf{A} = \mathbf{B} \tag{4.49}$$

where we define $\mathbf{A}$ as the *magnetic vector potential* (MVP). The MVP has units of webers/meter. The MVP is a measure of the magnetic flux. In integral form

$$\oint \mathbf{A} \cdot d\mathbf{l} = \psi_m \tag{4.50}$$

where ψ_m is equal to the flux passing through the area enclosed by the path of integration.

To see the connection between the MVP and the electric scalar potential, V, we consider Faraday's law.

$$\nabla \times \mathbf{E} = -\frac{\partial \mathbf{B}}{\partial t} \tag{4.51}$$

Figure 4.5 Vector potential and flux.

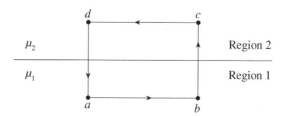

Using Equation (4.49), we have

$$\nabla \times \mathbf{E} = -\nabla \times \frac{\partial \mathbf{A}}{\partial t} \tag{4.52}$$

or

$$\nabla \times \left(\mathbf{E} + \frac{\partial \mathbf{A}}{\partial t} \right) = 0 \tag{4.53}$$

so that

$$\mathbf{E} = -\frac{\partial \mathbf{A}}{\partial t} - \nabla V \tag{4.54}$$

This is because for any scalar ϕ,

$$\nabla \times \nabla \phi = 0$$

To understand the interface conditions on the MVP, we consider Figure 4.5.

Consider the interface of material 1 and material 2 in Figure 4.5. If we integrate

$$\oint \mathbf{A} \cdot d\mathbf{l} = \psi_m \tag{4.55}$$

around the path *abcda*, then, as the sides *bc* and *da* approach zero length, the flux linked by the path goes to zero. Therefore, the tangential components of the magnetic vector potential must be continuous.

$$\hat{n} \times (\mathbf{A}_1 - \mathbf{A}_2) = 0 \tag{4.56}$$

4.1.5 The Curl–Curl Equation for the Magnetic Vector Potential

We will now develop the curl–curl equation for the MVP beginning with

$$\nabla \times \mathbf{A} = \mathbf{B} \tag{4.57}$$

where v is the magnetic reluctivity of the material. Taking the curl of both sides and using $v\mathbf{B} = \mathbf{H}$ we get

$$\nabla \times v\nabla \times \mathbf{A} = \nabla \times \mathbf{H} \tag{4.58}$$

From Maxwell's equations

$$\nabla \times v\nabla \times \mathbf{A} = \mathbf{J} \tag{4.59}$$

or

$$\nabla \times \nu \nabla \times \mathbf{A} = \sigma \mathbf{E} \tag{4.60}$$

Now in terms of the vector and scalar potentials,

$$\nabla \times \nu \nabla \times \mathbf{A} = \sigma \left(\frac{\partial \mathbf{A}}{\partial t} - \nabla V \right) \tag{4.61}$$

The curl–curl operator is usually replaced by the Laplacian operator. Applying the vector identity,

$$\nabla \times \nabla \times \mathbf{F} = \nabla \nabla \cdot \mathbf{F} - \nabla^2 \mathbf{F} \tag{4.62}$$

Equation (4.61) becomes

$$\nu \nabla^2 \mathbf{A} - \sigma \frac{\partial \mathbf{A}}{\partial t} = \nabla(\nu \nabla \cdot \mathbf{A}) + \sigma \nabla V \tag{4.63}$$

If ν is not constant, then

$$\nabla \times \nu \nabla \times \mathbf{A} = -\nu \nabla^2 \mathbf{A} + \nu \nabla \nabla \cdot \mathbf{A} + \nabla \nu \times \nabla \times \mathbf{A} \tag{4.64}$$

so

$$\nu \nabla^2 \mathbf{A} - \nabla \nu \times \nabla \times \mathbf{A} - \sigma \frac{\partial \mathbf{A}}{\partial t} = \nu(\nabla \nabla \cdot \mathbf{A} + \sigma \nabla V) \tag{4.65}$$

We will now look at some possible choices for the divergence.

4.1.6 Choosing the Divergence of A

In vector calculus, a quantity is fully defined if both the curl and the divergence are specified. In the case of the MVP, we are interested only in the curl, since this gives the flux density. The divergence is not usually used and can be independently selected to simplify the formulation. There are a number of possibilities and we will consider a few here. Consider the equation

$$\nabla^2 \mathbf{A} - \mu \sigma \frac{\partial \mathbf{A}}{\partial t} - \mu \epsilon \frac{\partial^2 \mathbf{A}}{\partial t^2} = \nabla \left[\nabla \cdot \mathbf{A} + \mu \sigma V + \mu \epsilon \frac{\partial V}{\partial t} \right] \tag{4.66}$$

A popular choice for the divergence of $\mathbf{A}$ is

$$\nabla \cdot \mathbf{A} = 0 \tag{4.67}$$

If the divergence $\mathbf{A}$ is zero, then

$$\int_S \nabla \cdot \mathbf{A} \, dS = 0 \tag{4.68}$$

Consider the material interface of Figure 4.6. If the divergence of the vector potential is zero, then the flux of the vector $\mathbf{A}$ which enters the disc from the bottom, must leave the disc through the top as the height of the disc approaches zero. This means that

$$\hat{n} \cdot (A_1 - A_2) = 0 \tag{4.69}$$

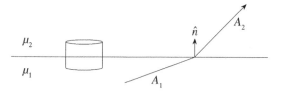

Figure 4.6 Vector potential at material boundary.

Therefore, we conclude that the normal component of **A** is continuous at a material interface. This choice of divergence is called the *Coulomb gauge*. We then have from Equation (4.65)

$$\nabla^2 \mathbf{A} - \nabla(\mu\sigma V) = -\mu \mathbf{J} \tag{4.70}$$

which is a popular choice for eddy current modeling.

If we go back to Equation (4.66)

$$\nabla^2 \mathbf{A} - \mu\sigma \frac{\partial \mathbf{A}}{\partial t} - \mu\epsilon \frac{\partial^2 \mathbf{A}}{\partial t^2} = \nabla \left[\nabla \cdot A + \mu\sigma V + \mu\epsilon \frac{\partial V}{\partial t} \right] \tag{4.71}$$

We can see that another possible choice for divergence is

$$\nabla \cdot \mathbf{A} = -\mu\epsilon \frac{\partial V}{\partial t} \tag{4.72}$$

Applying the same argument at a material interface to evaluate the flux leaving a closed surface, we get

$$\oint \mathbf{A} \cdot d\mathbf{S} = \mathbf{S} \cdot \hat{n} \left(\mu_2\epsilon_2 - \mu_1\epsilon_1 \right) \frac{\partial V}{\partial t} \tag{4.73}$$

This gives

$$\hat{n} \cdot (A_1 - A_2) = (\mu_2\epsilon_2 - \mu_1\epsilon_1) \frac{\partial V}{\partial t} \tag{4.74}$$

We see then that, in this case, the normal component of **A** is not continuous. We can simplify the equation to

$$\nabla^2 \mathbf{A} - \mu\sigma \frac{\partial A}{\partial t} - \mu\epsilon \frac{\partial^2 \mathbf{A}}{\partial t^2} = \mu\sigma \nabla V \tag{4.75}$$

Yet another possible choice for divergence is

$$\nabla \cdot \mathbf{A} = -\mu\sigma V - \mu\epsilon \frac{\partial V}{\partial t} \tag{4.76}$$

From Maxwell's equations

$$\nabla \cdot \mathbf{E} = \frac{\rho}{\epsilon} \tag{4.77}$$

Therefore

$$\nabla \cdot \mathbf{E} = \nabla \cdot \left(-\frac{\partial \mathbf{A}}{\partial t} - \nabla V \right) = -\frac{\partial}{\partial t} \nabla \cdot \mathbf{A} - \nabla^2 V \tag{4.78}$$

We now have a set of coupled equations

$$\nabla^2 V - \mu\sigma\frac{\partial V}{\partial t} - \mu\epsilon\frac{\partial^2 V}{\partial t^2} = -\frac{\rho}{\epsilon}$$

$$\nabla^2 \mathbf{A} - \mu\sigma\frac{\partial \mathbf{A}}{\partial t} - \mu\epsilon\frac{\partial^2 \mathbf{A}}{\partial t^2} = 0 \tag{4.79}$$

We again have a discontinuity in the normal component of the MVP at the interface between two materials. We get

$$\hat{n} \cdot (\mathbf{A}_1 - \mathbf{A}_2) = -(\mu_1\sigma_1 - \mu_2\sigma_2)V - (\mu_1\epsilon_1 - \mu_2\epsilon_2)\frac{\partial V}{\partial t} \tag{4.80}$$

4.1.7 The A* Formulation

The $\mathbf{A} - V$ formulation that we just discussed is quite general and popular. This formulation uses four unknowns at each point, three vector components, and one scalar. To reduce the number of unknowns, we can sometimes use the $\mathbf{A}^*$ formulation. We begin by considering the diffusion equation for eddy currents.

$$\nabla \times \nu\nabla \times \mathbf{A} + \sigma\left(\frac{\partial \mathbf{A}}{\partial t} + \nabla V\right) = \mathbf{J}_0 \tag{4.81}$$

The continuity equation ($\nabla \cdot \mathbf{J} = 0$) leads to

$$\nabla \cdot \sigma\left(\frac{\partial \mathbf{A}}{\partial t} + \nabla V\right) = 0 \tag{4.82}$$

We define a *modified vector potential* $\mathbf{A}^*$ such that

$$\frac{\partial \mathbf{A}^*}{\partial t} = \frac{\partial \mathbf{A}}{\partial t} + \nabla V = 0 \tag{4.83}$$

This gives for Equation (4.81)

$$\nabla \times \nu\nabla \times \mathbf{A}^* + \sigma\frac{\partial \mathbf{A}^*}{\partial t} = \mathbf{J}_0 \tag{4.84}$$

We have now eliminated the scalar potential. This gives us fewer unknowns at each point. The $\mathbf{A}^*$ formulation is less general and can only be used in problems having uniform conductivity. The continuity condition $\nabla \cdot \mathbf{J} = \nabla \cdot \sigma\mathbf{A}^* = 0$ is satisfied only in a weak sense as is $\mathbf{J} \cdot \hat{n}$ on the boundary.

4.1.8 Electric Vector Potential, the T − Ω Formulation

We have seen that the continuity equation for the magnetic flux, $\nabla \cdot \mathbf{B} = 0$, allowed us to represent $\mathbf{B}$ using a vector potential (MVP) so that $\nabla \times \mathbf{A} = \mathbf{B}$. The continuity of current density, $\nabla \cdot \mathbf{J} = 0$, allows us to define a current vector potential, often referred to as the electric vector potential, $\mathbf{T}$.

$$\nabla \times \mathbf{T} = \mathbf{J} \tag{4.85}$$

Note that $\nabla \times \mathbf{H} = \mathbf{J}$ as well, so $\mathbf{H}$ and $\mathbf{T}$ can differ by the gradient of a scalar. Both quantities also have the same units, amperes/meter. Therefore,

$$\mathbf{H} = \mathbf{T} - \nabla\Omega \tag{4.86}$$

Using $\mathbf{E} = \rho\mathbf{J}$, and from Faraday's law

$$\nabla \times \mathbf{E} = -\frac{\partial \mathbf{B}}{\partial t} = \nabla \times \rho\nabla \times \mathbf{T} \tag{4.87}$$

Substituting for the magnetic flux density

$$\mathbf{B} = \mu\mathbf{H} = \mu(\mathbf{T} - \nabla\Omega) \tag{4.88}$$

we obtain

$$\nabla \times \rho\nabla \times \mathbf{T} + \mu\frac{\partial \mathbf{T}}{\partial t} - \mu\nabla\frac{\partial \Omega}{\partial t} = 0 \tag{4.89}$$

In current-free regions, we can find the magnetic field from the scalar potential.

$$\mathbf{H} = -\nabla\Omega \tag{4.90}$$

where Ω can be found from Laplace's equation

$$-\nabla \cdot \mu\nabla\Omega = 0 \tag{4.91}$$

As in the case of the MVP, the solution can be made unique by specifying the divergence. For example, we can use the Coulomb gauge, $\nabla \cdot \mathbf{T} = 0$. With this choice, Equation (4.77) becomes [47]

$$\nabla \times \rho\nabla \times \mathbf{T} - \nabla\rho\nabla \cdot \mathbf{T} + \mu\frac{\partial \mathbf{T}}{\partial t} - \mu\nabla\frac{\partial \Omega}{\partial t} = 0 \tag{4.92}$$

To find the relationship between the vector and scalar potential, we use $\nabla \cdot \mathbf{B} = 0$. From (4.76) we have

$$\nabla \cdot \mu(\mathbf{T} - \nabla\Omega) = 0 \tag{4.93}$$

We have presented some of the most widely-used formulations for eddy currents. The reader interested in more detail is referred to [46]. We will now apply some of these formulations to finite difference, finite element, and integral equation methods for eddy current problems.

5

Finite Differences

In Section 4.1, we found that electromagnetic phenomena are described by partial differential equations, with the dependent variables being either potentials or field values. One of the earliest attempts to solve these problems numerically was by the finite difference method. In the finite difference method, the partial derivatives are replaced by difference equations [48, 49]. Space is divided into finite difference cells and the equations are written for the unknown potential or field values at nodes. The partial differential equations are replaced by a set of simultaneous difference equations. The unknowns are computed on a finite set of points (the nodes). If we need the values at other locations, we must interpolate between the nodes.

In this section, we will introduce the concept of difference equations and apply it to the eddy current problem. We will also look at the application of boundary conditions and the treatment of non-homogeneous materials. We will then show that these difference equations can be represented by an equivalent circuit to find the solution to the eddy current problem.

5.1 Difference Equations

The basic approximations for the spatial derivatives with respect to x in the finite difference method are

$$f'(x_0) \approx \frac{f(x_0 + \Delta x) - f(x_0)}{\Delta x}, \quad \text{forward difference}$$

$$f'(x_0) \approx \frac{f(x_0) - f(x_0 - \Delta x)}{\Delta x}, \quad \text{backward difference} \qquad (5.1)$$

$$f'(x_0) \approx \frac{f(x_0 + \Delta x) - f(x_0 - \Delta x)}{2\Delta x}, \quad \text{central difference}$$

These values are illustrated in Figure 5.1.

Eddy Currents: Theory, Modeling, and Applications, First Edition.
Sheppard J. Salon, M. V. K. Chari, Lale T. Ergene, David Burow, and Mark DeBortoli.

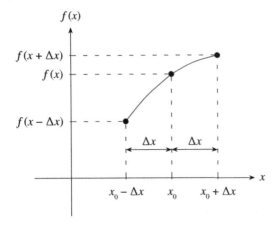

Figure 5.1 Forward, backward, and central differences.

The approximate second derivatives are found by using a central difference formula based on the first derivatives, where the first derivatives are estimated at the intermediate points $(x + \Delta x/2)$ and $(x - \Delta x/2)$. This gives

$$f''(x_0) = \frac{f'(x_0 + \Delta x/2) - f'(x_0 - \Delta x/2)}{\Delta x}$$

$$f''(x_0) = \frac{\dfrac{f(x_0 + \Delta x) - f(x_0)}{\Delta x} - \dfrac{f(x_0) - f(x_0 - \Delta x)}{\Delta x}}{\Delta x} \tag{5.2}$$

$$f''(x_0) = \frac{f(x_0 + \Delta x) - 2f(x_0) + f(x_0 - \Delta x)}{(\Delta x)^2}$$

We can find the truncation errors by expanding the function $f(x)$ around $x_0 + \Delta x$ and $x_0 - \Delta x$ in a Taylor series.

$$f(x_0 + \Delta x) = f(x_0) + \Delta x f'(x_0) + \frac{(\Delta x)^2}{2!} f''(x_0) + \cdots \tag{5.3}$$

$$f(x_0 - \Delta x) = f(x_0) - \Delta x f'(x_0) + \frac{(\Delta x)^2}{2!} f''(x_0) - \cdots \tag{5.4}$$

If we subtract Equation (5.4) from (5.3), we obtain

$$f(x_0 + \Delta x) - f(x_0 - \Delta x) = 2\Delta x f'(x_0) + O(\Delta x)^3 \tag{5.5}$$

where the symbol O means *order of*. Dividing Equation (5.5) by $2\Delta x$, we obtain the central difference formula (5.1). We see that the truncation error is of the order $(\Delta x)^2$.

Solving for the first derivatives in Equations (5.3) and (5.4) separately, we obtain the forward and backward difference formulas of Equation (5.1). We see from the Taylor formula that the truncation error associated with each of these is of the order of Δx.

If we add Equations (5.3) and (5.4), we obtain

$$f(x_0 + \Delta x) + f(x_0 - \Delta x) = 2f(x_0) + (\Delta x)^2 f''(x_0) + O(\Delta x)^4 \tag{5.6}$$

Solving Equation (5.6) for the second derivative, we obtain Equation (5.2) and see that the truncation error is of the order of $(\Delta x)^2$.

5.2 The Two-Dimensional Diffusion Equation

We will now focus on the two-dimensional diffusion equation for the magnetic vector potential. This has been well developed for problems in Cartesian and polar coordinates. In this particular case, where the fields have only two components and the current has only one component, the formulation simplifies and it is only necessary to solve for one component of the magnetic vector potential. The limitation of the two-dimensional analyses is that there is no variation of the fields or currents in the direction of the current.

Let us consider the diffusion equation for homogeneous materials in two-dimensional Cartesian coordinates.

$$v\nabla^2 A = v\left(\frac{\partial^2 A}{\partial x^2} + \frac{\partial^2 A}{\partial y^2}\right) = -J_0(x,y) + \sigma\frac{dA}{dt} \tag{5.7}$$

where v is the magnetic material property (reluctivity). Using the expansion for the second derivative in Equation (5.2), we obtain (see Figure 5.2)

$$\frac{\partial^2 A}{\partial x^2} = \frac{A_{i+1,j} - 2A_{i,j} + A_{i-1,j}}{(\Delta x)^2} + O(\Delta x)^2 \tag{5.8}$$

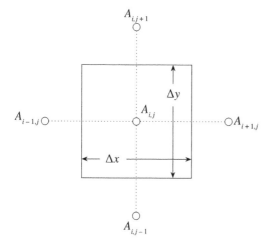

Figure 5.2 Finite difference cell and diagram.

and

$$\frac{\partial^2 A}{\partial y^2} = \frac{A_{i,j+1} - 2A_{i,j} + A_{i,j-1}}{(\Delta y)^2} + O(\Delta y)^2 \tag{5.9}$$

Substituting these into (5.7), multiplying through by $\Delta x \Delta y$, and adding, we obtain

$$\nu \left(\Delta y \frac{A_{i+1,j} - 2A_{i,j} + A_{i-1,j}}{\Delta x} + \Delta x \frac{A_{i,j+1} - 2A_{i,j} + A_{i,j-1}}{\Delta y} \right)$$
$$= J_0(x,y)\Delta x \Delta y - \sigma \frac{dA}{dt} \Delta x \Delta y \tag{5.10}$$

This equation is written for each node in the problem. This will give us a set of simultaneous equations. The set of equations is singular until the potential of at least one node is specified. The first term on the right-hand side of the equation is the source current density times the area of the finite difference cell. This is the total applied current in the cell. The second term is the eddy current in the cell.

5.2.1 Interfaces Between Materials

One of the reasons why we use numerical methods is the ease with which they handle complex material properties and irregular interfaces. Consider Figure 5.3, in which we have a five-point finite difference graph, including a boundary between different materials. Recall that the well-known interface condition of $B_{n1} = B_{n2}$ (continuous normal flux density) comes directly from application of Gauss' law for

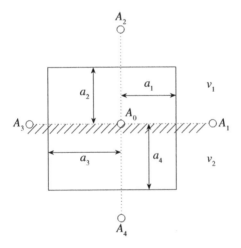

Figure 5.3 Interface between different materials.

the magnetic field. The interface condition $H_{t1} = H_{t2}$ (continuous tangential magnetic field) must also be satisfied. In the case of the finite difference method for the vector potential used here, the normal component of flux density must be continuous at the boundaries, as can be seen by noting that $B_y = \frac{(A_0 - A_1)}{\Delta x}$ is the normal flux density for both the first and fourth quadrants of the figure. This condition is exactly satisfied in the vector potential formulation. The tangential magnetic field, however, is only approximately continuous since this quantity involves A_2 and A_4. This discontinuity in tangential field will get smaller as the mesh size gets smaller and can be used as a measure of the error in the solution.

To find the finite difference formula for the interface, we use Ampere's law

$$\oint H \cdot d\ell = I_{\text{enc}} \tag{5.11}$$

Evaluating Equation (5.11) around the finite difference cell we find,

$$v_1 \frac{(A_0 - A_1) a_2}{a_1} \frac{a_2}{2} + v_1 \frac{(A_0 - A_2) a_1}{a_2} \frac{a_1}{2} + v_2 \frac{(A_0 - A_2) a_3}{a_2} \frac{a_3}{2} + v_2 \frac{(A_0 - A_3) a_2}{a_3} \frac{a_2}{2}$$
$$v_3 \frac{(A_0 - A_3) a_4}{a_3} \frac{a_4}{2} + v_3 \frac{(A_0 - A_4) a_3}{a_4} \frac{a_3}{2} + v_4 \frac{(A_0 - A_4) a_1}{a_4} \frac{a_1}{2} + v_4 \frac{(A_0 - A_1) a_4}{a_1} \frac{a_4}{2}$$
$$= \left(\frac{a_1 a_2}{4} J_1 + \frac{a_2 a_3}{4} J_2 + \frac{a_3 a_4}{4} J_3 + \frac{a_4 a_1}{4} J_4 \right)$$
$$- j\omega \left(\sigma_1 \frac{a_1 a_2}{4} + \sigma_2 \frac{a_2 a_3}{4} + \sigma_3 \frac{a_3 a_4}{4} + \sigma_4 \frac{a_4 a_1}{4} \right) A_0 \tag{5.12}$$

Combining terms we can write

$$\alpha_1(A_0 - A_1) + \alpha_2(A_0 - A_2) + \alpha_3(A_0 - A_3) + \alpha_4(A_0 - A_4) = I_0 - I_e \tag{5.13}$$

where

$$\alpha_1 = \frac{v_1 a_2 + v_4 a_4}{2 a_1}$$
$$\alpha_2 = \frac{v_1 a_1 + v_2 a_3}{2 a_2}$$
$$\alpha_3 = \frac{v_2 a_2 + v_3 a_4}{2 a_3}$$
$$\alpha_4 = \frac{v_4 a_1 + v_3 a_3}{2 a_4}$$

$$I_0 = \frac{1}{4} \left(a_1 a_2 J_1 + a_2 a_3 J_2 + a_3 a_4 J_3 + a_4 a_1 J_4 \right)$$
$$I_e = \frac{j\omega A_0}{4} \left(a_1 a_2 \sigma_1 + a_2 a_3 \sigma_2 + a_3 a_4 \sigma_3 + a_4 a_1 \sigma_4 \right) \tag{5.14}$$

Where I_0 is the applied current in the finite difference cell and I_e is the eddy current in the cell.

5.2.2 Dirichlet and Neumann Boundary Conditions

For the diffusion equation to have a unique solution, either the potential or the normal derivative of the potential must be specified at each point on the boundary. If the potential is specified, this is called a Dirichlet condition and the unknown nodal potential is eliminated. If the normal derivative is specified (Neumann boundary condition), then we apply the following procedure. Consider Figure 5.4. In this case, the potential at node (i,j) is unknown. If we consider the finite difference expansion for the potential at node (i,j), then we see that it involves the potential at point $(i+1,j)$, which is outside the domain of interest. We express the normal derivative at (i,j), as

$$\frac{\partial A}{\partial n} = \frac{\partial A}{\partial x} = \frac{A_{i+1,j} - A_{i-1,j}}{2\Delta x} \tag{5.15}$$

From (5.15), we solve for the potential at $(i+1,j)$.

$$A_{i+1,j} = 2\Delta x \frac{\partial A}{\partial n} + A_{i-1,j} \tag{5.16}$$

The finite difference expansion for the potential at node (i,j) now becomes (using Equation (5.10)),

$$v_x \frac{2\Delta x \frac{\partial A}{\partial n} - 2A_{i,j} + 2A_{i-1,j}}{(\Delta x)^2} + v_y \frac{A_{i,j+1} - 2A_{i,j} + A_{i,j-1}}{(\Delta y)^2} \tag{5.17}$$

where v_x and v_y are the material properties in the x and y directions, respectively. A special, but very common case is the homogeneous Neumann boundary condition where $\frac{\partial A}{\partial n} = 0$. In this case, the equipotential lines are perpendicular to

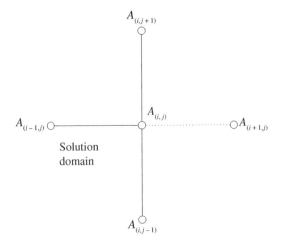

Figure 5.4 Boundary with normal derivative specified.

the surface. This is the case if, for example, we have a plane of symmetry where the potential on one side of the surface is the same as the potential on the other side of the surface. From Figure 5.4, we have $A_{i-1,j} = A_{i+1,j}$, and, setting $\frac{\partial A}{\partial n} = 0$ in Equation (5.17), we see that this condition is satisfied.

5.3 Time-Domain Solution of the Diffusion Equation

This section introduce some popular methods used in solving problems in the time domain.

$$\nabla^2 A - \mu\sigma\frac{dA}{dt} = -\mu J_0 \tag{5.18}$$

Considering the finite difference representation of the time derivative, we can use the forward, backward, or central difference formulas. To solve an ordinary differential equation (ODE), we need not only Dirichlet and/or Neumann boundary conditions but also initial conditions. From the initial conditions, we project forward in time by using an approximation for the time derivative. To illustrate the different methods, we will use the one-dimensional diffusion equation.

5.3.1 The Explicit Integration Scheme

If we use a central difference formula for the Laplacian terms and a backward difference formula for the time derivative, the finite difference formula is

$$\frac{A_{i-1}(t - \Delta t) - 2A_i(t - \Delta t) + A_{i+1}(t - \Delta t)}{(\Delta x)^2}$$

$$- \mu\sigma\frac{A_i(t) - A_i(t - \Delta t)}{\Delta t} = \mu J \tag{5.19}$$

Rearranging terms,

$$A_i(t) = \frac{\Delta t}{\mu\sigma(\Delta x)^2}(A_{i-1}(t - \Delta t) - 2A_i(t - \Delta t) + A_{i+1}(t - \Delta t))$$

$$+ A_i(t - \Delta t) - J\frac{\Delta t}{\sigma} \tag{5.20}$$

Defining

$$\alpha = \frac{\Delta t}{\mu\sigma(\Delta x)^2} \tag{5.21}$$

we obtain

$$A_i(t) = \alpha A_{i-1}(t - \Delta t) + (1 - 2\alpha)A_i(t - \Delta t)$$

$$+ \alpha A_{i+1}(t - \Delta t) - \alpha\mu(\Delta x)^2 J \tag{5.22}$$

This is called an *explicit* integration scheme. The advantage of an explicit method is that the solution at time, t, can be found by using the known values of the unknown at the previous time, $t - \Delta t$. Therefore, we avoid solving a large system of simultaneous equations. We *sweep* through the equations and use the results from the previous time step to obtain the solution at the next time step. Although this is certainly an advantage, the backward difference scheme is not used very frequently. We previously saw that the error associated with the backward difference calculation is $O(\Delta t)$. This means that very small time steps are required to obtain an accurate result. It may also be that, for large-time steps, the solution is unstable [50].

5.3.2 Implicit Integration Schemes

To avoid these very small time steps and improve accuracy, we usually rely on implicit integration schemes. These methods express the values of the unknowns at a particular time in terms of the unknowns at the previous time step and the current time step. The drawback of this process is that we must solve a system of simultaneous equations for the unknowns. This means that the method is more computationally intensive. The advantage is that we can use much larger time steps. These implicit methods are therefore much more popular.

To illustrate the process, we will use the one-dimensional diffusion equation for the magnetic vector potential. For the implicit schemes, we replace the Laplacian by averages of the central difference formulas at time t and time $t + \Delta t$. These are weighted averages, with the weight between 0 and 1, with weighting factor θ. These methods are sometimes called θ methods. Expanding the Laplacian

$$\frac{\partial^2 A}{\partial x^2} \approx \frac{A(x - \Delta x, t) - 2A(x, t) + A(x + \Delta x, t)}{(\Delta x)^2}(1 - \theta)$$
$$+ \frac{A(x - \Delta x, t + \Delta t) - 2A(x, t + \Delta t) + A(x + \Delta x, t + \Delta t)}{(\Delta x)^2}\theta \qquad (5.23)$$

The time derivative of A is approximated by a forward difference

$$\frac{\partial A}{\partial t} \approx \frac{A(x, t + \Delta t) - A(x, t)}{\Delta t} \qquad (5.24)$$

In terms of the parameter α

$$\alpha(1 - \theta)A(x - \Delta x, t) + (1 - 2\alpha + 2\alpha\theta)A(x, t)$$
$$+ \alpha(1 - \theta)A(x + \Delta x, t) + \alpha\theta A(x - \Delta x, t + \Delta t)$$
$$- (2\alpha\theta + 1)A(x, t + \Delta t) + \alpha\theta A(x + \Delta x, t + \Delta t) = 0 \qquad (5.25)$$

For $\theta = 0$, we have a backward difference and the equations are the same as in the explicit method. For $\theta = 1$, we have a forward difference equation. For $\theta = 0.5$, we have the Crank–Nicholson method.

5.4 Equivalent Circuit Representation for Finite Difference Equations

We will now show that the finite difference equations have an equivalent circuit representation [26, 27, 58]. We will illustrate this with the two-dimensional expression for the magnetic vector potential

$$\frac{1}{\mu}\nabla^2 A = -J_0 + \sigma\frac{dA}{dt} \tag{5.26}$$

Consider the resistive-capacitive network of Figure 5.5. Summing the currents leaving node 0, using Kirchhoff's current law, we find

$$\frac{v_0 - v_1}{R_1} + \frac{v_0 - v_2}{R_2} + \frac{v_0 - v_3}{R_3} + \frac{v_0 - v_4}{R_4} + C\frac{dv_0}{dt} = I_0 \tag{5.27}$$

We can see that Equation (5.26) has the same form as Equation (5.10). We can make this equivalent with the following substitutions.

$$
\begin{aligned}
R_1 &= R_3 = \frac{\mu\Delta x}{\Delta y} \\
R_2 &= R_4 = \frac{\mu\Delta y}{\Delta x} \\
v_i &= A_i \\
C &= \sigma\Delta x\Delta y \\
I_0 &= -J_0(\Delta x\Delta y)
\end{aligned}
\tag{5.28}
$$

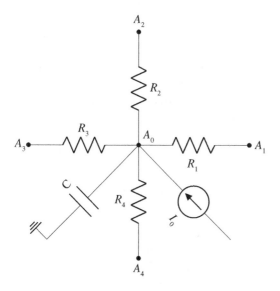

Figure 5.5 R–C circuit for the vector potential eddy current problem.

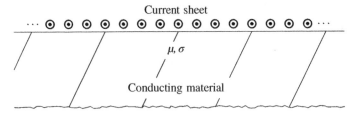

Current sheet

μ, σ

Conducting material

Figure 5.6 One-dimensional diffusion problem.

Therefore, in this equivalent circuit for the MVP, the nodal voltages are the z components of the MVP, the resistances represent permeances normal to the direction of the flux, the source current is the total input current to the finite difference cell, and the currents in the resistors are the ampere-turns (magneto-motive force) magnetizing that section of the cell [51, 52].

The capacitor connected to ground represents the eddy current path and current flowing through the capacitor, is the z directed eddy current in the cell. The total current in the cell is the sum of the eddy current and the source current. Since the ground or neutral point is included in the circuit, it is not necessary to set a value of potential in order to obtain a non-singular system of equations. This is not the case with magneto-static problems in which not setting a reference value of potential would result in the system being ill-conditioned.

It is interesting that this circuit reflects the observed behavior of skin effect problems. Let us look at the one-dimensional problem of a current sheet and a linear conducting material as illustrated in Figure 5.6. The finite difference equivalent circuit is shown in Figure 5.7. We know from Chapter 1.1, that as the frequency increases, the skin depth decreases. Considering the equivalent circuit, as the frequency increases, the reactances of the capacitors decrease and more current leaves the network near the surface. This results in a smaller skin depth. As the conductivity increases, the capacitors get larger and again the eddy currents flow more at the surface. We also see that if the permeability increases, the resistors in the equivalent circuit get larger, limiting the depth to which the current can penetrate. Similarly, as either the frequency decrease, the conductivity decreases, or the permeability decreases, the currents can penetrate farther into the material.

5.4.1 Numerical Example Using a Ladder Network Solution

We will now look at the result of a one-dimensional example using the ladder network. This example has been chosen for a number of reasons. First, there is a simple analytical solution. There are also a number of practical applications for this case. We can also demonstrate how the results of the numerical computation can be used to find quantities such as losses, effective resistance and inductance,

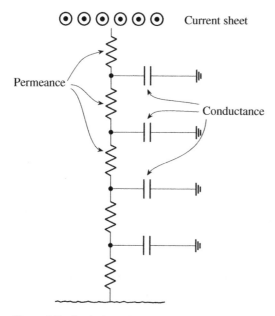

Figure 5.7 Equivalent circuit for the one-dimensional diffusion problem.

and electromagnetic force. We can use this equivalent circuit for steady-state ac (complex phasor analysis) as well as instantaneous (time domain) analysis. The circuit is also easily modified to include nonlinear and non-homogeneous materials.

One method of computing global quantities such as loss, resistance and reactance, is by means of the Poynting vector. For the complex or phasor case, the complex Poynting vector is defined as

$$P = E \times H^*$$
(5.29)

Since we are using the magnetic vector potential in the formulation, the electric field is found as

$$E = j\omega A$$
(5.30)

The magnetic field is

$$H = \frac{1}{\mu}(\nabla \times A)$$
(5.31)

With these, we can compute the Poynting vector.

We can also use stored energy as an option to find impedance. For the linear example above, the energy in the magnetic field is

$$W_m = \frac{1}{2}\int_V B \cdot H \, dv$$
(5.32)

The magnetic permeance can be found as

$$P = \frac{2W_m}{I^2} \tag{5.33}$$

In terms of the vector potential

$$W_m = \frac{1}{2} \int_V \nabla \times A \cdot H \, dv \tag{5.34}$$

From a vector identity, we have

$$\nabla \cdot (A \times H) = (\nabla \times A) \cdot H - (\nabla \times H) \cdot A \tag{5.35}$$

Now, using $\nabla \times H = J$, we have

$$W_m = \frac{1}{2} \int_V J \cdot A \, dv + \frac{1}{2} \int_V \nabla \cdot (A \times H) \, dv \tag{5.36}$$

$$W_m = \frac{1}{2} \int_V J \cdot A \, dv + \frac{1}{2} \oint_S A \times H \cdot dS$$

As we allow the surface in the second integral to go to infinity, we note that $A \propto 1/r$ and $H \propto 1/r^2$ while $S \propto r^2$. We conclude that this integral vanishes as $r \to \infty$.

This leaves

$$W_m = \frac{1}{2} \int_V J \cdot A \, dv \tag{5.37}$$

Consider the slot in Figure 5.8. The example here uses a one-dimensional ladder network that can represent a rectangular slot in a motor or a section of a semi-infinite slab of uniform isotropic conductor. The problem is excited with a current sheet located at $y = 0$. In this particular case, this is equivalent to applying the total current to the conductor, since current injected at the first node will leave the network through the grounded capacitors. The current distribution will then be determined by the network solution. This example also gives us an opportunity

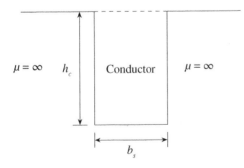

Figure 5.8 One-dimensional example.

to illustrate the calculation of some useful quantities, such as losses, resistance, inductance, energy and force.

In the numerical example which follows, we explore the solution in the frequency domain (phasor analysis) and then in the time domain with steady-state sinusoidal and step function inputs. The conductor is copper and the conductivity is $\sigma = 5.8 \times 10^7 \, \mathrm{S\,m^{-1}}$. The width of the conductor is 0.01 m and the height is 0.10 m. In this one-dimensional example, the width is not important as there is no variation of any quantity in the x or z directions. The width will be important of course in the evaluation of losses, resistance, and forces. The depth is chosen so that for power frequencies ($50 - 60\,\mathrm{Hz}$) the skin depth is small compared to the depth of the conductor and a simple analytical formula can be used to validate the finite difference solution.

In the first example, we apply a current sheet as the excitation at $y = 0$, which corresponds to the top of the conductor. This example is done in the complex mode so the solution is steady-state sinusoidal and the variables are complex numbers. We solve directly for the magnetic vector potential at the nodes (corresponding to the nodal voltage). The flux density, only in the x direction, is found by taking the curl of the vector potential. In this case, the numerical derivative is approximated as the difference of the voltage at two adjacent nodes divided by the distance between the nodes. The eddy current corresponds to the capacitor current, $j\omega AC$. All other quantities can now be found from these values as will be illustrated below.

The analytical solution for this problem has been given in Section 2.2. The solution for the current density is

$$J(y) = J_0 e^{-\frac{(1+j)y}{\delta}} \tag{5.38}$$

where

$$J_0 = H_0 \frac{1+j}{\delta} \tag{5.39}$$

The value of H_0 is simply the total current divided by the conductor width. We know that the real and imaginary components of the eddy current should be equal at the surface ($y = 0$). Also we know that the magnitude of the eddy current density will decay exponentially with y, so that the value at one skin depth will be $1/e$ times that at the surface and the phase will lag the phase at the surface by one radian. This is illustrated in Figure 5.9 in which we see the current density as a function of depth. Point A is at the surface and point B is located one skin depth below the surface. A comparison of the finite difference solution and the analytical expression of Equation (5.38) shows that the ladder network gives an extremely good approximation of the solution. There are two curves in Figure 5.10.

We can now use this example to illustrate some of the *post-processing* that can be done with the solution. We may require local quantities such as the flux density

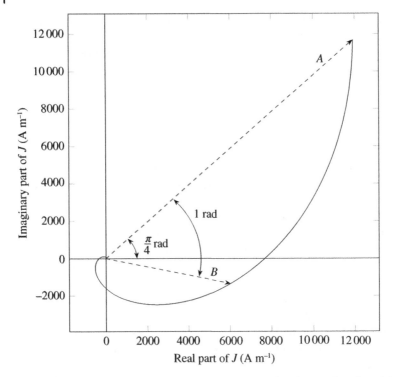

Figure 5.9 Real and imaginary components of current density as a function of depth.

or loss density at a specific location, or we may be interested in global quantities such as resistance, inductance, and force. One quantity of interest is the loss (often referred to as Joule loss). One method of determining the loss is to integrate the loss density over the surface of the conductor. In our example, we assume the depth is 1 m (axial direction). The loss in each of the finite difference cells is then the square of the cell current magnitude times the cell resistance. The cell resistance is

$$R_c = \frac{1}{b_s h_{\text{cell}} \sigma} \quad \Omega \text{m}^{-1} \tag{5.40}$$

where b_s is the cell width and h_{cell} is the cell height.

In the example used here, the total current is 1.0 A. Adding up the loss in all of the cells gives 0.1987×10^{-3} W. Since the total current is 1.0 A, the effective resistance is $0.1987 \times 10^{-3} \, \Omega$. We can compare this result to the analytical expression we found in Section 2.2 (high-frequency limit).

$$R_{ac} = \frac{1}{\sigma \delta b_s} \tag{5.41}$$

This result is also $R_{ac} = 0.1987 \times 10^{-3} \, \Omega$.

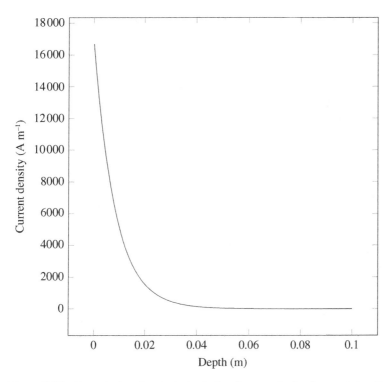

Figure 5.10 Numerical and analytical solution for magnitude of current density at different depths.

Alternatively, we can use the Poynting vector discussed previously. The equation is repeated here.

$$P = E \times H^*$$ (5.42)

Recall that the Poynting vector gives the complex power density crossing a surface. The surface of interest is at the top of the conductor, so $b_s \times 1.0$. Evaluating the Poynting vector at the surface of our finite difference model, we obtain $PV \times b_s = (0.00019867 + j0.00019871)$VA.

This is $P + jQ$ and we need only divide by the current squared to get the resistance and inductive reactance. The resistance agrees with the value we have obtained above. The inductive reactance is then $X = 0.0001987\,\Omega$. Our previous discussion of the one-dimensional plate gave us the result that the resistance and reactance should be equal, which is the case.

The eddy currents (or skin effect) will cause the ac resistance to be higher than dc resistance by restricting the area that the current passes through. The inductance is also affected. By restricting the area that the flux must pass through, the inductance will be reduced by the same factor. Recalling that the skin depth is proportional to the inverse square root of the frequency ($\delta \propto 1/\sqrt{\omega}$) we see how the resistance and reactance change by the same factor as a function of frequency. For example, if we double the frequency, the resistance increases by $\sqrt{2}$ and the inductance decreases by the same amount. To find the reactance, we multiply the inductance by frequency so the resistance and reactance change by the same factor. In the example above, if we make the frequency 120 Hz, we obtain $R + jX = 0.000281 + j0.000281$ which is the 60 Hz result times $\sqrt{2}$.

We can also use the magnetic energy to find the inductance. As shown above, with a magnetic vector potential solution, it is convenient to find the magnetic energy using

$$W_m = \frac{1}{2}\int_V J \cdot A \, dv \tag{5.43}$$

The result for the 60 Hz case is that the stored energy is 2.635×10^{-7} J. Since the current is 1.0 A, the inductance is twice the value of the stored energy or 5.27×10^{-7} H. Multiplying this by the angular frequency we obtain the same reactance as found above.

The redistribution of the current due to the skin effect, not only changes the resistance and inductance, but the electromagnetic force on the conductor. In the case of force on nonmagnetic conductors, we can use the Lorentz formula

$$dF = J \times Bd\ell \tag{5.44}$$

This local force density is then integrated over the conductor cross section. In the ladder network, we shall add the contributions of each cell. For the 60 Hz case, we obtain a force of 6.22×10^{-5} N downward. We can check the use of this local approach by evaluating the Maxwell Stress at the top of the conductor. The normal force (to the top surface) is then

$$F_n = \frac{B_n^2 - B_t^2}{2\mu_0}b_s \tag{5.45}$$

The result is 6.28×10^{-5} N or less than a 1% difference.

5.4.2 Time-Domain Example for the Finite Difference Network

The same model can be used for the transient or time-domain solution. In this case, the excitation is a current sheet which is sinusoidal in time. To compare it to the phasor analysis for the steady-state sinusoidal case, we must let the solution go for several cycles until steady state is reached. A central difference method, which was described above, was used to find the time-domain solution. The current density near the surface and one skin depth down is shown in Figure 5.11. We note that the

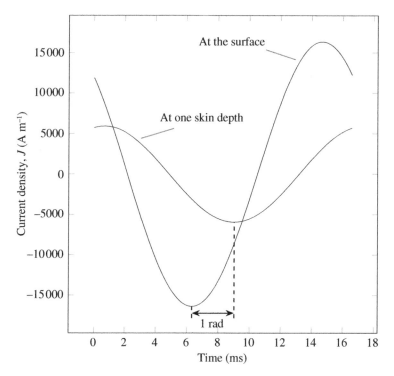

Figure 5.11 Current density at surface and at one skin depth with sinusoidal excitation.

phase angle of the current density is 45° at the surface which agrees with theory. The current density at a depth δ is lagging the current density at the surface by one radian. If we compare the peak values of the current density we see that the ratio is $1/e$ as expected.

Another interesting example is the diffusion of current produced by a step-function of current. This solution has already been discussed in Section 1.10 and is repeated here for convenience. For a semi-infinite conductor, initially unexcited, with an applied surface magnetic field of H_0, the current density as a function of space and time is

$$J(x,t) = \frac{2H_0}{\sqrt{4\pi Dt}} e^{-\frac{x^2}{4Dt}} \tag{5.46}$$

where

$$D = \frac{1}{\mu\sigma} \tag{5.47}$$

The finite difference model represents a slab of copper ($\mu = \mu_0$) which is excited by a step function current sheet at the surface. The conductor in the model is 1.0 m

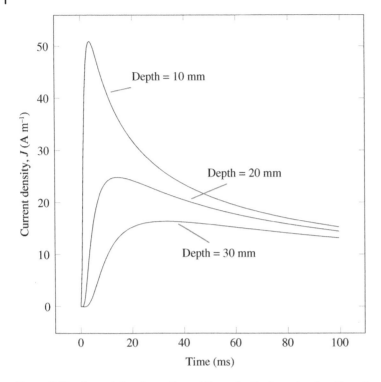

Figure 5.12 Current density vs. time at three depths for a step function of current.

deep and is divided into 1000 cells. The results of the model and the closed-form solution are shown in Figure 5.12 at three different locations below the surface of the conductor. As we move down from the surface, the peak current decreases and the time to peak increases which agrees with the theoretical predictions. The concept of treating this case as an $R - C$ ladder network also gives some insight into the phenomenon. At the application of a step function of current to the network, the first capacitor will take most of the current. As it charges and the voltage increases, current will be driven down the network and the second capacitor will begin to charge. This time delay is related to the diffusion time constant introduced previously. Since the total current in the conductor must equal the Ampere-turns in the current sheet, the capacitors further from the surface will have less current and that current will reach a maximum at a later time.

We have seen that the finite difference method is well-adapted to solving eddy current problems both in the time and frequency domain. For the two-dimensional MVP formulation, we developed an $R - C$ network which also gives physical insight into the eddy current distribution. The examples given were for Cartesian coordinates but the equivalent circuit is valid for polar coordinates as well. This circuit cannot be used in three-dimensions, since the vector potential will have multiple components. The finite difference method is extremely general and can be used to solve problems with complicated geometry and boundary conditions, as well as nonlinear and nonhomogeneous material properties. We will now turn our attention to the finite element method.

6

Finite Elements

6.1 Finite Elements

The finite element method is a numerical method which was first applied in structural and continuum mechanics in the 1960s and then became popular in electromagnetic and thermal analysis. The method can be applied to complex geometries having nonlinear and anisotropic materials. It can handle diffusion problems, like eddy currents, as well as wave problems. The method is based on two solution approaches: the variational method (also called Ritz's Method) and the Galerkin method. In the variational finite element method, an electromagnetic partial differential equation (PDE) is written in terms of an energy related functional. The approach yields a solution by minimizing this functional with respect to the unknown field variables or potentials. The finite element method using the Galerkin approach, uses the PDEs multiplied by a weighting function and integrated over the problem domain. The process then also minimizes an error function to find the unknown variables.

The method requires that the domain of the problem domain be discretized or divided into elements which completely cover the solution domain. Boundary conditions and sources can then be applied which makes the problem well-posed.

The steps involved in the finite element method are as follows:

- Define the electromagnetic problem by PDEs.
- Obtain a functional formulation for the PDEs in terms of the variational or the Galerkin approach.
- Subdivide the problem domain into finite elements.
- Start with a trial solution using the nodal values of the elements and interpolation functions.

Eddy Currents: Theory, Modeling, and Applications, First Edition.
Sheppard J. Salon, M. V. K. Chari, Lale T. Ergene, David Burow, and Mark DeBortoli.
© 2024 The Institute of Electrical and Electronics Engineers, Inc. Published 2024 by John Wiley & Sons, Inc.

- Minimize the functional with respect to each of the nodal potentials.
- Solve the set of algebraic equations for the nodal values.
- Post-process the nodal solution to find useful parameters.

6.2 The Variational Method

The variational method is one of the most practical and accurate methods for solving boundary value problems [46]. In this method, the PDE describing the problem is expressed as an energy-related formula which is called a functional. The solution to the PDE is based on the minimization of the functional.

In this section, the derivation of the functional and its variational form will be presented. A functional is an operator on a class of functions. For example, if we define ϕ to be a set of functions of two variables (x, y) subject to the condition $\phi = f(s)$ on S (where S is the surface bounding the region R) and which is continuously differentiable, then any quantity, such as $\mathcal{F}$, that takes a specific numerical value corresponding to each function in the set is said to be a functional on the set ϕ.

Here, the extremum of a functional with one independent variable will be considered in the form given below

$$\mathcal{F} = \int_a^b f(x, u, u')dx \tag{6.1}$$

where $u' = \frac{du}{dx}$, and the boundary conditions are

$$u(a) = A, \quad u(b) = B \tag{6.2}$$

where a, b, A, and B are constants. We will assume that f has continuous second-order derivatives with respect to its three arguments and require that the unknown function $u(x)$ possesses two derivatives everywhere in (a, b). A family of admissible functions that includes $u(x)$ is of the form

$$u(x) + \epsilon\eta(x) \tag{6.3}$$

where $\eta(x)$ is an arbitrary twice-differentiable function that vanishes at the end points (a, b). Thus,

$$\eta(a) = \eta(b) = 0 \tag{6.4}$$

Also, ϵ is a parameter that is a constant for any one function in the set but varies from one function to another.

If $u(x)$ is replaced by $u(x) + \epsilon\eta(x)$, then the integral $\mathcal{F}$ assumes the form (after dropping the terms in the parentheses for convenience)

$$\mathcal{F} = \int_a^b f(x, u + \epsilon\eta, u' + \epsilon\eta') \, dx \tag{6.5}$$

It is apparent that the maximum or minimum value (extremum) of $F(\epsilon)$ occurs when $\epsilon = 0$, that is, when the variation of u is zero. Hence it follows that

$$\frac{dF}{d\epsilon} = 0 \tag{6.6}$$

when $\epsilon = 0$.

Substituting for $F(\epsilon)$ from Equation (6.5) into Equation (6.6) and using the chain rule of differentiation

$$\frac{dF}{d\epsilon} = \int_a^b \left(\frac{\partial f(x, u + \epsilon\eta, u' + \epsilon\eta')}{\partial(u + \epsilon\eta)} \eta + \frac{\partial f(x, u + \epsilon\eta, u' + \epsilon\eta')}{\partial(u' + \epsilon\eta')} \eta' \right) dx \tag{6.7}$$

Setting $\epsilon = 0$, the equation becomes

$$\frac{dF}{d\epsilon} = \int_a^b \left(\frac{\partial f}{\partial u} \eta(x) + \frac{\partial f}{\partial u'} \eta'(x) \right) dx = F'(0) = 0 \tag{6.8}$$

The second term of the definite integral in Equation (6.8) can be transformed by integration by parts as follows:

$$\int_a^b \left(\frac{\partial f}{\partial u'} \eta'(x) \right) dx = \frac{\partial f}{\partial u'} \eta(x) \Big|_a^b - \int_a^b \left(\frac{\partial}{\partial x} \frac{\partial f}{\partial u'} \eta(x) \right) dx \tag{6.9}$$

Because $\eta(a) = \eta(b) = 0$, the first term on the right vanishes. Equation (6.9) then reduces to the form

$$\int_a^b \left(\frac{\partial f}{\partial u'} \eta'(x) \right) dx = - \int_a^b \left(\frac{\partial}{\partial x} \frac{\partial f}{\partial u'} \eta(x) \right) dx \tag{6.10}$$

Substituting (6.10) into (6.8),

$$F'(0) = - \int_a^b \left(\frac{\partial}{\partial x} \frac{\partial f}{\partial u'} - \frac{\partial f}{\partial u} \right) \eta(x) \, dx \tag{6.11}$$

The integrand of this function must be zero for any arbitrary function $\eta(x)$ in the neighborhood of $u(x)$, so that

$$\frac{\partial}{\partial x} \left(\frac{\partial f}{\partial u'} \right) - \left(\frac{\partial f}{\partial u} \right) = 0 \tag{6.12}$$

Equation (6.12) is called the *Euler–Lagrange* equation of the functional (6.1), subject to boundary conditions (6.2).

We will now present the functional for time-varying fields. We will consider two methods: the residual excitation method and the Poynting vector method.

Let us first consider the linear Poisson equation.

$$v\nabla^2 \phi_0 = f_0 \tag{6.13}$$

where ϕ_0 is the true potential solution, f_0 is the forcing function, and v is a physical parameter. If ϕ is an approximate solution, then a residue will be obtained as follows

$$v\nabla^2 \phi = f_0 - R \tag{6.14}$$

The right-hand side of (6.14) is expressed as an equivalent source density, f_1. We now multiply both sides by ϕ_0 and integrate the result over the entire region. We obtain

$$F = \int \phi_0 R \, dV = \int (\phi_0 v \nabla^2 \phi_0 - \phi_0 f_1) \, dV \tag{6.15}$$

The integral on the right-hand side of (6.15) yields the residual energy in the system, which we call the functional F. The minimization of this energy-related functional yields the true solution to the field problem defined by (6.13). This integral on the right-hand side of Equation (6.15) can be transformed by vector identities as follows:

$$\int v \phi_0 \nabla^2 \phi_0 \, dV = \int \nabla \cdot (v \phi_0 \nabla \phi_0) \, dV - \int v |\nabla \phi_0|^2 \, dV \tag{6.16}$$

Substituting (6.16) into (6.15),

$$F = -\int \phi_0 f_1 \, dV + \int \nabla \cdot (v \phi_0 \nabla \phi_0) \, dV - \int v |\nabla \phi_0|^2 \, dV \tag{6.17}$$

Using the divergence theorem, the second integral of (6.17) is transformed into a surface integral so that

$$F = -\int \phi_0 f_1 \, dV + \oint (v \phi_0 \nabla \phi_0) \, dS - \int v |\nabla \phi_0|^2 \, dV \tag{6.18}$$

Minimization of the Euler–Lagrange equation (6.12) results in the solution of the PDE.

We now consider the problem beginning with the Poynting vector. The functional is a scalar quantity, and its minimum value is the precondition for obtaining the true solution to scalar and vector potential field problems. The Poynting vector can be used to find a functional formulation for electromagnetic fields. Using a power balance, we have

$$-\oint_S (E \times H) \cdot dS = \int_V J \cdot E \, dV + \frac{\partial}{\partial t} \int_V \left(\frac{\epsilon}{2} E^2 + \frac{\mu}{2} H^2 \right) dV \tag{6.19}$$

We will neglect the energy term associated with the electrostatic field and the displacement currents. The equation then becomes

$$-\oint_S (E \times H) \cdot dS = \int_V J \cdot E \, dV + \frac{\partial}{\partial t} \int_V \frac{\mu}{2} H^2 \, dV \tag{6.20}$$

Defining the vector potential, A, such that

$$B = \nabla \times A \tag{6.21}$$

and using the relationship

$$\nabla \times E = -\frac{\partial B}{\partial t} \tag{6.22}$$

we obtain

$$\nabla \times E = -\frac{\partial}{\partial t}(\nabla \times A) \qquad (6.23)$$

or

$$E = -\frac{\partial A}{\partial t} \qquad (6.24)$$

First let us assume that the current density, J, does not vary with time.

$$\frac{\partial}{\partial t}(J \cdot A) = J \cdot \frac{\partial A}{\partial t} \qquad (6.25)$$

Thus from Equations (6.24) and (6.25)

$$\int_V J \cdot E \, dV = -\frac{\partial}{\partial t} \int_V J \cdot A \, dV \qquad (6.26)$$

Substituting Equation (6.26) into (6.20)

$$\oint_S (E \times H) \cdot dS = -\frac{\partial}{\partial t} \left(\int_V J \cdot A \, dV - \int_V \frac{\mu}{2} H^2 \, dV \right) \qquad (6.27)$$

Equation (6.27) can be rewritten in terms of the reluctivity v ($v = \frac{1}{\mu}$) so that

$$H = vB = v\nabla \times A \qquad (6.28)$$

and

$$-\oint_S (E \times H) \cdot dS = -\frac{\partial}{\partial t} \left(\int_V J \cdot A \, dV - \int \frac{v}{2}(\nabla \times A)^2 \, dV \right) \qquad (6.29)$$

Assuming that A has only a z-directed component A_z, we have, for the two-dimensional problem,

$$(\nabla \times A)^2 = |\nabla A|^2 = \left(\frac{\partial A}{\partial x} \right)^2 + \left(\frac{\partial A}{\partial y} \right)^2 \qquad (6.30)$$

Substituting for $(\nabla \times A)^2$ from (6.30) into (6.29) and integrating both sides of Equation (6.29) with respect to time, we have

$$\int \left(-\oint_S (E \times H) \cdot dS \right) dt = -\int_V J \cdot A \, dV + \int_V \frac{v}{2} \left(\left(\frac{\partial A}{\partial x} \right)^2 + \left(\frac{\partial A}{\partial y} \right)^2 \right) dV \qquad (6.31)$$

which is the energy flowing out of the surface enclosing the volume of the field region [46].

We now look at the energy functional for the diffusion problem. We have seen that electromagnetic field problems, including eddy current effects, can be modeled by the diffusion equation in terms of a magnetic vector potential A, source current density J_s, and scalar potential function, ϕ as

$$\nabla \times v\nabla \times A + \sigma\frac{\partial A}{\partial t} - \sigma\nabla\phi = J_s \qquad (6.32)$$

In addition to Equation (6.32), the vector and scalar potentials are related by a zero divergence condition on the current density vector. The resulting equation is

$$\nabla \cdot \sigma \frac{\partial A}{\partial t} - \nabla \cdot \sigma \nabla \phi = 0 \tag{6.33}$$

Integrating Equation (6.33) over time, assuming σ is constant,

$$\sigma \nabla \cdot A - \sigma \int \nabla^2 \phi \, dt = 0 \tag{6.34}$$

Equations (6.32) and (6.34) fully define the time-dependent eddy current problem. For two-dimensional problems with a single component of vector potential, the divergence of A is automatically zero, and therefore there is no need for a scalar potential function, ϕ.

The functional for the diffusion equation (6.32) and gauge condition (6.34) are obtained as follows [46]

$$F = \int_V \left(A \cdot \nabla \times v\nabla \times A + \sigma A \cdot \frac{\partial A}{\partial t} \right.$$
$$\left. -\sigma A \cdot \nabla \phi + \sigma \phi \nabla \cdot A - \sigma \phi \nabla^2 \phi - 2A \cdot J_s \right) dV \tag{6.35}$$

Using vector identities (Green's theorem and the divergence theorem), substituting ϕ for the term containing time in (6.35), replacing the integrals containing $\nabla \times v\nabla \times A$ and Laplace's equation in ϕ by volume and surface integrals, and rearranging terms, after considerable algebra, the energy-related functional for the diffusion equation is obtained as

$$F = \int_V \left(v(\nabla \times A)^2 + \sigma A \cdot \frac{\partial A}{\partial t} - \sigma A \cdot \nabla \phi + \sigma \phi \nabla \cdot A \right.$$
$$\left. + \ \sigma \nabla \phi \cdot \nabla \phi - 2A \cdot J_s \right) dV - \oint_S A \cdot (n \times v\nabla \times A) \, dS$$
$$- \oint \phi \frac{\partial \phi}{\partial n} \, dS \tag{6.36}$$

The steady-state time-harmonic diffusion equation and the related functional are obtained by representing the vector potential A, the scalar potential ϕ, and the source current density J_s as time-harmonic functions or phasors. Thus

$$A = |A|e^{j\theta_A}$$
$$\phi = |\phi|e^{j\theta_\phi}$$
$$J_s = |J_s|e^{j\theta_{J_s}} \tag{6.37}$$

Substituting for A, ϕ, and J_s from Equation (6.37) into Equations (6.32) and (6.34), the steady-state diffusion equation and the divergence condition on A [46] are obtained as Equations (6.38) and (6.39) respectively.

$$\nabla \times v\nabla \times A + j\omega\sigma A - \sigma\nabla\phi = J_s \tag{6.38}$$

$$\sigma\nabla \cdot A - \frac{\sigma}{j\omega}\nabla \cdot \nabla\phi = 0 \tag{6.39}$$

From the generalized expression for the functional of Equation (6.36), we have

$$\mathcal{F} = \langle \psi^T | \mathcal{L}\psi \rangle - 2\langle \psi^T | f \rangle \tag{6.40}$$

Making the following substitutions from Equations (6.38) and (6.39) in Equation (6.40)

$$\psi = \begin{pmatrix} A \\ \phi \end{pmatrix}$$

$$f = J_s \tag{6.41}$$

$$\mathcal{L}\psi = \begin{pmatrix} \nabla \times \nu\nabla \times A + j\omega\sigma A - \sigma\nabla\phi \\ \sigma\nabla \cdot A - \frac{\sigma}{j\omega}\nabla \cdot \nabla\phi \end{pmatrix} \tag{6.42}$$

The functional for the steady-state diffusion equation is obtained as

$$\mathcal{F} = \int_V \Big(A \cdot \nabla \times \nu\nabla \times A + j\omega\sigma A^2 - \sigma A \cdot \nabla\phi$$

$$+ \sigma\phi\nabla \cdot A - \frac{\sigma\phi}{j\omega}\nabla \cdot \nabla\phi - 2A \cdot J_s \Big) dV = 0 \tag{6.43}$$

Using the vector identities

$$A \cdot \nabla \times \nu\nabla \times A = \nu(\nabla \times A)^2 - \nabla \cdot [A \times \nu\nabla \times A] \tag{6.44}$$

$$-\frac{j\sigma}{\omega}\phi\nabla \cdot \nabla\phi = \frac{j\sigma}{\omega}(\nabla\phi)^2 - \frac{j\sigma}{\omega}\nabla \cdot (\phi\nabla\phi) \tag{6.45}$$

and substituting them in the expression for the functional of Equation (6.43), we obtain

$$\mathcal{F} = \int_V \Big(\nu(\nabla \times A)^2 + j\omega\sigma A^2 - \sigma A \cdot \nabla\phi - \sigma\phi\nabla \cdot A + \frac{j\sigma}{\omega}(\nabla\phi)^2 - 2A \cdot J_s \Big) dV$$

$$- \int_V \nabla \cdot (A \times \nu\nabla \times A) \, dV - \int_V \frac{j\sigma}{\omega}\nabla \cdot (\phi\nabla\phi) \, dV \tag{6.46}$$

The last two volume integrals on the right-hand side of Equation (6.46) are transformed into surface integrals using the divergence theorem. This gives

$$\mathcal{F} = \int_V \Big(\nu(\nabla \times A)^2 + j\omega\sigma A^2 - \sigma A \cdot \nabla\phi - \sigma\phi\nabla \cdot A + \frac{j\sigma}{\omega}(\nabla\phi)^2 - 2A \cdot J_s \Big) dV$$

$$- \oint_S A \cdot (n \times \nu\nabla \times A) \, dS - \oint_S \frac{j\sigma}{\omega}\phi\nabla\phi \cdot dS \tag{6.47}$$

Equation (6.47) represents the functional for the generalized steady-state diffusion equation.

6.2.1 Finite Element Discretization

The problem domain is divided into a number of finite elements. These elements can be first-order or higher-order elements depending on the approximating function. These elements are the finite element mesh. The elements will contain nodes,

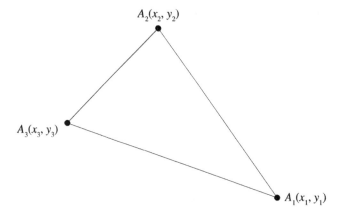

Figure 6.1 One triangular element.

which can be at the vertices, along the edges and in the interior of the elements. For example, a two-dimensional surface can be discretized into linear triangular elements. In this case, the nodes will be at the triangle vertices. An example of the polynomial for this approximating function can be illustrated by using a first-order triangular element as shown in Figure 6.1 with its nodal coordinates (x_1, y_1), (x_2, y_2), and (x_3, y_3). The linear approximating function for the potential can be written as shown in Equation (6.48).

$$A = a + bx + cy \tag{6.48}$$

By substituting each coordinate for the vertices in this potential equation, a set of simultaneous equations is obtained for the potentials at each node.

$$A_1 = a + bx_1 + cy_1$$
$$A_2 = a + bx_2 + cy_2$$
$$A_3 = a + bx_3 + cy_3 \tag{6.49}$$

These equations can be expressed in matrix form as

$$\begin{pmatrix} A_1 \\ A_2 \\ A_3 \end{pmatrix} = \begin{pmatrix} 1 & x_1 & y_1 \\ 1 & x_2 & y_2 \\ 1 & x_3 & y_3 \end{pmatrix} \begin{pmatrix} a \\ b \\ c \end{pmatrix} \tag{6.50}$$

We can obtain the a, b, and c constants for this element by solving Equation (6.50).

$$\begin{pmatrix} a \\ b \\ c \end{pmatrix} = \begin{pmatrix} 1 & x_1 & y_1 \\ 1 & x_2 & y_2 \\ 1 & x_3 & y_3 \end{pmatrix}^{-1} \begin{pmatrix} A_1 \\ A_2 \\ A_3 \end{pmatrix} \tag{6.51}$$

The inverse of the coefficient matrix is equal to

$$\begin{pmatrix} 1 & x_1 & y_1 \\ 1 & x_2 & y_2 \\ 1 & x_3 & y_3 \end{pmatrix}^{-1} = \frac{1}{2S} \begin{pmatrix} a_1 & b_1 & c_1 \\ a_2 & b_2 & c_2 \\ a_3 & b_3 & c_3 \end{pmatrix}^T \tag{6.52}$$

where S is the area of the triangle. So,

$$\begin{pmatrix} a \\ b \\ c \end{pmatrix} = \frac{1}{2S} \begin{pmatrix} a_1 & b_1 & c_1 \\ a_2 & b_2 & c_2 \\ a_3 & b_3 & c_3 \end{pmatrix}^T \begin{pmatrix} V_1 \\ V_2 \\ V_3 \end{pmatrix} \tag{6.53}$$

where

$$a_1 = x_2 y_3 - x_3 y_2$$
$$a_2 = x_3 y_1 - x_1 y_3$$
$$a_3 = x_1 y_2 - x_2 y_1$$
$$b_1 = y_2 - y_3$$
$$b_2 = y_3 - y_1 \tag{6.54}$$
$$b_3 = y_1 - y_2$$
$$c_1 = x_3 - x_2$$
$$c_2 = x_1 - x_3$$
$$c_3 = x_2 - x_1$$

Now the a, b, and c values can be substituted into Equation (6.48) to obtain the potential.

$$A = \frac{1}{2S}[(a_1 + b_1 x + c_1 y)A_1 + (a_2 + b_2 x + c_2 y)A_2 + (a_3 + b_3 x + c_3 y)A_3] \tag{6.55}$$

We can also use the shape functions α_1, α_2, and α_3, which are polynomial weighting functions, to obtain the approximation polynomials.

$$\alpha_1 = \frac{1}{2S}[a_1 + b_1 x + c_1 y]$$
$$\alpha_2 = \frac{1}{2S}[a_2 + b_2 x + c_2 y]$$
$$\alpha_3 = \frac{1}{2S}[a_3 + b_3 x + c_3 y] \tag{6.56}$$

The potential can be written by using these shape functions as

$$A = \sum_{i=1}^{n} \alpha_i A_i \tag{6.57}$$

where n is the number of nodes in the element. The sum of the shape functions of any element will always be equal to one.

$$\alpha_1 + \alpha_2 + \alpha_3 = 1 \tag{6.58}$$

These shape functions will be used to calculate the coefficient matrix for each element. Considering triangular element 1, the coefficient matrix for this element can be expressed by using Equation (6.59). This matrix is also called the stiffness matrix.

$$[K^1] = \begin{pmatrix} k^1_{11} & k^1_{12} & k^1_{13} \\ k^1_{21} & k^1_{22} & k^1_{23} \\ k^1_{31} & k^1_{32} & k^1_{33} \end{pmatrix} \tag{6.59}$$

The each element of this stiffness matrix is calculated separately by using the shape functions.

$$k^1_{ij} = \left[\int_J \nabla \alpha_i \nabla \alpha_y \, dS \right] \tag{6.60}$$

6.2.2 Global Matrix Assembly

The global stiffness matrix is created by assembling the element stiffness matrices. Here, an assembly example will be shown for two triangular elements given in Figure 6.2. Nodal potentials and stiffness matrices of each elements are given by Equations (6.61)–(6.64) respectively [46].

$$[A]^1 = \begin{pmatrix} A_1 \\ A_2 \\ A_3 \end{pmatrix} \tag{6.61}$$

$$[K]^1 = \begin{pmatrix} k^1_{11} & k^1_{12} & k^1_{13} \\ k^1_{21} & k^1_{22} & k^1_{23} \\ k^1_{31} & k^1_{32} & k^1_{33} \end{pmatrix} \tag{6.62}$$

$$[A]^2 = \begin{pmatrix} A_4 \\ A_5 \\ A_6 \end{pmatrix} \tag{6.63}$$

$$[K]^2 = \begin{pmatrix} k^2_{11} & k^2_{12} & k^2_{13} \\ k^2_{21} & k^2_{22} & k^2_{23} \\ k^2_{31} & k^2_{32} & k^2_{33} \end{pmatrix} \tag{6.64}$$

After taking derivatives of the shape functions and considering the a, b, and c constants for the related elements, each element of the coefficient matrix can be expressed in a general form in Equation (6.65) and matrix form in Equation (6.66).

$$k_{ij} = \frac{\mu}{4S}(b_i b_j + c_i c_j) \quad i = 1, 2, 3, \quad j = 1, 2, 3 \tag{6.65}$$

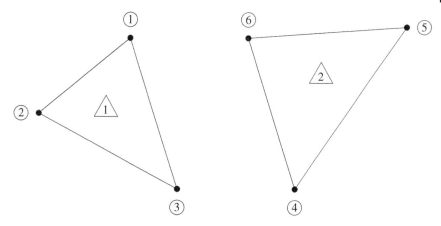

Figure 6.2 Two separate triangular elements.

$$[K]^e = \frac{\mu}{4S} \begin{pmatrix} b_1^2 + c_1^2 & b_1 b_2 + c_1 c_2 & b_1 b_3 + c_1 c_3 \\ b_1 b_2 + c_1 c_2 & b_2^2 + c_2^2 & b_2 b_3 + c_2 c_3 \\ b_1 b_3 + c_1 c_3 & b_2 b_3 + c_2 c_3 & b_3^2 + c_3^2 \end{pmatrix} \tag{6.66}$$

The energy equation for a triangular element is

$$W = \frac{\mu}{2} [A]_s^T [K]_s [A]_s \tag{6.67}$$

The vector potential and coefficient matrix are given as follows:

$$[A]_s^T = \begin{pmatrix} A_1 & A_2 & A_3 & A_4 & A_5 & A_6 \end{pmatrix} \tag{6.68}$$

$$[K]_s = \begin{bmatrix} [K]^1 & 0 \\ 0 & [K]^2 \end{bmatrix} \tag{6.69}$$

The nodes 1 and 6, and 3 and 4 are connected as shown in Figure 6.3. The assembled potential and system matrix equations are

$$[A]_c = \begin{pmatrix} A_1 \\ A_2 \\ A_3 \\ A_5 \end{pmatrix} \tag{6.70}$$

$$[K]_c = \begin{pmatrix} k_{11} & k_{12} & k_{13} & k_{15} \\ k_{21} & k_{22} & k_{23} & k_{25} \\ k_{31} & k_{32} & k_{33} & k_{35} \\ k_{51} & k_{52} & k_{53} & k_{55} \end{pmatrix} \tag{6.71}$$

The K_c matrix is the global coefficient matrix and is symmetric, sparse, and singular. It shows the connection between the nodes. In the above example, the global coefficient matrix with all sub elements is given in Equation (6.72).

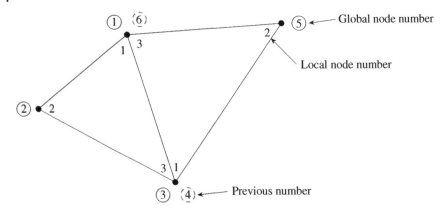

Figure 6.3 Two adjacent triangular elements.

$$[K]_c = \begin{pmatrix} k_{11}^1 + k_{33}^2 & k_{12}^1 & k_{13}^1 + k_{31}^2 & k_{32}^2 \\ k_{21}^1 & k_{22}^1 & k_{23}^1 & 0 \\ k_{31}^1 + k_{13}^2 & k_{32}^1 & k_{33}^1 + k_{11}^2 & k_{12}^2 \\ k_{32}^2 & 0 & k_{21}^2 & k_{22}^2 \end{pmatrix} \tag{6.72}$$

6.2.3 Numerical Examples: Conducting Plates

The finite element method is now applied to an example that we have solved in closed-form in Section 2.3 and with the finite difference method in Section 5.1. The example is a conducting plate in a sinusoidally time-varying, 60 HZ solenoidal field. The finite element mesh is made of first-order triangles and the excitation is a surface current sheet. The material is copper with conductivity $\sigma = 5.8 \times 10^7 \, \mathrm{Sm}^{-1}$ and permeability $\mu_0 = 4\pi \times 10^{-7} \, \mathrm{Hm}^{-1}$. The geometry is shown in Figure 6.4.

The mesh used for the example is a regular mesh constructed of triangles. The mesh chosen is especially easy to check. If we take a square and draw the two

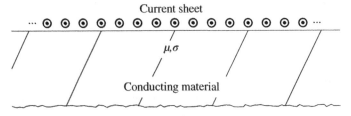

Figure 6.4 Current sheet and copper plate.

Figure 6.5 Mesh consisting of four-element, five-node blocks.

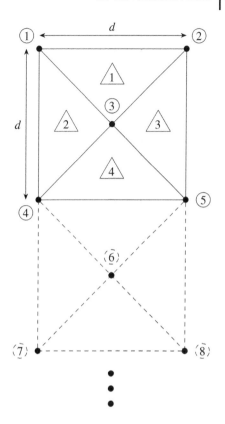

diagonals as shown in Figure 6.5, we obtain a regular pattern including four triangles (elements) and five vertices (nodes).

As illustrated in the figure, element 1 has nodes (1,3,2), element 2 has nodes (1,4,3), element 3 has nodes (2,3,5), and element 4 has nodes (3,4,5). The node ordering is in the counterclockwise direction. Table 6.1 shows the cyclic ordering of the nodes for each element.

Table 6.1 Elements and node numbering for the uppermost four-element block.

Element	i	j	k
1	1	3	2
2	1	4	3
3	2	3	5
4	3	4	5

The global finite element matrix consists of two parts: the element stiffness matrix (reluctance) and the mass matrix (conductance). The elements of the stiffness matrix are given by

$$[K] = \frac{1}{4\mu_0 S} \begin{bmatrix} b_i^2 + c_i^2 & b_i b_j + c_i c_j & b_i b_k + c_i c_k \\ & b_j^2 + c_j^2 & b_j b_k + c_j c_k \\ \text{symmetric} & & b_k^2 + c_k^2 \end{bmatrix} \tag{6.73}$$

We recall that

$$b_i = y_j - y_k$$
$$c_i = x_k - x_j \tag{6.74}$$

with the other indices varying in cyclic order. The values of these parameters for elements in the uppermost block are given in Table 6.2.

The S matrix for element 1 is then

$$[K]_1 = \frac{1}{\mu_0} \begin{bmatrix} 0.5 & -0.5 & 0 \\ -0.5 & 1 & -0.5 \\ 0 & -0.5 & 0.5 \end{bmatrix} \tag{6.75}$$

Following the same process for elements 2–4 and combining the resulting elemental stiffness matrices, we obtain the 5×5 stiffness matrix for the uppermost 4-element block:

$$[K] = \frac{1}{\mu_0} \begin{bmatrix} 1 & 0 & -1 & 0 & 0 \\ 0 & 1 & -1 & 0 & 0 \\ -1 & -1 & 4 & -1 & -1 \\ 0 & 0 & -1 & 1 & 0 \\ 0 & 0 & -1 & 0 & 1 \end{bmatrix} \tag{6.76}$$

The T (conductance) matrix for each element is given by

$$[T] = \frac{j\omega\sigma S}{12} \begin{bmatrix} 2 & 1 & 1 \\ 1 & 2 & 1 \\ 1 & 1 & 2 \end{bmatrix} \tag{6.77}$$

Table 6.2 Values of b, c, and S (area) for elements in the uppermost block.

Element	x_i	x_j	x_k	y_i	y_j	y_k	b_i	b_j	b_k	c_i	c_j	c_k	S
1	0	$\frac{d}{2}$	d	0	$\frac{-d}{2}$	0	$\frac{-d}{2}$	0	$\frac{d}{2}$	$\frac{d}{2}$	$-d$	$\frac{d}{2}$	$\frac{d^2}{4}$
2	0	0	$\frac{d}{2}$	0	$-d$	$\frac{-d}{2}$	$\frac{-d}{2}$	$\frac{-d}{2}$	d	$\frac{d}{2}$	$\frac{-d}{2}$	0	$\frac{d^2}{4}$
3	d	$\frac{d}{2}$	d	0	$\frac{-d}{2}$	$-d$	$\frac{d}{2}$	$-d$	$\frac{d}{2}$	$\frac{d}{2}$	0	$\frac{-d}{2}$	$\frac{d^2}{4}$
4	$\frac{d}{2}$	0	d	$\frac{-d}{2}$	$-d$	$-d$	0	$\frac{-d}{2}$	$\frac{d}{2}$	d	$\frac{-d}{2}$	$\frac{-d}{2}$	$\frac{d^2}{4}$

and the 5×5 conductance matrix for each 4-element block is then

$$[T] = \frac{j\omega\sigma S}{12} \begin{bmatrix} 4 & 1 & 2 & 1 & 0 \\ 1 & 4 & 2 & 0 & 1 \\ 2 & 2 & 8 & 2 & 2 \\ 1 & 0 & 2 & 4 & 1 \\ 0 & 1 & 2 & 1 & 4 \end{bmatrix} \tag{6.78}$$

As explained earlier, the global matrix is assembled by adding the contributions of each individual element. In the present example, we have the 5×5 matrix for the uppermost 4-element block. We then work downward, adding each successive 4-element block, noting that we are adding 3 new nodes each time: the 2 nodes at the bottom and the one at the center of the new block. The 2 nodes at the top of the new block are the same as the 2 nodes at the bottom of the previous block. It is then a straightforward process to form the global matrix. The row corresponding to each node will ultimately include contributions from all elements containing that node (2 elements for the nodes at the top and bottom, and 4 elements for all others).

To completely define the problem, we need a forcing or excitation function. For first-order triangular elements containing source current density, J, we apply one-third of the current within an element to each of the three nodes of the element to generate the element's contribution to the forcing function vector. In this example, the excitation is not provided by a region of current density but rather by an x-directed external field H_0 at the surface of the conductor. The source of this field can be viewed as a z-directed sheet of current of linear density H_0 amperes per meter at the surface of the conductor. The total excitation current in this problem is then H_0 times the width of the surface, d. Since there are 2 nodes at the surface, this current is equally divided and half is injected in each node. The input vector then is all zeros except for terms 1 and 2, which are identical and equal to one half the surface current:

$$\{F\} = \begin{bmatrix} H_0 d/2 & H_0 d/2 & 0 & 0 & \cdots & 0 \end{bmatrix}^T \tag{6.79}$$

We are now ready to solve the global matrix equation

$$[K + T]\{A\} = \{F\} \tag{6.80}$$

for the values of the vector potential at each node $\{A\}$. Since this problem is two-dimensional, the vector potential is purely z directed. The system was solved for the case of a 60 HZ applied field of $H_0 = 1.0\,\text{Am}^{-1}$, using 300 layers of 4-element blocks of side dimension $d = 0.283\,\text{mm}$ (i.e. one-thirtieth of a skin depth) making a total domain depth of 10 skin depths.

Once the model is solved for the values of the vector potential at the nodes, we can post-process these values to obtain flux densities and current densities. We

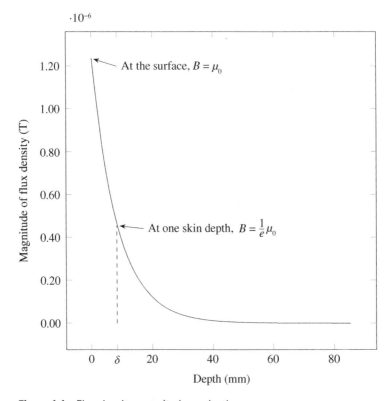

Figure 6.6 Flux density magnitude vs. depth.

note that since the vector potential has only a z component, then from $\mathbf{B} = \nabla \times \mathbf{A}$ we have

$$B = B_x = \frac{\partial A_z}{\partial y}$$

Therefore, the flux density in the region between two nodes is the difference of the nodal vector potentials divided by the y-distance between the two nodes.

The magnitude of the flux density as a function of depth, normalized to the flux density at the surface, $B_0 = \mu_0 H_0$, is given in Figure 6.6. At a depth of one skin depth, the magnitude of the calculated flux density is $1/e$ times the value at the surface, consistent with closed-form theory.

To find the current density, we first recall that $\mathbf{J} = \sigma \mathbf{E}$. Using Faraday's Law and the definition of magnetic vector potential

$$\nabla \times \mathbf{E} = -\frac{\partial \mathbf{B}}{\partial t} \qquad \mathbf{B} = \nabla \times \mathbf{A}$$

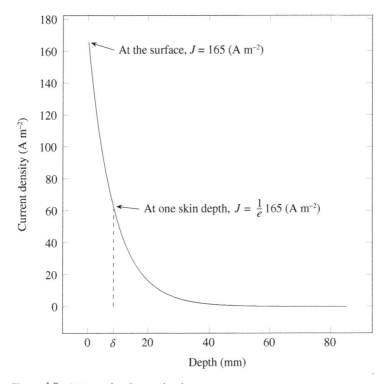

Figure 6.7 Current density vs. depth.

which, in the absence of electric scalar potential, can be combined to yield

$$E = -\frac{\partial A}{\partial t} \tag{6.81}$$

Substituting $j\omega$ for the time derivative, we have

$$E = -j\omega A \tag{6.82}$$

and using the resulting expression for **E** in the relationship for **J** above, we obtain for our example problem

$$J_z = -j\omega\sigma A_z \tag{6.83}$$

Figure 6.7 shows the current density magnitude as a function of depth in the conductor. As shown in Figure 6.7, the values of current density at points separated by one skin depth have the ratio $1/e$.

We can also check the phase angles of these variables. We plot the phase angles of the flux density and current density in Figure 6.8. As expected from $\mathbf{B} = \mu_0\mathbf{H}$, the flux density at the surface has a phase angle of zero, consistent with the applied **H** field. We further expect, from (2.27), that the phase shifts linearly with depth

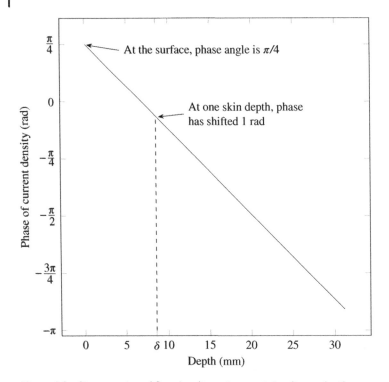

Figure 6.8 Phase angles of flux density and current density vs. depth.

and that the phase will shift by one radian as we move one skin depth into the material. The phase of the current density at the surface lags the magnetic field by 45° which agrees with (2.29).

As further verification, we can calculate the total current in the slab by integrating the current density over the area. The current density within each element can be found from (6.83), using the average of the vector potentials at the three vertices of the element. This is multiplied by the area of the element to obtain the total current in that element. The process is repeated for each element:

$$I_{eddy} = -j\omega \sum_{elm} \sigma_{elm} \left(\frac{A_{i,elm} + A_{j,elm} + A_{k,elm}}{3} \right) \Delta_{elm} = -284 \times 10^{-6} \text{A} \quad (6.84)$$

which is equal in magnitude and opposite in sign to the input current found from $H_0 d$.

To evaluate global quantities, we can start by finding the total apparent power P in the slab by integrating the Poynting vector, $\mathbf{E} \times \mathbf{H}^*$ over a surface Γ surrounding the region

$$P = \oint_{\Gamma} (\mathbf{E} \times \mathbf{H}^*) \cdot d\Gamma \quad (6.85)$$

In this one-dimensional problem, the contributions from the vertical sides cancel, and the fields at the bottom boundary should be negligible since the problem is assumed to be infinite in this direction. The surface integral, therefore, reduces to a line integral of $\mathbf{E} \times \mathbf{H}^*$ along the upper boundary of the domain only, multiplied by the depth of the problem in the z direction (1 m). The magnetic field at the surface has been specified as purely x directed. The electric field is purely z directed and therefore the Poynting vector is y directed along the upper surface.

The electric field is found from (6.82)

$$E = E_z = -j\omega A_z \tag{6.86}$$

Using vector potential at nodes on the upper boundary of the problem, we find that $E_{z0} = (-2.02 - j2.02) \times 10^{-6}\ Vm^{-1}$ at the upper boundary. Combining this with the specified surface magnetic field of $H_0 = (1.0 + j0.0)\,Am^{-1}$, the apparent power into the region through this surface is found as

$$P = E_{z0}H_0d \cdot 1\,m = (-575 - j575) \times 10^{-12}\ VA \tag{6.87}$$

The negative value indicates that power is flowing downward through the upper surface and into the domain of the problem, which is as expected.

We obtain the resistance and reactance by dividing the apparent power by the square of the current:

$$Z = \frac{P}{|I|^2} = (0.007 + j0.007)\Omega \tag{6.88}$$

We can check this result by estimating the resistance using a current path of length 1 m and cross-sectional area δd, since the slab is much deeper than the skin depth, and the current is largely contained within the skin depth:

$$R = \frac{1\,m}{\sigma\delta d} = 0.007\Omega \tag{6.89}$$

which is consistent with (6.88). We also observe that the resistance and reactance are equal, which is consistent with closed-form expressions for an infinite slab.

Another check on the results is a comparison with the closed-form solution to the problem of a finite depth plate. Consider the problem in Figure 6.9. The material properties are the same here but the depth of the plate is 2δ. The center line is an axis of symmetry. The solution for a plate of thickness $2b$ is repeated here in Equations (6.90) and (6.91).

$$B_x = B_0 \frac{\cosh(y\sqrt{1+j}/\delta)}{\cosh(b\sqrt{1+j}/\delta)} \tag{6.90}$$

$$J_z = -\frac{B_0}{\mu\delta}(1+j)\left(\frac{\sinh(y/\delta)\cos(y/\delta) + j\cosh(y/\delta)\sin(y/\delta)}{\cosh(b/\delta)\cos(b/\delta) + j\sinh(b/\delta)\sin(b/\delta)} \right) \tag{6.91}$$

In Figure 6.10, we see the results of the closed-form solution for the flux density magnitude and the finite element results which agree extremely well (There are

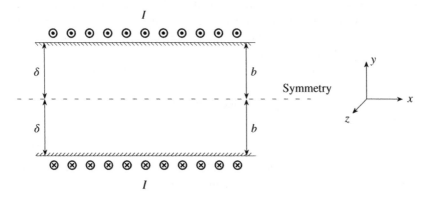

Figure 6.9 Finite depth plate example.

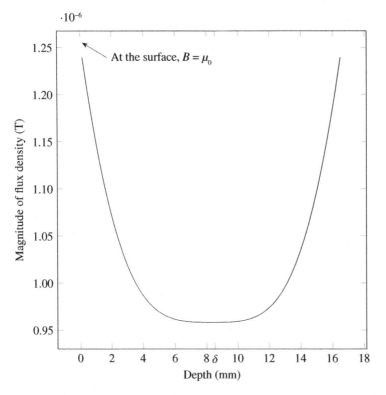

Figure 6.10 Comparison between FEA results and closed-form solution for flux density magnitude vs. depth.

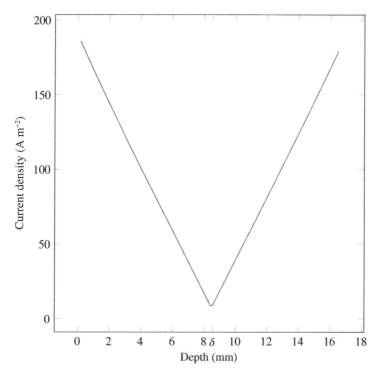

Figure 6.11 Comparison between FEA results and closed-form solution for current density magnitude vs. depth.

two curves plotted). In Figure 6.11, we compare the results for the current density magnitude.

6.2.4 Equivalent Circuit for Two-Dimensional Finite Element Eddy Current Analysis

We have seen in Section 5.1 that we can have an equivalent circuit representation for the two-dimensional finite difference formulation of the eddy current problem. In a similar way, Carpenter has developed an equivalent circuit for the finite element formulation [54].

As described above, for first-order triangular elements (see Figure 6.12), the linear interpolation is described by

$$A(x, y) = \frac{1}{2\Delta} \sum_{i=1}^{3} \left(a_i + b_i x + c_i y \right) A_i \tag{6.92}$$

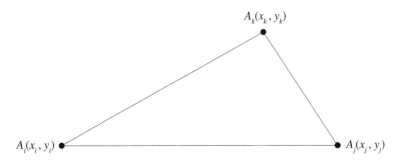

Figure 6.12 First-order triangular element.

The interpolation coefficients are defined in terms of the triangle vertex coordinates.

$$a_i = x_k y_j - x_j y_k$$
$$b_i = y_k - y_j$$
$$c_i = x_j - x_k \tag{6.93}$$

In the first-order element, since the vector potential varies linearly, the current density also varies linearly, so that

$$J(x, y) = \frac{1}{2\Delta} \sum_{i=1}^{3} \left(a_i + b_i x + c_i y \right) J_i \tag{6.94}$$

As in the finite difference equivalent circuit, the *resistive* part of the circuit represents the permeance in a magnetic circuit where the current in that branch is equivalent to the MMF (ampere-turns).

These circuit elements are evaluated as

$$R_k = \frac{\mu \ell_{ij}}{w_k} \tag{6.95}$$

The equivalent height is defined as

$$w_k = \frac{h_k}{2} \left(1 - \frac{\ell_{ik} \ell_{jk}}{h_k^2} \right) \tag{6.96}$$

as shown in Figure 6.13. We can progress by analogy by considering the finite element matrix equations derived above. There are two terms in the element equation, one referring to the magnetic flow (network permeances) and one to the electric current flow (network conductance). We treat these equations as an admittance network with the magnetic vector potential as the nodal voltage, just as we did for the finite difference equivalent circuit in Section 5.1.

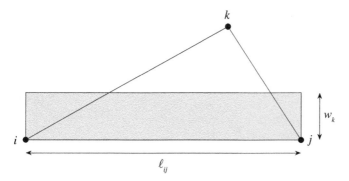

Figure 6.13 Definitions for Equation (6.95).

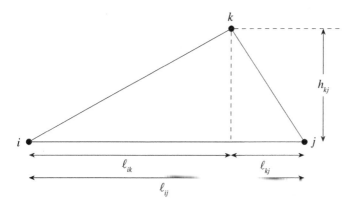

Figure 6.14 Triangular finite element for equivalent circuit.

The equations are repeated here for convenience. The stiffness matrix is given by

$$\mathbf{S} = \frac{1}{4\mu\Delta} \begin{bmatrix} b_i^2 + c_i^2 & b_i b_j + c_i c_j & b_i b_k + c_i c_k \\ b_i b_j + c_i c_j & b_j^2 + c_j^2 & b_j b_k + c_j c_k \\ b_i b_k + c_i c_k & b_j b_k + c_j c_k & b_k^2 + c_k^2 \end{bmatrix} \tag{6.97}$$

Referring to Figure 6.14, we can write the b and c terms in Equation (6.97) in terms of the indicated distances.

As an example, consider the S_{12} term, which will represent the equivalent reluctance between nodes 1 and 2. We have

$$S_{12} = \frac{1}{4\mu\Delta} \left(b_1 b_2 + c_1 c_2 \right) \tag{6.98}$$

where

$$b_1 = y_3 - y_2 = h_{23}$$
$$b_2 = y_1 - y_3 = -h_{23}$$
$$c_1 = x_2 - x_3 = \ell_{32}$$
$$c_2 = x_3 - x_1 = \ell_{13} \tag{6.99}$$

Therefore,

$$S_{12} = -h_{23}^2 + \ell_{32}\ell_{13} \tag{6.100}$$

The equivalent permeance is

$$P = \frac{4\mu\Delta}{-h_{23}^2 + \ell_{32}\ell_{13}} \tag{6.101}$$

The element area can be found as

$$\Delta = \frac{1}{2}\left((x_2 - x_1)(y_3 - y_1) - (y_2 - y_1)(x_3 - x_1)\right) = \frac{\ell_{12}h_{23}}{2} \tag{6.102}$$

We can write the permeance in the more traditional form by using the path length divided by an equivalent area.

$$P = \frac{2\mu\ell_{12}H_{12}}{-h_{23}^2 + \ell_{32}\ell_{13}} = \frac{\mu\ell_{12}}{w_{eq}} \tag{6.103}$$

where w_k is found (see Figure 6.13) as

$$w_{eq} = \frac{h_{23}}{2}\left(1 - \frac{\ell_{13}\ell_{32}}{h_{23}}\right) \tag{6.104}$$

By interpreting the finite element system equations as a network admittance matrix, we can find the network elements by inspection. Recall that in network theory, we form the admittance matrix by the following algorithm. On the diagonal, we have the sum of all admittances connected to the node.

$$Y_{ii} = \sum_{k=1}^{N} y_{ik} \tag{6.105}$$

On the off-diagonal, we have the negative of the admittance connected between nodes. For the off-diagonal terms, we have

$$Y_{ij} = -y_{ij} \tag{6.106}$$

It follows that for nodes that are not connected, the admittance is zero. The admittance matrix is then a sparse symmetric matrix as is the finite element matrix. (In our notation, the upper case Y refers to the matrix element, while the lower case y is the circuit admittance.) Following Carpenter's approach, we can infer two networks, one for the flux and one for the current. The flux network

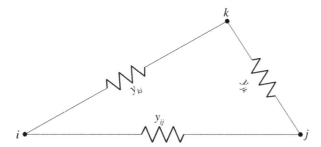

Figure 6.15 Flux network of triangular element.

can now be described as a set of permeances connecting the three nodes of the element. This is illustrated in Figure 6.15. By applying the rules of the admittance matrix formation, the reader can verify that we obtain the magnetic flux matrix of Equation (6.97).

In the same way, we can use the conductance matrix to infer a network representing the electric current.

Recall that the **T** matrix (conductance) is given by

$$\mathbf{T} = \frac{j\omega\sigma\Delta}{12}\begin{bmatrix} 2 & 1 & 1 \\ 1 & 2 & 1 \\ 1 & 1 & 2 \end{bmatrix} \tag{6.107}$$

The elements of the **T** matrix, interpreted as an admittance matrix, give the circuit elements. The off-diagonal terms are the negatives of the direct circuit connections between the nodes. For example, the (1,2) connection is a negative capacitance.

$$c_{12} = -\frac{j\omega\sigma\Delta}{12} \tag{6.108}$$

The diagonal term is the sum of all admittances connected to the node, and we know that each node has 2 negative capacitances given by Equation (6.108). By subtracting these from the diagonal, we obtain the connection to ground equal to

$$c_g = \frac{j\omega\sigma\Delta}{3} \tag{6.109}$$

This is illustrated in Figure 6.16. Once again, we see the elements in the network add up to the elements in the admittance matrix.

If we compare the finite element equivalent circuit with the finite difference equivalent circuit, we notice that the flux carrying elements (reluctances) are the same for the case of a square grid which was used in the previous example. Several terms in the matrix (those opposite a right angle) vanish in the finite element matrix.

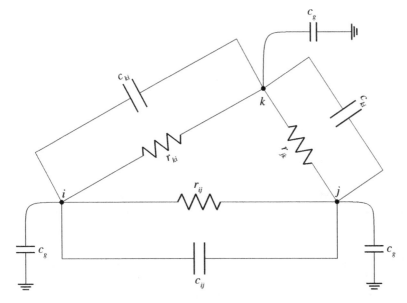

Figure 6.16 Network representing the electric current.

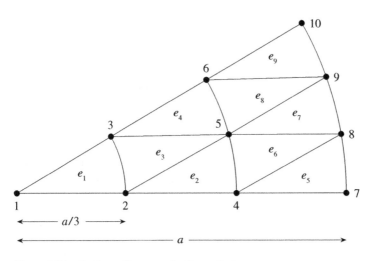

Figure 6.17 Section of long conducting cylinder.

Following Carpenter [54], we look at the case of a long cylindrical conductor carrying current in the axial direction as a numerical example. Due to symmetry in the θ direction, we analyze 1/8 of the problem or a 45° section. The model has 9 first-order triangular elements and 10 nodes as shown in Figure 6.17.

The assembly of the 9 elements results in a 10×10 nodal admittance matrix. The circuit elements are the permeances between the nodes, the negative capacitances between the nodes, and the positive capacitances to ground represent the conductance network in the case of an infinitely long wire. In this example, we are interested in the impedance of the conductor. This can easily be found from the magnetic vector potential solution as will be described below. Of interest here is the relatively few elements that are required for a reasonable solution. In this case, we have the exact solution in terms of Bessel functions as described in Section 3.1. We can compare the resistance and reactance per unit depth. If the skin depth is much larger than the conductor radius, we expect a solution in which the current is approximately uniformly distributed and the resistance approaches the dc resistance. As the skin depth gets smaller, we expect that the solution accuracy diminishes. Carpenter finds good results even with one first-order element per skin depth. After that the error is not acceptable. This is indeed the case. The finite element method minimizes the error in the global energy, and quantities such as resistance and inductance can be found with relatively few elements. If we wanted an accurate description of the current vs. depth however, the best we can get with this model is a linear variation over each element.

The inputs to the model are ac current sources (with phase angle of zero) injected at nodes 7–10. Physically, this would represent a current sheet at the boundary, and the return current is forced into the conductor. The total current in the conductor is 1 A, therefore, 1/8 A is injected into the model. The current at nodes 7 and 10 is 1/48 A. On nodes 8 and 9, we inject 1/24 A. In order to find the impedance per unit length of the conductor, we can continue with the circuit analogy. Since we have 1 A in the conductor, the complex impedance will be equal to the voltage drop per meter in the conductor. If we take the vector potential solution at the conductor surface, then the electric field is

$$E_z = j\omega A_z \tag{6.110}$$

In the specific model, the vector potential on the surface should be a constant. In practice, nodes 7 and 10 have the same vector potential and nodes 8 and 9 have the same vector potential but they are very slightly different. In this case, we have averaged them. It is interesting to compare this with the Poynting vector method used previously.

$$P = E_z \times H_\theta \tag{6.111}$$

Recall that the Poynting vector integrated over the surface of the conductor gives the real and reactive power flow into the conductor. Since we have 1 A, this will also be the resistance and inductive reactance. The electric field at the surface is given by Equation (6.110) above. The magnetic field is found by application of

Ampere's law, integrated around the conductor. Since we have 1 A in the conductor, for a conductor of radius a,

$$H_\theta = \frac{1}{2\pi a} \tag{6.112}$$

We now integrate the Poynting vector around the conductor. Since the Poynting vector is constant, we simply multiply by the circumference. This gives

$$\oint P \cdot d\ell = \frac{1}{2\pi a} \times j\omega A_z \times 2\pi a = j\omega A_z \tag{6.113}$$

We can supply a physical interpretation of the currents in the equivalent circuit. The current injection at the perimeter of the model (nodes 7–10) are physical axial currents representing the current sheet at the edge of the conductor. The current through the positive capacitors to ground represents the physical axial eddy currents. The currents between the nodes through the resistor and parallel negative capacitor can be thought of as *ampere-turn* flow. These currents are in the plane of the problem, while the real currents are all axial. These currents produce a potential difference between the nodes they connect and this (vector) potential difference is equal to the magnetic flux per unit depth, crossing the line connecting the two nodes. All of the physical current injected at the surface leaves the network through the grounded capacitors.

As a check on the results, we ran the model at a very low frequency. In this case, we should obtain the dc resistance. The dc resistance for our example is $R_{dc} = 5.488 \times 10^{-5}\,\Omega\text{m}^{-1}$. The model gives $R_{dc} = 5.603 \times 10^{-5}\,\Omega\text{m}^{-1}$. The difference is 2%. Most of this is explained by the area of the model being slightly less than the theoretical cylinder. The perimeter of the model is made of four straight-line segments. This reduces the area of the model by 1.2%. Correcting for this area difference, the results differ by less than 1%. In the case of dc, we have a very simple solution to the current and flux distribution. The current is uniform, which agrees with the model result, and the flux density is zero at the center and increases linearly to

$$B_\theta = \frac{\mu I}{2\pi a} \tag{6.114}$$

at the surface $r = a$. The circuit, which is based on first-order elements, gives constant flux density in each element. If we compare the theoretical flux density and the model flux density (from Figure 6.18), we see that the model does give a relatively accurate representation of the flux density variation. The plot shows the real part of the flux density, but in this case, the flux density and current density have negligible imaginary parts.

Even when the element size is equal to the skin depth, the model gives reasonable results. For the example above, that corresponds to a frequency of 393.06 HZ. The skin depth is then $\delta = 0.00333\,\text{m}$, or 1/3 of the conductor radius. Figure 6.19 shows the real part of the flux density for the model and for the exact solution.

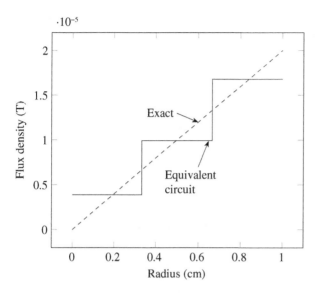

Figure 6.18 Exact and equivalent circuit flux density at dc.

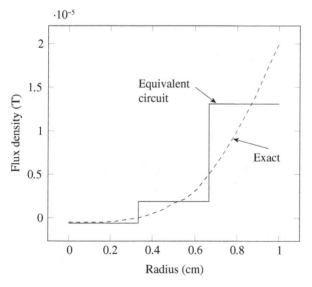

Figure 6.19 Exact and equivalent circuit flux density at $\delta = a/3$.

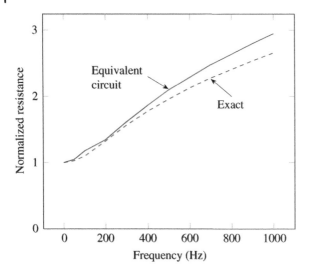

Figure 6.20 Exact and equivalent circuit normalized resistance vs. frequency.

The model gives very good results if we use it to find the ac resistance and internal reactance. We have already found the resistance and reactance for the long conducting cylinder in Section 3.1 in terms of modified Bessel functions. We can compare the result for the resistance as a function of frequency. In the specific example, we have the radius of the conductor equal to 0.01 m. The conductivity is $\sigma = 5.8 \times 10^7 \, \text{Sm}^{-1}$, and the permeability is μ_0. In order to make the results more general, we plot the resistance divided by the dc resistance of the conductor. This is plotted as a function of frequency (Figure 6.20).

We see that with just a few elements, the equivalent circuit gives reasonable results for the effective resistance, even when the first-order element is larger than the skin depth.

6.3 Axisymmetric Finite Element Eddy Current Formulation with Magnetic Vector Potential

There are many practical applications of eddy currents in axisymmetric geometries. Some of these have been illustrated in Chapter 3. As in the eddy current analysis for Cartesian coordinates, we can describe the problem using only one component of the magnetic vector potential. Both the MVP, A, and the current density, J, have only θ components.

$$A = A_\theta$$

$$J = J_\theta \tag{6.115}$$

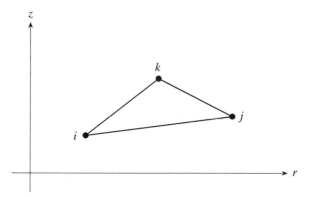

Figure 6.21 Axisymmetric first-order triangle.

Our functional becomes

$$\mathcal{F} = \frac{1}{2}\iint_R \left\{ v\left[(\nabla \times A_\theta) \cdot (\nabla \times A_\theta) \right] - 2A_\theta \cdot J_\theta + j\omega\sigma A_\theta^2 \right\} dR$$
$$+ \oint_C A_\theta \left(\frac{\partial A_\theta}{\partial n} + \frac{A_\theta}{r} \cdot \frac{\partial r}{\partial n} \right) dc \tag{6.116}$$

Writing this in cylindrical coordinates, we obtain

$$\mathcal{F} = 2\pi \iint_R \left(\frac{vr}{2}\left[\left(\frac{\partial A_\theta}{\partial r}\right)^2 + \left(\frac{\partial A_\theta}{\partial z}\right)^2 \right] + vA_\theta \frac{\partial A_\theta}{\partial r} + \frac{vA_\theta^2}{2r} - J_\theta r A_\theta \right) dr\, dz$$
$$+ 2\pi \iint_R \frac{j\omega\sigma A_\theta^2 r}{2} dr\, dz + \oint_C A_\theta \left(\frac{\partial A_\theta}{\partial n} + \frac{A_\theta}{r} \cdot \frac{\partial r}{\partial n} \right) dc \tag{6.117}$$

The line integral term is normally set to zero, except in the case where the finite elements region is coupled to a different solution technique [46], such as a closed-form solution or an integral equation solution. In these cases, the line integral term can be used to ensure continuity at the boundary. We now minimize $\mathcal{F}$ over the problem domain. As in the two-dimensional case, we divide the region into triangular elements as shown in Figure 6.21.

The shape functions for the first-order elements are

$$\zeta_i = (a_i + b_i z + c_i r)/2\Delta$$
$$\zeta_j = (a_j + b_j z + c_j r)/2\Delta \tag{6.118}$$
$$\zeta_k = (a_k + b_k z + c_k r)/2\Delta$$

where

$$a_i = \begin{vmatrix} z_j & z_k \\ r_j & r_k \end{vmatrix}$$

$$b_i = r_j - r_k$$
$$c_i = z_k - z_j \tag{6.119}$$

and the indices are cyclic. For each element we have

$$A = \sum_{i,j,k} A_i \zeta_i = \sum_{i,j,k} \frac{(a_i + b_i z + c_i r)A_i}{2\Delta} \tag{6.120}$$

We minimize the functional by setting the first derivative of (6.117) to zero.

$$\frac{\delta F}{\delta A_i} = 2\pi \iint_R \left\{ vr \left[\left(\frac{\partial A}{\partial r} \right) \frac{\partial}{\partial A_i} \left(\frac{\partial A}{\partial r} \right) + \left(\frac{\partial A}{\partial z} \right) \frac{\partial}{\partial A_i} \left(\frac{\partial A}{\partial z} \right) \right] \right.$$
$$+ v \left(\frac{\partial A}{\partial A_i} \right) \left(\frac{\partial A}{\partial r} \right) + vA \frac{\partial}{\partial A_i} \left(\frac{\partial A}{\partial r} \right) + \frac{vA}{r} \left(\frac{\partial A}{\partial A_i} \right) + j\omega\sigma rA \left(\frac{\partial A}{\partial A_i} \right)$$
$$\left. - Jr \left(\frac{\partial A}{\partial A_i} \right) \right\} \, dr \, dz = 0 \tag{6.121}$$

Substituting (6.120) into (6.121)

$$\sum_{i=1}^{n} 2\pi \iint_R \left\{ vr \left[\left(\frac{\partial \zeta_i}{\partial r} \right) \left(\frac{\partial \zeta_j}{\partial r} \right) + \left(\frac{\partial \zeta_i}{\partial z} \right) \left(\frac{\partial \zeta_j}{\partial z} \right) \right] [A_i] \right.$$
$$+ v \left[\zeta_i \left(\frac{\partial \zeta_i}{\partial r} \right) + (\zeta_i)^T \left(\frac{\partial \zeta_i}{\partial r} \right)^T \right] [A_i]$$
$$+ v \left[\frac{\zeta_i \zeta_j}{r} \right] [A_i] + j\omega\sigma r \left[\zeta_i \zeta_j \right] [A_i]$$
$$\left. - Jr\zeta_i \right\} \, dR = 0 \tag{6.122}$$

In matrix form, Equation (6.122) becomes

$$v[S][A] + v[D][A] + v[E][A] + j\omega\sigma[T][A] = [T][J] \tag{6.123}$$

The matrices $[S]$, $[D]$, $[E]$, and $[T]$ have elements

$$S_{ij} = \iint r \left[\left(\frac{\partial \zeta_i}{\partial r} \right) \left(\frac{\partial \zeta_j}{\partial r} \right) + \left(\frac{\partial \zeta_i}{\partial z} \right) \left(\frac{\partial \zeta_j}{\partial z} \right) \right] \, dr \, dz$$

$$D_{ij} = \iint \left[\zeta_i \left(\frac{\partial \zeta_j}{\partial r} \right) + \zeta_j \left(\frac{\partial \zeta_i}{\partial r} \right) \right] \, dr \, dz$$

$$E_{ij} = \iint \frac{1}{r} \zeta_i \zeta_j \, dr \, dz$$

$$T_{ij} = \iint r \zeta_i \zeta_j \, dr \, dz \tag{6.124}$$

Note that the third term goes to infinity as r approaches 0. To overcome this, Konrad and Silvester [55] have utilized a change of variables. Let

$$A = \sqrt{r}\phi$$
$$J = \sqrt{r}\psi \tag{6.125}$$

Substituting these into the matrix equation (6.121) we obtain

$$
S_{ij} = \iint r^2 \left[\left(\frac{\partial \zeta_i}{\partial r} \right) \left(\frac{\partial \zeta_j}{\partial r} \right) + \left(\frac{\partial \zeta_i}{\partial z} \right) \left(\frac{\partial \zeta_j}{\partial z} \right) \right] dr \, dz
$$

$$
D_{ij} = \iint \frac{3}{2} r \left[\zeta_i \left(\frac{\partial \zeta_i}{\partial r} \right) + \zeta_i \left(\frac{\partial \zeta_i}{\partial r} \right) \right] dr \, dz
$$

$$
E_{ij} = \iint \frac{9}{4} \zeta_i \zeta_j \, dr \, dz
$$

$$
T_{ij} = \iint r^2 \zeta_i \zeta_j \, dr \, dz \tag{6.126}
$$

We see that the singularity in Equation (6.124) has now disappeared in (6.126). For first-order elements, the matrices in Equation (6.126) become

$$
\zeta_m = \frac{a_m + b_m z + c_m r}{2\Delta}, \quad m = i, j, k
$$

$$
\frac{\partial \zeta_m}{\partial r} = \frac{c_m}{2\Delta}, \quad m = i, j, k
$$

$$
\frac{\partial \zeta_m}{\partial z} = \frac{b_m}{2\Delta}, \quad m = i, j, k \tag{6.127}
$$

r is written as a linear combination of the element vertex values.

$$
r = \sum_{m=i,j,k} \zeta_m r_m \tag{6.128}
$$

Substituting,

$$
S_{ij} = 2\pi \iint (\zeta_i r_i + \zeta_j r_j + \zeta_k r_k)^2 \left[\frac{(b_i b_j + c_i c_j)}{4\Delta^2} \right] dr \, dz \tag{6.129}
$$

We now use

$$
dr \, dz = 2\Delta d\zeta_i d\zeta_j \tag{6.130}
$$

and

$$
\zeta_i + \zeta_j + \zeta_k = 1 \tag{6.131}
$$

where the limits of ζ_i are 0 to 1, and the limits of ζ_j are 0 to $(1 - \zeta_i)$. Substituting this into Equation (6.129).

$$
S_{ij} = \frac{2\pi v}{2\Delta} \int_0^1 \int_0^{1-\zeta_i} \left\{ \zeta_i^2 (r_i - r_k)^2 + \zeta_j^2 (r_j - r_k)^2 + r_k^2 \right.
$$

$$
+ 2\zeta_i \zeta_j (r_i - r_k)(r_j - r_k) + 2r_k \zeta_j (r_j - r_k)
$$

$$
\left. + 2r_k \zeta_i (r_i - r_k) \right\} [b_i b_j + c_i c_j] \, d\zeta_1 d\zeta_j \tag{6.132}
$$

We now evaluate the integrals. The first integral becomes

$$
\frac{2\pi v}{2\Delta} \int_0^1 \int_0^{1-\zeta_i} \zeta_i^2 (r_i - r_k)^2 \, d\zeta_i \, d\zeta_j = \frac{\pi v}{12\Delta} (r_i - r_k)^2 \tag{6.133}
$$

The other terms are evaluated [46] in the same way. We obtain

$$[S] = \frac{\pi v}{12}(r_i^2 + r_j^2 + r_k^2 + r_i r_j + r_j r_k + r_k r_i)$$

$$\times \begin{bmatrix} b_i^2 + c_i^2 & b_i b_j + c_i c_j & b_i b_k + c_i c_k \\ & b_j^2 + c_j^2 & b_j b_k + c_j c_k \\ \text{symmetric} & & b_k^2 + c_k^2 \end{bmatrix} \tag{6.134}$$

For the elements of the $[D]$ matrix

$$D_{ij} = 4\pi\Delta \iint \frac{3}{2}(r_i\zeta_i + r_j\zeta_j + r_k\zeta_k)\left[\zeta_i\frac{\partial\zeta_j}{\partial r} + \zeta_j\frac{\partial\zeta_i}{\partial r}\right] d\zeta_i\, d\zeta_j \tag{6.135}$$

Multiplying the terms, we get

$$D_{ij} = 6\pi\Delta \iint \left[r_i\zeta_i^2\left(\frac{\partial\zeta_j}{\partial r}\right) + r_j\zeta_i\zeta_j\left(\frac{\partial\zeta_j}{\partial r}\right) + r_k\zeta_i\zeta_k\left(\frac{\partial\zeta_j}{\partial r}\right)\right] \partial\zeta_i\, \partial\zeta_j$$

$$+ \left[r_i\zeta_i\zeta_j\left(\frac{\partial\zeta_i}{\partial r}\right) + r_j\zeta_j^2\left(\frac{\partial\zeta_i}{\partial r}\right) + r_k\zeta_j\zeta_k\left(\frac{\partial\zeta_i}{\partial r}\right)\right] d\zeta_i\, d\zeta_j \tag{6.136}$$

Integrating ζ_i from 0 to 1 and ζ_j from 0 to $(1 - \zeta_i)$ gives

$$D_{ij} = \frac{\pi}{8}\{c_i r_j + c_j r_i - c_k(r_i + r_j + r_k)\} \tag{6.137}$$

Therefore, D becomes

$$D = \frac{v\pi}{8} \begin{bmatrix} 2c_i(2r_i + r_j + r_k) & c_i r_j + c_j r_i & -c_j(r_i + r_j + r_k) \\ & -c_k(r_i + r_j + r_k) & +c_i r_k + c_k r_i \\ & 2c_j(r_i + 2r_j + r_k) & -c_i(r_i + r_j + r_k) \\ & +c_j r_k + c_k r_j & \\ \text{symmetric} & & 2c_k(r_i + r_j + 2r_k) \end{bmatrix} \tag{6.138}$$

The elements of the T matrix are found as

$$T_{ij} = j2\pi\omega\sigma \iint r^2\phi\left(\frac{\partial\phi}{\partial\phi_k}\right)$$

$$= j2\pi\omega\sigma \iint 2\Delta(\zeta_i r_i + \zeta_j r_j + \zeta_k r_k)^2 \sum \zeta_i\zeta_j\phi_i\, d\zeta_i\, d\zeta_j \tag{6.139}$$

Integrating over the same limits, the first row of T becomes

$$j\omega\sigma T_{ii}, T_{ij}, T_{ik} = \frac{j2\pi\omega\sigma\Delta}{180}\left[\left(12r_i^2 + 6r_i r_j + 6r_i r_k + 2r_j^2 + 2r_j r_k + 2r_k^2,\right.\right.$$

$$(3r_i^2 + 4r_i r_j + 2r_i r_k + 3r_j^2 + r_k^2),$$

$$\left.\left. 3r_i^2 + 2r_i r_j + 4r_i r_k + r_j^2 + 2r_j r_k + 3r_k^2\right)\right] \tag{6.140}$$

For the second row of the matrix, we need only the jj and jk terms since the matrix is symmetric.

$$j\omega\sigma T_{jj} = \frac{j2\pi\omega\sigma\Delta}{180}(2r_i^2 + 6r_i r_j + 2r_i r_k + 12r_j^2 + 6r_j r_k + 2r_k^2)$$

$$j\omega\sigma T_{kj} = \frac{j2\pi\omega\sigma\Delta}{180}(r_i^2 + 2r_ir_j + 2r_ir_k + 3r_j^2 + 4r_jr_k + 3r_k^2) \tag{6.141}$$

The T_{kk} term is

$$j\omega\sigma T_{kk} = \frac{j2\pi\omega\sigma\Delta}{180}(2r_i^2 + 2r_ir_j + 6r_ir_k + 2r_j^2 + 6r_jr_k + 12r_k^2) \tag{6.142}$$

For the E matrix, we have

$$E_{ij} = \frac{9\pi\nu}{2}\iint \phi\frac{\partial\phi}{\partial\phi_k} = \frac{9\pi\nu}{2}\int_0^1\int_0^{1-\zeta_i} 2\Delta(\zeta_i^2, \zeta_i\zeta_j, \zeta_i\zeta_k)[\phi]\, d\zeta_i\, d\zeta_j \tag{6.143}$$

Integrating each term of the equation gives

$$\frac{9\pi\nu}{2}\iint \phi\left(\frac{\partial\phi}{\partial\phi_k}\right) = \frac{3\pi\nu\Delta}{8}[2,1,1,][\phi] \tag{6.144}$$

Integrating over i, j, and k, we obtain

$$[E] = \frac{3\pi\nu\Delta}{8}\begin{bmatrix} 2 & 1 & 1 \\ 1 & 2 & 1 \\ 1 & 1 & 2 \end{bmatrix}\begin{pmatrix} \phi_i \\ \phi_j \\ \phi_k \end{pmatrix} \tag{6.145}$$

7

Integral Equations

In Chapters 5 and 6, we have presented the finite difference and finite element methods. These are *differential* formulations. These techniques approximate the differential equation over small regions, the finite difference cell, or finite element. The shape functions or approximating functions are not required to be solutions to the differential equation. Also, the element shape functions are defined only inside an individual element. This is referred to as *local support*. A different numerical approach to electromagnetic problems is to use integral equations. In this formulation, the field or potential is approximated by a series of functions, which are usually exact solutions of the differential equations, and have *global support*. Their actions are over the entire problem domain. We shall see, however, that while the solutions to the governing differential equations are exact, the boundary conditions may be only approximately satisfied. This is in contrast to the finite element method (FEM), in which the Dirichlet boundary conditions are satisfied exactly but the operator equation is satisfied only approximately.

We will present a number of integral equations approaches to the eddy current problem in this chapter.

7.1 Surface Integral Equation Method for Eddy Current Analysis

In two-dimensional eddy current analysis, we have found it useful to use the magnetic vector potential, A, as we only need a single component. We will now develop a numerical method based on the vector potential in which a conductor is divided into two-dimensional *patches*, on which we will assume that the vector potential, and therefore the current density, is constant. The result will be surface integral equations for either the vector potential or the current density.

We recall that

$$J = \sigma E \tag{7.1}$$

Eddy Currents: Theory, Modeling, and Applications, First Edition.
Sheppard J. Salon, M. V. K. Chari, Lale T. Ergene, David Burow, and Mark DeBortoli.
© 2024 The Institute of Electrical and Electronics Engineers, Inc. Published 2024 by John Wiley & Sons, Inc.

Recall that the electric field has two parts. One is the induced electric field (Faraday's law) and the other part can be described by the gradient of a scalar. So we may write J as

$$J = -\sigma \frac{dA}{dt} - \sigma \nabla V \tag{7.2}$$

The magnetic vector potential in two-dimensions, produced by an axial current distribution, can be found as

$$A = \frac{\mu}{2\pi} \iint J \ln r \, dS \tag{7.3}$$

We then obtain an expression for the current density in terms of an integral over the current distribution.

$$J = \frac{\mu\sigma}{2\pi} \iint \frac{\partial J}{\partial t} \ln \left((x - \xi)^2 + (y - \eta)^2\right)^{1/2} d\xi \, d\eta + \sigma E_0 \tag{7.4}$$

For the case of sinusoidally time-varying fields, we can replace the time derivative with $j\omega$ to obtain

$$J = \frac{j\omega\mu\sigma}{2\pi} \iint J(\xi, \eta) \ln \left((x - \xi)^2 + (y - \eta)^2\right)^{1/2} d\xi \, d\eta + \sigma E_0 \tag{7.5}$$

This last equation is known as a Fredholm integral equation of the second kind, in which the unknown, J, is both under the integral sign and outside the integral. We now divide the surface into small patches over which we can assume that the current density is constant. Since the current density is constant over each element, we can take the current density out of the integral.

$$J_i = \frac{j\omega\mu\sigma}{2\pi} \sum_{i=1}^{N} J_i \iint \ln \left((x - \xi)^2 + (y - \eta)^2\right)^{1/2} d\xi \, d\eta + \sigma E_0 \tag{7.6}$$

In the evaluation of Equation (7.6), we can use the distance between the center points of the patches for all terms in the series except for the self-term ($i = j$). For this term, the distance between the source point and the field point is zero, and the logarithm goes to infinity. The integral is finite, however, and can be evaluated either numerically or in closed-form for different shapes. For a square element, the geometric mean distance (GMD) to itself is $0.44705a$ where a is the dimension of a side of the square. Rosa and Grover [56] give an approximation for the GMD for rectangular ($a \times b$) shapes that is useful in evaluating this term.

$$\text{GMD} = 0.2235(a + b) \tag{7.7}$$

The formula is used as follows. If, for example, the ratio of a and b is 0.5, then GMD $= 0.2235(1 + 0.5)a = 0.3354a$. For $b = 0.25a$, we have GMD $= 0.2796a$. If b is $1/10$ of a then GMD $= 0.2459a$.

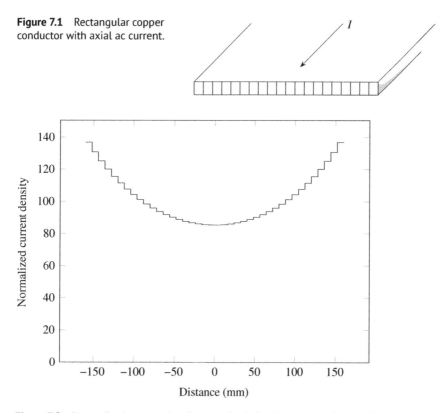

Figure 7.1 Rectangular copper conductor with axial ac current.

Figure 7.2 Normalized current density magnitude for thin rectangular conductor.

The following example will illustrate the use of this technique. Consider a rectangular conductor of dimensions $0.004\,\text{m} \times 0.16\,\text{m}$. The conductor is made of copper with $\sigma = 5.8 \times 10^7\,\text{Sm}^{-1}$ and has permeability $\mu_0 = 4\pi \times 10^{-7}\,\text{Hm}^{-1}$. We will apply an electric field of $1.0\,\text{Vm}^{-1}$ at 60 Hz in the z direction (see Figure 7.1).

The conductor is then divided into 40 square elements ($a = 0.004$) and the system of equations described by Equation (7.6) was formed using the value of $0.44705a$ for the diagonal term and using the distance between center points for the off diagonals. In this case, we need only solve for 20 terms due to symmetry. The results are shown for the magnitude and phase angle of the current density in Figures 7.2 and 7.3. In Figure 7.2, we see the magnitude of the normalized current density. The total current in the bar was found and then the individual current density values in each element were divided by the average current density and multiplied by 100. This gives us the percent normalized current density. We also note that the *staircase* shape of the curve results from the model giving constant values of current density over the element.

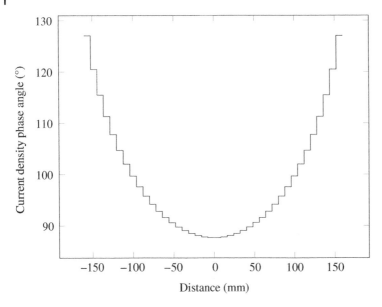

Figure 7.3 Current density phase for thin rectangular conductor.

In this problem, the current density only varies in one direction, so we will consider a second example of a square copper conductor, $0.04\,m \times 0.04\,m$, with an applied electric field of $1.0\,Vm^{-1}$ at $60\,Hz$. In this case, the current density is a function of both x and y. The conductor is divided into a 40×40 mesh. In this case, we have 1600 unknown values of current density. In Figure 7.4, we see the magnitude of the current density over the conductor. As expected, we observe that the current density is highest near the surface.

We compare the results to a finite element solution shown in Figure 7.5. We see that the results match very well.

We note here some properties of the surface integral equation method. First, we see that the system matrix described by Equation (7.6) is a full matrix. Recall that the system matrices we obtained for the finite difference and finite element methods were sparse. This results in longer computation times for populating the matrix and solving the system of equations. On the other hand, we see that it was not necessary to mesh or in any other way describe the surrounding air region. The logarithm function, which is the kernel of the integral, is the analytical solution for flux linkage of a long current filament in two-dimensions. The far-field and open boundary conditions are automatically included in the formulation. The finite element model that was used for Figure 7.5, used significantly more elements and had significantly more unknowns. Furthermore, a special technique involving a

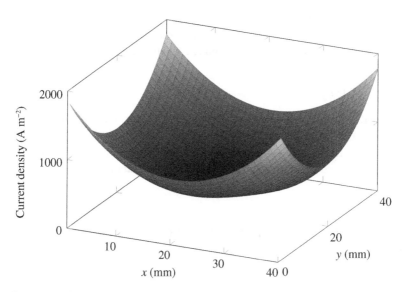

Figure 7.4 Current density magnitude for square conductor by integral equations.

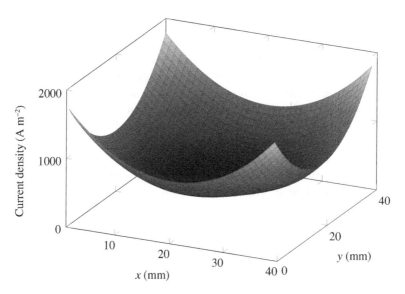

Figure 7.5 Current density magnitude for square conductor by finite elements.

transformation on the outer boundary to simulate the infinite space surrounding the conductor was used.

The reader may notice that the system matrix for the integral equation method has a striking resemblance to a network model. The inductance of a loop made of parallel filaments is

$$L_{i,j} = \frac{\mu_0}{2\pi} \ln d_{ij} \tag{7.8}$$

where $d_{i,j}$ is the distance between the two parallel filaments. We can interpret this surface integral method for the current density as a set of coupled circuits with resistance and self and mutual inductances. In fact, referring back to Section 1.11, where we introduced the coupled circuit model, we see that the set of equations is the same.

7.2 Boundary Element Method for Eddy Current Analysis

In Section 7.1 we introduced the surface integral method. In this analysis, the surface of the regions with current were divided into small patches, and a set of simultaneous equations was developed for the potential at each of the segments. In the boundary element method, we also divide the conductor into small segments but only on the boundary of the region. There are no unknowns on the inside of the regions. The interior local variables, such as current density or field, are found in the post-processing stage in terms of the values on the boundary. The boundary element method, therefore, results in a very small set of equations when compared to finite difference, finite element, or even surface integrals techniques.

Typical eddy current problems have regions that contain conducting materials, but there may be regions that are nonconducting or contain sources. These regions are described by Laplace's or Poisson's equation as opposed to the eddy current region, which is described by the diffusion equation.

In this section, we will develop the boundary element relations for the Laplace or the diffusion equation in two-dimensions. Consider the region and boundary shown in Figure 7.6. We begin with

$$\nabla^2 \phi = f(x, y) \tag{7.9}$$

Consider now the function $G(r)$, which satisfies

$$\nabla^2 G = \delta(r) \tag{7.10}$$

where G is the Green's function and $\delta(r)$ is defined by

$$\delta(r) = \begin{cases} 0 & \text{for } r \neq 0 \\ \infty & \text{for } r = 0 \end{cases} \tag{7.11}$$

Figure 7.6 Two-dimensional region and surface.

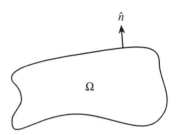

The *delta function* has the property that

$$\int_{-\infty}^{\infty} \delta(r)\, dr = 1 \tag{7.12}$$

where $r = \sqrt{(\xi - x)^2 + (\eta - y)^2}$, and (ξ, η) and (x, y) are the source and field points, respectively.

Integrating the difference of Equation (7.9) times G and Equation (7.10) multiplied by ϕ over the region of interest, Ω, we obtain

$$\iint_{\Omega} (G\nabla^2 \phi - \phi \nabla^2 G) d\Omega = \iint_{\Omega} fG\, d\Omega \tag{7.13}$$

The function G, defined by Equation (7.10), is the Green's function or the potential due to an infinite line source located at $r = 0$. G can be found as the solution of

$$\frac{1}{r}\frac{\partial}{\partial r} r \frac{\partial G}{\partial r} = \delta(r) \tag{7.14}$$

The solution of this equation is

$$G = -\frac{1}{2\pi} \ln(r) \tag{7.15}$$

Applying Green's theorem to the left-hand side of Equation (7.13), we obtain

$$\int_{\ell} \left(G\frac{\partial \phi}{\partial n} - \phi\frac{\partial G}{\partial n} \right) d\ell = \iint_{\Omega} fG\, d\Omega \tag{7.16}$$

The kernel of the line integral in Equation (7.16) contains a singularity when the source and field points coincide. The integral, however, is finite. To evaluate the line integral, we divide it into two parts, one which is singularity-free and the other which contains the singularity as shown in Figure 7.7. We can then evaluate

$$\int_{\ell} \left(G\frac{\partial \phi}{\partial n} - \phi\frac{\partial G}{\partial n} \right) d\ell + \lim_{\Delta\ell \to 0} \int_{\Delta\ell} \left(G\frac{\partial \phi}{\partial n} - \phi\frac{\partial G}{\partial n} \right) d\ell = \iint_{\Omega} fG\, d\Omega \tag{7.17}$$

The contour $\Delta\ell$, shown in Figure 7.7, is a semicircle of radius ϵ with center on the singularity. On the contour $\Delta\ell$ we have $\frac{\partial}{\partial n} = -\frac{\partial}{\partial r}$ and $d\ell = \epsilon d\theta$. Using these

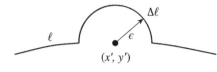

Figure 7.7 Integrating around the singularity.

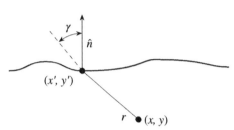

Figure 7.8 Definition of boundary terms.

relationships, the integral over $\Delta \ell$ becomes

$$\lim_{\Delta \ell \to 0} \int_{\Delta \ell} \left(G \frac{\partial \phi}{\partial n} - \phi \frac{\partial G}{\partial n} \right) d\ell = \frac{1}{2\pi} \lim_{\epsilon \to 0} \int_0^{2\pi} \left(\frac{\partial \phi}{\partial r} \epsilon \ln \epsilon - \phi \right) d\theta = -\phi$$

(7.18)

Referring to Figure 7.8, we see that

$$\frac{\partial r}{\partial n} = \cos \gamma$$

(7.19)

and

$$\frac{\partial G}{\partial n} = \frac{\partial G}{\partial r} \frac{\partial r}{\partial n} = -\frac{\cos \gamma}{2\pi r}$$

(7.20)

Substituting into Equation (7.17),

$$\phi(x, y) = \frac{1}{2\pi} \int_\ell \left(\phi \frac{\cos \gamma}{r} - \frac{\partial \phi}{\partial n} \ln r \right) d\ell + \frac{1}{2\pi} \int \int_\Omega f \ln r \, d\Omega$$

(7.21)

Equation (7.21) is an expression for the potential at any position in the region in terms of the potential and its normal derivative at the boundaries, plus a contribution due to the sources in the region. This demonstrates that by knowing the values of the potential and its normal derivative on the boundary, we can completely describe the potential inside the region.

If the sources are known, as is frequently the case, the surface integral in Equation (7.21) can be evaluated numerically or sometimes analytically. In an important class of problems, the surface integral can be transformed into another boundary integral and the potential can be expressed entirely in terms of values at the boundary. To see this, let us assume that the forcing function is harmonic, that is, it is a solution of Laplace's equation.

$$\nabla^2 f = 0$$

(7.22)

Figure 7.9 Integration for a field point located on the boundary.

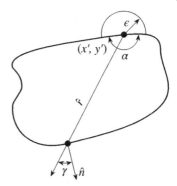

An important example would be a constant source. We now define a function, g, that satisfies

$$\nabla^2 g = G \tag{7.23}$$

We see that the last term of Equation (7.21) is of the form

$$\iint_\Omega fG\,d\Omega \tag{7.24}$$

Equation (7.24) is equal to the integral of (7.23) times f minus the integral of (7.22) (which is zero) times g. Thus,

$$\iint_\Omega fG\,d\Omega = \iint_\Omega (f\nabla^2 g - g\nabla^2 f)\,d\Omega \tag{7.25}$$

We use Green's theorem again to obtain

$$\iint_\Omega fG\,d\Omega = \int_\ell \left(f\frac{\partial g}{\partial n} - g\frac{\partial f}{\partial n} \right) d\ell + \lim_{\Delta\ell \to 0}\int_{\Delta\ell} \left(f\frac{\partial g}{\partial n} - g\frac{\partial f}{\partial n} \right) d\ell \tag{7.26}$$

As before, we can integrate around the singularity at $r = 0$ by distorting the boundary into a circular arc. We then take the limit as the radius of the circle goes to zero.

We can evaluate g in Equation (7.23) by solving

$$\frac{1}{r}\frac{\partial}{\partial r} r \frac{\partial g}{\partial r} = -\frac{1}{2\pi}\ln r \tag{7.27}$$

The solution for g is

$$g = \frac{r^2}{8\pi}(1 - \ln r) \tag{7.28}$$

The last integral in Equation (7.26) now becomes

$$\frac{1}{8\pi}\lim_{\epsilon\to 0}\int_0^{2\pi} \left(f\epsilon^2(2\ln\epsilon - 1) + \frac{\partial f}{\partial r}\epsilon^3(1 - \ln\epsilon) \right) d\theta = 0 \tag{7.29}$$

and we see that

$$\frac{\partial g}{\partial n} = \frac{\partial g}{\partial r}\frac{\partial r}{\partial n} = \frac{r}{8\pi}(1 - 2\ln r)\cos\gamma \tag{7.30}$$

Substituting Equations (7.29) and (7.30) into (7.26), and the result into Equation (7.21), we obtain

$$\phi(x, y) = \frac{1}{2\pi} \int_\ell \left(\phi \frac{\cos \gamma}{r} - \frac{\partial \phi}{\partial n} \ln r \right) d\ell$$
$$- \frac{1}{8\pi} \int_\ell \left(\frac{\partial f}{\partial n} r^2 (1 - \ln r) - fr(1 - 2\ln r) \cos \gamma \right) d\ell \tag{7.31}$$

The singularity in this last integral is removed by integrating around the singular point along a circular arc and taking the limit as the radius of the arc goes to zero. So we have

$$\lim_{\Delta\ell \to 0} \int_{\Delta\ell} \left(\frac{\partial f}{\partial n} r^2 (1 - \ln r) - f \frac{\partial}{\partial r} r^2 (1 - \ln r) \right) d\ell \tag{7.32}$$

Finally, we obtain an expression for the potential at any point (x', y') on the contour as in Figure 7.9,

$$\alpha\phi(x', y') + \int_\ell \left(\frac{\partial \phi}{\partial n} \ln r - \phi \frac{\cos \gamma}{r} \right) d\ell$$
$$= \frac{1}{4} \int_\ell \left(\frac{\partial f}{\partial n} r^2 (1 - \ln r) - fr(1 - 2\ln r) \cos \gamma \right) d\ell \tag{7.33}$$

Note that the integral should not be evaluated at the singularity because the (finite) contribution to the integral has already been evaluated. Equation (7.33) contains two unknowns, the potential and its normal derivative on the contour. Once these are known, Equation (7.31) can be used to find the potential at any point in the region.

7.2.1 $T - \Omega$ Boundary Element Eddy Current Formulation in Two-Dimensions

If we consider the set of problems in Cartesian coordinates with long uniform conductors and applied axial fields, we then will have eddy currents circulating in the (x, y) plane and a magnetic field in the z direction. We have considered this class of problems in Chapters 2 and 3 for rectangular and circular conductors. For these problems, we can use a $T - \Omega$ formulation [46]. The vector T is called the *current vector potential* and behaves similarly to the magnetic vector potential, A. As contours of equal magnetic vector potential are flux lines in two-dimensions, lines of constant current vector potential are the current flow lines (streamlines in a fluid analogy). Also, since the flux density is found by taking the curl of the magnetic vector potential, the current density is found by taking the curl of the current vector potential,

$$\nabla \times T = J \tag{7.34}$$

The reader will note the similarity between this and

$$\nabla \times H = J \tag{7.35}$$

In fact, the units of H and T are the same, and since the two vectors have the same curl, they can differ only by the gradient of a scalar. T offers more flexibility in the application of boundary conditions and that has made it more useful in numerical computation.

Similarly to the magnetic vector potential problem, T has only a z component for the two-dimensional case. Using $\nabla \cdot T = 0$, the governing differential equation is

$$\frac{1}{\sigma} \nabla^2 T - j\omega\mu T = j\omega\mu H_0 \tag{7.36}$$

where H_0 is the applied external field.

For homogeneous media, we can write Equation (7.36) as

$$\nabla^2 T - \alpha^2 T = \alpha^2 H_0 \tag{7.37}$$

where

$$\alpha^2 = j\omega\mu\sigma \tag{7.38}$$

Consider the Green's function that satisfies the equation

$$\nabla^2 G - \alpha^2 G = \delta(\xi - x, \eta - y) \tag{7.39}$$

where (x, y) are the field points and (ξ, η) are the source points. Multiplying equation (7.39) by T and Equation (7.37) by G, subtracting, and integrating over the region R, we obtain

$$\int_R T\delta(\xi - x, \eta - y)\,d\xi\,d\eta = \alpha^2 \int_R H_0 G\,d\xi\,d\eta + \int_R (T\nabla^2 G - G\nabla^2 T)\,d\xi\,d\eta \tag{7.40}$$

Using Green's theorem on the last term of Equation (7.40) gives

$$T(x, y) = \alpha^2 \int_R H_0 G\,d\xi d\eta + \int_C \left(T\frac{\partial G}{\partial n} - G\frac{\partial T}{\partial n} \right) dC \tag{7.41}$$

To evaluate the Green's function, we consider Equation (7.39), which becomes

$$\frac{\partial^2 G}{\partial r^2} + \frac{1}{r}\frac{\partial G}{\partial r} - \alpha^2 G = \delta(\xi - x, \eta - y) \tag{7.42}$$

where

$$r = \sqrt{(\xi - x)^2 + (\eta - y)^2} \tag{7.43}$$

The solution of Equation (7.42) is

$$G(\xi, \eta; x, y) = CI_0(\alpha r) + DK_0(\alpha r) \tag{7.44}$$

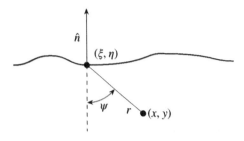

Figure 7.10 Boundary definitions.

where I_0 and K_0 are modified Bessel functions of the first and second kinds of order zero, respectively. From the behavior of these functions at infinity, we deduce that $C = 0$. The constant D is found to be $\frac{1}{2\pi}$ by integrating over a small disk centered at $r = 0$.

The term $\frac{\partial G}{\partial n}$ is evaluated as

$$\frac{\partial G}{\partial n} = \frac{\partial G}{\partial r}\frac{\partial r}{\partial n} \tag{7.45}$$

where

$$\frac{\partial G}{\partial r} = -\frac{\alpha}{2\pi}K_1(\alpha r) \tag{7.46}$$

and

$$\frac{\partial r}{\partial n} = -\cos\psi \tag{7.47}$$

where K_1 is the modified Bessel function of the first kind of order 1. The boundary terms are defined in Figure 7.10.

Using Equations (7.46) and (7.47) we obtain

$$T(x,y) = \frac{\alpha^2}{2\pi}\int_R H_0 K_0(\alpha r)\,d\xi\,d\eta + \int_C T K_1(\alpha r)\cos\psi\,dC - \frac{1}{2\pi}\int_C \frac{\partial T}{\partial n}K_0(\alpha r)\,dC \tag{7.48}$$

As shown before, the surface integral involving the source can be transformed into a line integral if we limit H_0 to the class of functions that satisfy Laplace's equation. Otherwise we must integrate over the region R. If we have

$$\nabla^2 H_0 = 0 \tag{7.49}$$

then we define a function g such that

$$\nabla^2 g = K_0(\alpha r) \tag{7.50}$$

The first integral in Equation (7.48) can be written as

$$\int_R H_0 K_0(\alpha r)\,dR = \int_R (H_0\nabla^2 g - g\nabla^2 H_0)\,dR \tag{7.51}$$

Figure 7.11 Integrating around a singularity.

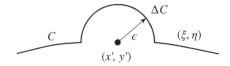

Applying Green's theorem, we get

$$\int_R H_0 K_0(\alpha r)\, dR = \int_C \left(H_0 \frac{\partial g}{\partial n} - g \frac{\partial H_0}{\partial n} \right) dC \tag{7.52}$$

The solution to Equation (7.50) is

$$g = -\frac{1}{\alpha^2} K_0(\alpha r) \tag{7.53}$$

We also have

$$\frac{\partial g}{\partial n} = -\frac{K_1(\alpha r)\cos\psi}{\alpha} \tag{7.54}$$

Substituting Equations (7.53) and (7.54) into Equation (7.52) gives

$$\int_R H_0 K_0(\alpha r)\, dR = -\frac{1}{\alpha} \int_C H_0 K_1(\alpha r)\, dR + \frac{1}{\alpha^2} \int_C \frac{\partial H_0}{\partial n} K_0(\alpha r)\, dC \tag{7.55}$$

Substituting this result into Equation (7.48)

$$T(x,y) = -\frac{\alpha}{2\pi} \int_C H_0 K_1(\alpha)\cos\psi\, dC + \frac{1}{2\pi} \int_C \frac{\partial H_0}{\partial n} K_0(\alpha r)\, dC$$
$$+ \frac{\alpha^2}{2\pi} \int_C T K_1(\alpha r)\cos\psi\, dC - \frac{1}{2\pi} \int_C \frac{\partial T}{\partial n} K_0(\alpha r)\, dC \tag{7.56}$$

We see then that the potential at any point in the region is expressed entirely in terms of values on the boundary. As before, we must remove the singularities that occur on the boundary. Referring to Figure 7.11, we integrate around a singularity along an arc of a circle. We allow (x,y) to approach the point (x',y') on the boundary. Therefore T may be written as

$$T(x,y) = -\frac{\alpha}{2\pi} \int_{C-\Delta C} H_0 K_1(\alpha)\cos\psi\, dC - \frac{\alpha}{2\pi} \int_{\Delta C} H_0 K_1(\alpha)\cos\psi\, dC$$
$$+ \frac{1}{2\pi} \int_{C-\Delta C} \frac{\partial H_0}{\partial n} K_0(\alpha r)\, dC + \frac{1}{2\pi} \int_C \frac{\partial H_0}{\partial n} K_0(\alpha r)\, dC$$
$$+ \frac{\alpha^2}{2\pi} \int_{C-\Delta C} T K_1(\alpha r)\cos\psi\, dC + \frac{\alpha^2}{2\pi} \int_C T K_1(\alpha r)\cos\psi\, dC$$
$$- \frac{1}{2\pi} \int_{C-\Delta C} \frac{\partial T}{\partial n} K_0(\alpha r)\, dC$$
$$- \frac{1}{2\pi} \int_C \frac{\partial T}{\partial n} K_0(\alpha r)\, dC \tag{7.57}$$

All singularities have been isolated in the three integrals around ΔC. Using the asymptotic property of K_0, we have

$$K_0(\alpha r) \rightarrow -\ln r \tag{7.58}$$

As $r \rightarrow 0$ in these integrals, we get:

For the first integral

$$\lim_{\epsilon \to 0} -\frac{\alpha}{2\pi} \int_{\Delta C} H_0 K_1(\alpha r) \cos \psi \, dC = \frac{1}{2\pi} \int_{\Delta C} H_0 \frac{\partial}{\partial n} K_0(\alpha r) \, dC$$

$$= \frac{1}{2\pi} \int_{\Delta C} H_0 \frac{\partial}{\partial n} (-\ln r) \, dC$$

$$= \frac{1}{2\pi} \int_0^\pi H_0 \frac{1}{\epsilon} (\epsilon \, d\theta) = \frac{H_0}{2} \tag{7.59}$$

For the second integral, we get

$$\lim_{\epsilon \to 0} \int_{\Delta C} \frac{\partial H_0}{\partial n} K_0(\alpha r) \, dC = \int_{\Delta C} O\left(\epsilon \ln \epsilon\right) dC = 0 \tag{7.60}$$

For the third integral, we get

$$\lim_{\epsilon \to 0} \frac{\alpha}{2\pi} \int_{\Delta C} T K_1(\alpha r') \cos \psi \, dC = \frac{1}{2\pi} \int_0^\pi T \frac{1}{\epsilon} (\epsilon \, d\theta) = \frac{T(x', y')}{2} \tag{7.61}$$

And for the fourth integral, we get

$$\lim_{\epsilon \to 0} -\frac{1}{2\pi} \int_{\Delta C} \frac{\partial T}{\partial n} K_0(\alpha r) \, dC = \int_{\Delta C} O\left(\epsilon \ln \epsilon\right) dC = 0 \tag{7.62}$$

We can now express T at the boundary as

$$T(x', y') = -H_0 - \alpha \int_C H_0 K_1(\alpha r) \cos \psi \, dC$$

$$+ \frac{\alpha}{\pi} \int_C T K_1(\alpha r) \cos \psi \, dC$$

$$- \frac{1}{\pi} \int_C \frac{\partial T}{\partial n} K_0(\alpha r) \, dC \tag{7.63}$$

We now can write a system of simultaneous algebraic equations by assuming that T and $\frac{\partial T}{\partial n}$ are constant on the straight-line boundary elements (pulse functions). Writing the integrals in Equation (7.63) as summations, we get

$$T_i = -\pi H_0 - \alpha \sum_{\substack{j=1 \\ j \neq i}}^N H_0 K_1(\alpha r_{ij}) \cos \psi_{ij} \Delta C_j$$

$$+ \frac{\alpha}{\pi} \sum_{\substack{j=1 \\ j \neq i}}^N T_j K_1(\alpha r_{ij}) \cos \psi_{ij} \Delta C_j$$

Table 7.1 Definition of c and d.

k	c_k	d_k
0	2.0	0.8456
1	0.6667	0.5041
2	0.100	0.1123

$$-\frac{1}{\pi}\sum_{\substack{j=1 \\ j\neq i}}^{N}\frac{\partial T_j}{\partial n}K_0(\alpha r_{ij})\Delta C_j + \frac{\alpha T_i}{\pi}\int_i K_1(\alpha r)\cos\psi\, dC$$

$$-\frac{1}{\pi}\int_i K_0(\alpha r)\, dC \tag{7.64}$$

The two integrals in Equation (7.64) contain singularities. The first integral is zero because the direction cosine is zero. The second integral has been evaluated by Luke [41] and is

$$\frac{1}{\pi}\int_i K_0(\alpha r)\cos\psi\, dC = \frac{2}{\alpha\pi}\left(\sum_{k=0}^{2}d_k\beta^{2k+1} - \ln\beta\sum_{k=0}^{2}c_k\beta^{2k+1}\right) \tag{7.65}$$

where $\beta = \alpha\Delta C_i/4$ and the coefficients c_k and d_k are defined in Table 7.1.

Thus, using Equation (7.64), we obtain the set of simultaneous equations

$$(J)\{T\} + (K)\left(\frac{\partial T}{\partial n}\right) = \{F\} \tag{7.66}$$

At each point on the boundary either T or $\frac{\partial T}{\partial n}$ must be specified in order to make the problem well posed.

7.2.2 Example Problem

As an example of this formulation, we look at a square copper conductor of dimension 10 mm on a side. A uniform magnetic field $H_0 = 1.0 + j0.0\,\mathrm{Am}^{-1}$ is applied in the z direction. Because of symmetry, we will solve for only one-fourth of the geometry. The boundary conditions are $T = 0$ on the conductor boundary and $\frac{\partial T}{\partial n} = 0$ on the two lines of symmetry. If the normal derivative is zero, then there is no current component normal to the boundary. We see that either T or $\frac{\partial T}{\partial n}$ is specified at each point on the boundary. In this case, the boundary was divided into 72 elements and Equation (7.66) was solved. Equation (7.56) was then used to find T in the conductor. The results for the real part of the current vector potential

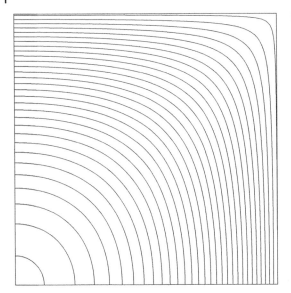

Figure 7.12 Real part of *T*.

are shown in Figure 7.12. Recall that we interpret lines of equipotential as indicating the direction of the current. We see the expected pattern with current parallel to the outer boundaries and perpendicular to the lines of symmetry. We also note that the current density is higher near the outer edges (lines closer together).

We have seen that the boundary element method is a very efficient process, resulting in a very small set of equations, yet capturing all of the physics of the problem. A drawback of the boundary element method is that it is difficult to apply in problems with nonhomogeneous or nonlinear material characteristics. In these cases, the Green's function is either not known or difficult to find, and methods that discretize the entire region are usually applied.

7.3 Integral Equations for Three-Dimensional Eddy Currents

We have seen examples of integral equations in which we divide a surface or boundary into small elements or patches. In general, we can extend the integral equation method into three-dimensions, in which case we divide the relevant space into small volumetric elements [5, 11, 44].

For the development of the integral equations, the following assumptions are made:

- The problem is linear and isotropic.
- Material properties such as permeability and conductivity are constant.

- Only non-ferromagnetic conducting materials are considered.
- The field is quasi-stationary and displacement currents are neglected.

From Maxwell's equations of Ampere's and Faraday's laws excluding displacement we have currents

$$\nabla \times H = J_e + J_s \tag{7.67}$$

$$\nabla \times E = -\frac{\partial B}{\partial t} \tag{7.68}$$

where	J_e	eddy current density	$A\,m^{-2}$
	J_s	eddy current free source current density	$A\,m^{-2}$
	H	magnetic field	$A\,m^{-1}$
	B	magnetic flux density	(T)
	E	electric field	$V\,m^{-1}$
	t	time	(s)

Using a vector potential function, A, such that B equals the curl of A and the constitutive relations, we obtain the eddy current diffusion equation as follows:

$$B = \nabla \times A \tag{7.69}$$

$$J_e = \sigma E \tag{7.70}$$

$$\nabla \times H = \nabla \times \frac{1}{\mu} B = \nabla \cdot \frac{1}{\mu} \times A = \sigma E + J_s \tag{7.71}$$

$$\nabla \times E = -\frac{\partial B}{\partial t} = -\frac{\partial}{\partial t} \nabla \cdot A \tag{7.72}$$

or

$$E = -\frac{\partial A}{\partial t} - \nabla \phi \tag{7.73}$$

where ϕ is the scalar potential function which acts as a gauge on A.

Substituting Equation (7.72) into Equation (7.71), the differential form of the diffusion equation is obtained as

$$\nabla \times \frac{1}{\mu} \nabla \times A = -\sigma \frac{\partial A}{\partial t} - \sigma \nabla \phi + J_s \tag{7.74}$$

The vector potential can also be split up as

$$A = A_e + A_s \tag{7.75}$$

where A is the total vector potential, A_e is the vector potential due to eddy currents, and A_s is the vector potential due to source currents. In integral form, A_e and A_s can be expressed in terms of the respective current densities J_e and J_s as

$$A_e = \frac{\mu_0}{4\pi} \int \frac{J_e}{|r'-r|} dV + \frac{\mu_0}{4\pi} \int \frac{\sigma E}{|r'-r|} dV = -\frac{\mu_0}{4\pi} \int \frac{\sigma \frac{\partial A}{\partial t} + \sigma \nabla \phi}{|r'-r|} dV \tag{7.76}$$

$$A_s = \frac{\mu_0}{4\pi} \int \frac{J_s}{|r'-r|} dV \tag{7.77}$$

where r' is the radius to the observation point from the origin and r is the radius to the source point in Cartesian coordinates such that

$$|r'-r| = \sqrt{(x'-x)^2 + (y'-y)^2 + (z'-z)^2}.$$

From Equations (7.76) and (7.77), the total vector potential, A, is obtained in integral form as

$$A = A_e + A_s = -\frac{\mu_0}{4\pi} \int \frac{\sigma \frac{\partial A}{\partial t} + \sigma \nabla \phi}{|r'-r|} dV + A_s \tag{7.78}$$

Equation (7.78) is a Fredholm integral equation of the second kind. Since the vector potential is not unique, we must set its divergence to zero by applying the Coulomb gauge, $\nabla \cdot A = 0$. Taking the divergence of Equation (7.78) and setting it to zero by applying the Coulomb gauge, we obtain

$$\nabla \cdot A = 0 = -\frac{\mu_0}{4\pi} \int \nabla \cdot \frac{\sigma \frac{\partial A}{\partial t} + \sigma \nabla \phi}{|r'-r|} dV + \nabla \cdot A_s \tag{7.79}$$

Because the divergence of the vector potential due to eddy current free source currents must be zero, the second term on the right-hand side of Equation (7.79) vanishes. We shall now expand the integral on the right-hand side of Equation (7.79) term by term.

From Green's theorem, we know for isotropic conductivity σ,

$$\nabla \cdot \left(\sigma \frac{\frac{\partial A}{\partial t}}{|r'-r|} \right) = \sigma \nabla \left(\frac{1}{|r'-r|} \right) \cdot \frac{\partial A}{\partial t} + \sigma \frac{\frac{\partial \nabla \cdot A}{\partial t}}{|r'-r|} \tag{7.80}$$

Once again, by the application of the Coulomb gauge, the second term on the right-hand side of Equation (7.80) vanishes, yielding

$$\nabla \cdot \left(\sigma \frac{\frac{\partial A}{\partial t}}{|r'-r|} \right) = \sigma \nabla \left(\frac{1}{|r'-r|} \right) \cdot \frac{\partial A}{\partial t} \tag{7.81}$$

Also, by the application of Green's theorem

$$\nabla \cdot \left(\frac{\nabla \phi}{|r'-r|} \right) = \nabla \phi \cdot \nabla \left(\frac{1}{|r'-r|} \right) + \left(\frac{\nabla \cdot \nabla \phi}{|r'-r|} \right) \tag{7.82}$$

Since, by using the Coulomb gauge, $\nabla \cdot A = 0$, then $\nabla \cdot \nabla\phi$ must necessarily be zero. Therefore, Equation (7.82) reduces to the form

$$\nabla \cdot \left(\frac{\nabla\phi}{|r' - r|} \right) = \nabla\phi \cdot \nabla \left(\frac{1}{|r' - r|} \right) \tag{7.83}$$

Substituting Equations (7.81) and (7.83) into (7.79), we have

$$-\frac{\sigma\mu_0}{4\pi} \int \frac{\partial A}{\partial t} \cdot \nabla \left(\frac{1}{|r' - r|} \right) dV - \frac{\sigma\mu_0}{4\pi} \int \nabla\phi \cdot \nabla \left(\frac{1}{|r' - r|} \right) dV = 0 \tag{7.84}$$

Equations (7.78) and (7.84) must be simultaneously solved to obtain the solution for the vector potential, A, and scalar potential, ϕ, which are the unknowns.

Restating the two equations, we have

$$A = A_e + A_s = -\frac{\mu_0}{4\pi} \int \frac{\sigma \frac{\partial A}{\partial t} + \sigma\nabla\phi}{|r' - r|} dV + A_s \tag{7.85}$$

and

$$-\frac{\sigma\mu_0}{4\pi} \int \frac{\partial A}{\partial t} \cdot \nabla \left(\frac{1}{|r' - r|} \right) dV - \frac{\sigma\mu_0}{4\pi} \int \nabla\phi \cdot \nabla \left(\frac{1}{|r' - r|} \right) dV = 0 \tag{7.86}$$

For time-harmonic cases where $A = |A|\, e^{j\omega t}$ and $\phi = |\phi|\, e^{j\omega t}$, Equations (7.85) and (7.86) reduce to the form

$$A = -\frac{\mu_0\sigma}{4\pi} \int \frac{(j\omega A + \nabla\phi)}{|r' - r|} dV + A_s \tag{7.87}$$

$$-\frac{\sigma\mu_0}{4\pi} \int j\omega A \cdot \nabla \left(\frac{1}{|r' - r|} \right) dV - \frac{\sigma\mu_0}{4\pi} \int \nabla\phi \cdot \nabla \left(\frac{1}{|r' - r|} \right) dV = 0 \tag{7.88}$$

Further, in Cartesian coordinates,

$$\nabla \left(\frac{1}{|r' - r|} \right) = \frac{(x' - x)\hat{u}_x + (y' - y)\hat{u}_y + (z' - z)\hat{u}_z}{\left[(x' - x)^2 + (y' - y)^2 + (z' - z)^2 \right]^{3/2}} \tag{7.89}$$

Decomposing Equations (7.87) and (7.88) with Equation (7.89) into their respective Cartesian components, we have, after rearranging terms

$$A_{sx} = A_x + \frac{\sigma\mu_0}{4\pi} \int \frac{j\omega A_x + \frac{\partial\phi}{\partial x}}{|r' - r|} dV \tag{7.90}$$

$$A_{sy} = A_y + \frac{\sigma\mu_0}{4\pi} \int \frac{j\omega A_y + \frac{\partial\phi}{\partial y}}{|r' - r|} dV \tag{7.91}$$

$$A_{sz} = A_z + \frac{\sigma \mu_0}{4\pi} \int \frac{j\omega A_z + \frac{\partial \phi}{\partial z}}{|r' - r|} dV \qquad (7.92)$$

$$\frac{\sigma \mu_0}{4\pi} \int \frac{\left(j\omega A_x + \frac{\partial \phi}{\partial x}\right)(x' - x)}{|r' - r|^3} dV + \frac{\sigma \mu_0}{4\pi} \int \frac{\left(j\omega A_y + \frac{\partial \phi}{\partial y}\right)(y' - y)}{|r' - r|^3} dV$$

$$+ \frac{\sigma \mu_0}{4\pi} \int \frac{\left(j\omega A_z + \frac{\partial \phi}{\partial z}\right)(z' - z)}{|r' - r|^3} dV = 0 \qquad (7.93)$$

where $|r' - r|$ $\sqrt{(x' - x)^2 + (y' - y)^2 + (z' - z)^2}$

A_{sx}, A_{sy}, A_{sz} are vector potentials due to eddy current free source currents

A_x, A_y, A_z are unknown vector potentials

$\frac{\partial \phi}{\partial x}, \frac{\partial \phi}{\partial y}, \frac{\partial \phi}{\partial z}$ are derivatives of the unknown scalar potential ϕ

Using the summation convention, we can expand r, x, y, z, the radius vector, the vector potentials, and the derivatives of the scalar potential in terms of shape functions inside the volume integrals of the above equations, and obtain the following expressions:

$$r = \sum_{i=1}^{N} \zeta_i r_i, \quad x = \sum_{i=1}^{N} \zeta_i x_i, \quad y = \sum_{i=1}^{N} \zeta_i y_i, \quad z = \sum_{i=1}^{N} \zeta_i z_i \qquad (7.94)$$

where ζ_i's are the shape functions (see Appendix F), i is the element node and N is the number of nodes in the element.

Therefore,

$$A_{sx} = \delta_{k\ell} A_x + \frac{\sigma \mu_0}{4\pi} \int \frac{j\omega \sum_{i=1}^{N} \zeta_i A_{xi} + \sum_{i=1}^{N} \frac{\partial \zeta_i}{\partial x} \phi_i}{\left| \sum_{i=1}^{N} \zeta_i r'_i - \sum_{i=1}^{N} \zeta_i r_i \right|} dV \qquad (7.95)$$

$$A_{sy} = \delta_{k\ell} A_y + \frac{\sigma \mu_0}{4\pi} \int \frac{j\omega \sum_{i=1}^{N} \zeta_i A_{yi} + \sum_{i=1}^{N} \frac{\partial \zeta_i}{\partial y} \phi_i}{\left| \sum_{i=1}^{N} \zeta_i r'_i - \sum_{i=1}^{N} \zeta_i r_i \right|} dV \qquad (7.96)$$

$$A_{sz} = \delta_{k\ell} A_z + \frac{\sigma \mu_0}{4\pi} \int \frac{j\omega \sum_{i=1}^{N} \zeta_i A_{zi} + \sum_{i=1}^{N} \frac{\partial \zeta_i}{\partial z} \phi_i}{\left| \sum_{i=1}^{N} \zeta_i r'_i - \sum_{i=1}^{N} \zeta_i r_i \right|} dV \qquad (7.97)$$

$$\frac{\sigma\mu_0}{4\pi}\int\frac{\left(j\omega\sum_{i=1}^{N}\zeta_i A_{xi}+\sum_{i=1}^{N}\frac{\partial\zeta_i}{\partial x}\phi_i\right)\left(\sum_{i=1}^{N}\zeta_i x_i'-\sum_{i=1}^{N}\zeta_i x_i\right)}{\left|\sum_{i=1}^{N}\zeta_i r_i'-\sum_{i=1}^{N}\zeta_i r_i\right|^3}$$

$$+\frac{\sigma\mu_0}{4\pi}\int\frac{\left(j\omega\sum_{i=1}^{N}\zeta_i A_{xi}+\sum_{i=1}^{N}\frac{\partial\zeta_i}{\partial x}\phi_i\right)\left(\sum_{i=1}^{N}\zeta_i x_i'-\sum_{i=1}^{N}\zeta_i x_i\right)}{\left|\sum_{i=1}^{N}\zeta_i r_i'-\sum_{i=1}^{N}\zeta_i r_i\right|^3}$$

$$+\frac{\sigma\mu_0}{4\pi}\int\frac{\left(j\omega\sum_{i=1}^{N}\zeta_i A_{xi}+\sum_{i=1}^{N}\frac{\partial\zeta_i}{\partial x}\phi_i\right)\left(\sum_{i=1}^{N}\zeta_i x_i'-\sum_{i=1}^{N}\zeta_i x_i\right)}{\left|\sum_{i=1}^{N}\zeta_i r_i'-\sum_{i=1}^{N}\zeta_i r_i\right|^3}=0 \quad (7.98)$$

where δ_{kl} is the Kronecker delta function ($= 1$ for $k = \ell$; $= 0$ for $k \neq \ell$) and k, ℓ are element numbers. With the above Equations (7.95) through (7.98), a matrix can be formed as shown in Equation (7.99).

$$\begin{bmatrix} A & 0 & 0 & D \\ 0 & B & 0 & E \\ 0 & 0 & C & F \\ G & H & I & J \end{bmatrix}\begin{bmatrix} A_x \\ A_y \\ A_z \\ \phi \end{bmatrix}=\begin{bmatrix} A_{sx} \\ A_{sy} \\ A_{sz} \\ 0 \end{bmatrix} \quad (7.99)$$

where A and D are from Equation (7.95), B and E are from (7.96), C and F are from (7.97) and G, H, I and J are from (7.98). The evaluation of the isoparametric element is shown in Appendix F.

Part III

Applications

8

Induction Heating

8.1 Simplified Induction Heating Analysis

Brown et al. [59] present a simplified approach to understand the principles of induction heating. First consider the problem of a wire carrying current I located a distance h above a conducting half-space (the load) as shown in Figure 8.1. We have presented a complete solution to this problem in Section 2.10, but for now, we will assume the skin depth to be very small, so that the current distribution in the conducting space can be considered a surface current sheet.

The field produced at the surface by the wire is given as

$$H_1 = \frac{I}{2\pi r} \tag{8.1}$$

where $r = \sqrt{h^2 + x^2}$. Using the method of images

$$H = 2H_1 \cos(\phi) = 2H_1 \frac{h}{r} \tag{8.2}$$

Using the current sheet approximation, the tangential field at the surface is equal to the surface current density.

$$J_s = H = \frac{I}{\pi} \frac{h}{h^2 + x^2} = \frac{I}{\pi h} \frac{1}{\left(1 + \left(\frac{x}{h}\right)^2\right)} \tag{8.3}$$

Figure 8.2 shows the surface current density for different values of height, h, for a filament conductor with 1 A. As we expect, if the conductor is closer to the surface, the current density just beneath the conductor is higher and falls off quickly.

As a check, we find the total current in the conducting plate by integrating the surface current density to infinity.

$$\int_{-\infty}^{\infty} J_s \, dx = \frac{hI}{\pi} \int_{-\infty}^{\infty} \frac{dx}{h^2 + x^2} = \frac{2I}{\pi} \tan^{-1} \left(\frac{x}{h}\right) \Big|_0^{\infty} = I \tag{8.4}$$

To find the losses in the load, consider the incremental section shown in Figure 8.3.

Eddy Currents: Theory, Modeling, and Applications, First Edition.
Sheppard J. Salon, M. V. K. Chari, Lale T. Ergene, David Burow, and Mark DeBortoli.
© 2024 The Institute of Electrical and Electronics Engineers, Inc. Published 2024 by John Wiley & Sons, Inc.

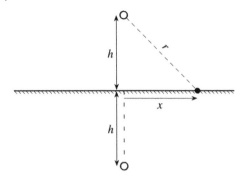

Figure 8.1 Long wire parallel to conducting half-plane.

The incremental resistance of this section is

$$dR = \frac{1}{\sigma \delta dx} \tag{8.5}$$

To find the total loss in the sheet, we integrate the loss density over the surface.

$$dW_L = (J \cdot dx)^2 \, dR = \frac{J^2 dx}{\sigma \delta} \tag{8.6}$$

Therefore

$$W_L = \frac{1}{\sigma \delta} \int_{-\infty}^{\infty} J^2(x) dx = \frac{2h^2 I^2}{\pi^2 \sigma \delta} \int_{0}^{\infty} \frac{dx}{\left(h^2 + x^2\right)^2} \tag{8.7}$$

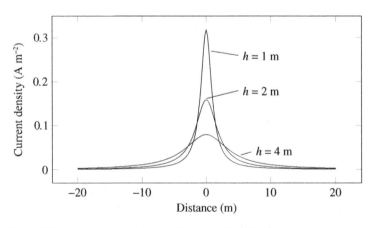

Figure 8.2 Surface current density for wire of height h above a conducting plane.

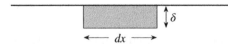

Figure 8.3 Incremental conducting section on surface.

Integrating

$$\int_0^\infty \frac{dx}{\left(h^2 + x^2\right)^2} = \frac{1}{2h^3} \tan^{-1}\left(\frac{x}{h}\right)\Big|_0^\infty = \frac{\pi}{4h^3} \tag{8.8}$$

Therefore

$$W_L = \frac{I^2}{2\pi\sigma\delta h} \tag{8.9}$$

We will now consider the losses in the cylindrical conductor. We will assume that the conductor has a circular cross section and radius a. We will first assume that the conductor is sufficiently far away that the current distribution is not affected by the currents in the conducting plate. We will then consider the current redistribution produced by the currents in the conducting sheet. In the first case, the current will be uniformly distributed around the conductor and is contained in the skin depth as shown in Figure 8.4. A more accurate representation can be found in Section 3.2 in terms of Bessel functions, but we have already shown that for the case in which the skin depth is much smaller than the radius, the present assumption is valid.

The power dissipated in the conductor is therefore

$$W_c = \frac{I^2}{2\pi a \sigma \delta} \tag{8.10}$$

We now consider the more general situation in which the cylindrical conductor is close to the load. In this case, the current distribution is not uniform around the conductor, and the current distribution in the load is affected by the nearby finite cylinder. We can use the method of images (see Figure 8.5) and replace the conductor by a filament located at a height above the conducting plate equal to (see Appendix D).

$$h' = \sqrt{h^2 - a^2} \tag{8.11}$$

Figure 8.4 Circular conductor and current distribution.

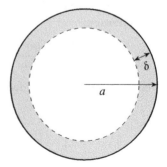

The current density in the load can now be studied by using Equation (8.3) and replacing h with h'. The new formula is

$$J_s = \frac{I}{\pi h} \frac{\sqrt{1-\left(\frac{a}{h}\right)^2}}{\left(1-\left(\frac{a}{h}\right)^2+\left(\frac{x}{h}\right)^2\right)} \tag{8.12}$$

The power dissipated in the load can now be found from Equation (8.9) by replacing h with h'. This gives

$$W = \frac{2h^2 I^2}{\pi^2 \sigma \delta} \int_0^\infty \frac{dx}{\left(h^2+x^2\right)^2} \tag{8.13}$$

The integral is evaluated by

$$\int_0^\infty \frac{dx}{\left(h^2+x^2\right)^2} = \frac{1}{2h^3} \tan^{-1}\left(\frac{x}{h}\right)\Big|_{x=0}^{x=\infty} = \frac{\pi}{4h^3} \tag{8.14}$$

So the power dissipated is

$$W_L = \frac{I^2}{2\pi\sigma\delta h} \tag{8.15}$$

The current distribution in the cylindrical conductor is found by adding the tangential components of field produced by the two sources as shown in Figure 8.5.

The current distribution is given by [59] as

$$J = \frac{I}{2\pi a} \frac{\sqrt{1-\left(\frac{a}{h}\right)^2}}{\left(1-\frac{a}{h}\cos\theta\right)} \tag{8.16}$$

In Figure 8.6, we see the current density around the conductor for 1 A as a function of the conductor height to radius ratio. As $h \to \infty$ the distribution is uniform as expected.

The loss in the conductor is now found by integrating the loss density around the perimeter.

$$W_c = 2\int_0^\pi \frac{(Ja\,d\theta)^2}{\sigma\delta a\,d\theta} \tag{8.17}$$

The integral can be evaluated as

$$\frac{\pi}{\left[1-\left(\frac{a}{h}\right)^2\right]^{3/2}} \tag{8.18}$$

Figure 8.5 Geometry to find the image source and current distribution around the conductor.

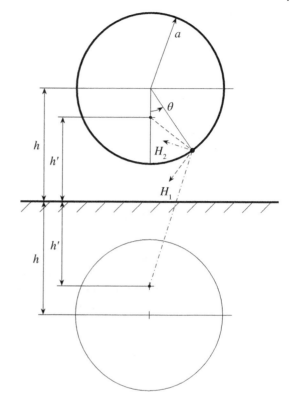

Figure 8.6 Surface current density around conductor as a function of h/a.

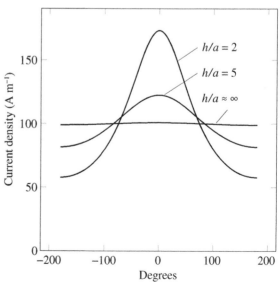

Therefore the conductor loss is

$$W_c = \frac{I^2}{2\pi a\sigma\delta}\left[1 - \left(\frac{a}{h}\right)^2\right]^{-1/2} \tag{8.19}$$

Calling the power dissipated in the load W_L and the power dissipated in the conductor W_c, we find the ratio of the two

$$\frac{W_L}{W_c} = \frac{a\delta_c}{h\delta_L} \tag{8.20}$$

We can then define the efficiency of the induction heating process as

$$\eta = \frac{W_L}{W_L + W_c} = \frac{1}{1 + \frac{h\delta_L}{a\delta_c}} \tag{8.21}$$

We will now look at a numerical example. We have a hollow copper circular conductor of radius 0.01 m and the conductor thickness is 0.002 m. The height of the conductor above the conducting plane is $h = 0.05$ m. The current density around the conductor, by Equation (8.16), is shown in Figure 8.7.

A finite element model was also constructed for this problem and the current density around the conductor is shown in Figure 8.8.

As we can see, the results are almost identical. This shows that the assumptions made in the simplified analysis is valid for this case. The losses in the conductor, by finite element analysis are shown in Figure 8.9.

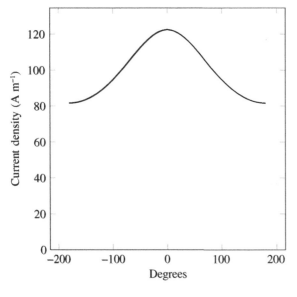

Figure 8.7 Surface current density around copper conductor of radius $a = 0.01$ m.

Figure 8.8 Surface current density around copper conductor by finite element analysis.

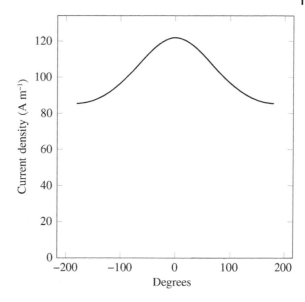

Figure 8.9 Loss density around copper conductor of radius $a = 0.01$ m.

8.2 Coupled Eddy Current and Thermal Analysis: Induction Heating

Induction heating of metals is an important industrial application. In induction heating, there is a coupling between the electrical, electromagnetic, and thermal phenomena. To illustrate this point, we consider the induction heating of magnetic steel. The application may be, for example, surface hardening for gears or axles. As we apply current to the induction coil, the steel work-piece saturates magnetically, as was explained in Section 2.6. This affects the skin depth and the impedance seen from the power supply, which in turn affects the current from the power supply. Another coupling is due to the effect of the temperature on the resistivity of the steel. During the induction heating process, the temperature is

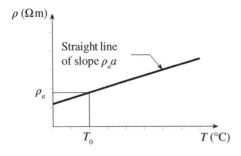

Figure 8.10 Resistivity vs. temperature.

usually rising very rapidly and the resistivity of the load is continuously changing. This dependence of resistivity on temperature is illustrated in Figure 8.10 where we see that the resistivity of the material has a linear relationship with temperature in the temperature range of interest.

As the temperature rises, not only does the electrical resistivity change, but the magnetic permeability changes as well. As the temperature rises, the permeability decreases. We, therefore, need a series of measured curves such as that shown in Figure 8.11.

Since the temperature distribution is highly nonuniform, the material properties involved in the electromagnetic computation vary locally. There are no closed-form or analytical solutions which can accurately account for this phenomenon. Numerical methods, like the finite element method, are ideal since they are able to represent the nonlinearity of the magnetic steel and the local variation of permeability and resistivity due to temperature effects.

All of these changes to the electrical material properties and possibly to the coil current, affect the losses and therefore the temperature rise. The thermal conductivity and specific heat of the work piece are also temperature dependent and will also be changing over the duration of the process. The thermal conductivity as a function of temperature is shown in Figure 8.12.

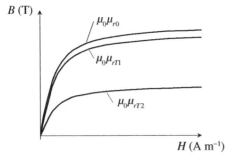

Figure 8.11 Magnetization curves vs. temperature.

Figure 8.12 Thermal conductivity vs. temperature.

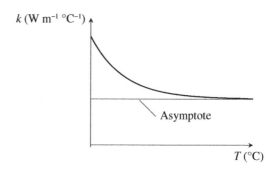

Figure 8.13 Specific heat vs. temperature.

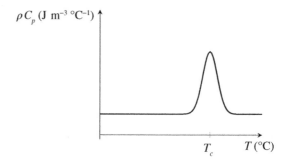

The specific heat as a function of temperature is shown in Figure 8.13. At one point, the material may go through a phase change (Curie point) where the specific heat increases dramatically. The analysis in this region often involves a variable time step since it is very easy to jump over this region if the time step is too large. As we approach the phase transition, the time step is shortened using the following process. The temperature is found at the end of a time step. We then go back to the beginning of the time step and using smaller time steps, arrive at the end of the original time step. If the results agree, we go on. If not, we again reduce the time step until we get agreement.

With this strong coupling between the electromagnetic and thermal phenomena, we must either solve the magnetic and thermal equations simultaneously or solve each set of equations separately and then exchange parameters (with possible iterations), to ensure that the electrical and thermal properties correspond to the up-to-date value of temperature and flux density. Time steps must be kept very small in order to capture these material property changes.

We usually opt for the second option and solve the electromagnetic and thermal equations separately and then iterate between the solutions until convergence is reached. The difficulty is that the thermal time constants are considerably longer than the electromagnetic time scales. In fact, we normally solve the

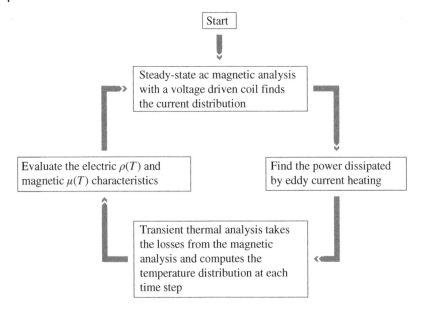

Figure 8.14 Flow chart of computation process.

electromagnetic eddy current problem using steady-state ac or phasor analysis, then transfer the losses to the time-domain thermal model. The thermal model gives the temperature distribution, which is then sent to the electromagnetic model so that the material properties can be adjusted. At each time step in the thermal analysis, there is an iterative exchange with the electromagnetic analysis until the process converges. This process is illustrated in the flow chart of Figure 8.14.

The eddy current analysis is typically done using a $T - \Omega$ or an $A - V$ formulation as described in Section 4.1. The local losses computed in this step are then transferred to the thermal model and the temperature is found. The local temperature distribution is then transferred back to the electromagnetic model and the material properties are updated.

In many cases involving induction heating, the skin depth is very small compared with the other dimensions in the problem. In these cases, it is very difficult to make a good finite element mesh. As was discusses in Section 6.1, we need several elements per skin depth for a good solution. The elements in this case would be very narrow and therefore have a poor aspect ratio. An alternative which is sometimes used is to employ the surface impedance concept. We have seen the idea of surface impedance applied to an electric machine rotor in Section 3.11. In this case, we replace the eddy current layer with a boundary that has the property that

$$\hat{n} \times E = Z_s \hat{n} \times (\hat{n} \times H) \tag{8.22}$$

The surface impedance Z_s is found as

$$Z_s = \frac{1+j}{\sigma\delta} \tag{8.23}$$

The evaluation of the temperature distribution, $T(x, y, z, t)$, in regions described by thermal conductivity k and specific heat ρC_p along with volumetric and/or surface heat flux input q is described by the Fourier heat diffusion equation (note that this is the same equation that describes the eddy current phenomena but the dependent variable in this case is the scalar temperature, T).

$$\rho C_p \frac{\partial T}{\partial t} = \nabla(k\nabla T) + q \tag{8.24}$$

The boundary conditions are given by

$$k\hat{n} \cdot \nabla T = h(T - T_a) + \epsilon C_n \left(T^4 - T_a^4\right) + q_s \tag{8.25}$$

where h is the heat transfer coefficient, ϵ is the radiation coefficient, T_a is the ambient temperature, and q_s the surface thermal flux. The initial temperature distribution $T(x, y, z, 0)$ must be known to make the problem well-posed. The thermal conductivity is also temperature dependent as is the specific heat as described above.

8.2.1 Example of Coupling the Magnetic and Thermal Problems

As an example, we will describe the models used in an induction heating problem. The process of coupling the eddy current problem with the thermal problem is illustrated in Figure 8.14. The method illustrated below uses a so-called "weak" coupling between the electromagnetic and thermal problems. The problems are solved separately and the material properties are continuously corrected to account for the field and temperature properties.

The computation proceeds as follows: A field problem is done using a set of material properties at the assumed starting temperature. The eddy current distribution and power is found in the lossy regions. This electromagnetic problem is solved as a steady-state sinusoidal problem, i.e. with complex phasor analysis. With these losses, the thermal problem is then solved in the time domain. Space is discretized using the finite element method and time is discretized by finite differences. The electromagnetic problem is solved again with all of the relevant electric and magnetic properties corrected for the temperature. This process is repeated, performing the eddy current and thermal calculations and correcting the material properties until convergence is reached. A given error criterion is met in which the material properties are consistent with both the electromagnetic and thermal solutions. We then move to the next time step and repeat the process. As one might imagine, depending on the complexity of the geometry and length of the calculation, the solution time can be quite long.

As an example, we consider the case of an axisymmetric billet of magnetic steel with resistivity $\rho = 25.0 \times 10^{-6}$ Ωm. The radius of the billet is 20.0 mm. The frequency of excitation is 50 kHz. The permeability is modeled by a series of effective permeability curves as discussed in Section 2.6. The steady-state eddy current problem was solved using the finite element method and the $A - V$ formulation discussed in Section 4.1. The iterative procedure described above was performed and the results after 3 s are displayed in the following set of figures.

In Figure 8.15, we see the flux distribution in the core, coils, and load regions computed at $t = 3$ s. Due the high frequency, high resistivity, and high permeability of the load, we can see that the skin depth is extremely small. In Figure 8.16, we see the power distribution in the load. Again, we note that the power is dissipated in a very thin region near the surface.

In Figure 8.17, we see the temperature distribution at $t = 3$ s. Even though the heat is flowing from the surface into the load, the temperature distribution shows that, for short times, the temperature rise remains limited to the region near the surface. In this way, the heat treatment process can affect the surface and not the interior regions.

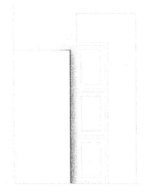

Figure 8.15 Flux distribution at $t = 3$ s showing source coils and load.

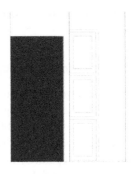

Figure 8.16 Loss density in the load at $t = 3$ s.

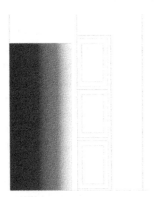

Figure 8.17 Temperature distribution in the load at $t = 3$ s.

9

Wattmeter

The watt-hour meter is a good example of an electro-mechanical device with moving conductors. Watt-hour meters are extremely complex due to their three-dimensional topology and required accuracy.

There are three main parts to an electro-mechanical watt-hour meter: the motor which produces the torque, the magnetic brake which retards the speed of the disc, and the register which counts the motor revolutions. The portions of the stator energized by the line voltage and load current are known as the voltage coil and the current coil respectively. The voltage coil is designed to be highly inductive while the current coil is highly resistive. Therefore, the current through voltage coil lags the line voltage by almost 90°. If one neglects saturation, the fluxes produced by these currents will be 90° out of phase. Torque can be produced by two alternating fluxes which have both time and a space displacement. These fluxes, which are out of phase by approximately 90°, induce an electromotive force (emf) on the conducting disc. Since the disc has its own impedance (mostly resistive), there will be circulating eddy currents on the disc, and these currents will be interacting with the fluxes which produce them. This interaction produces a torque in the system and disc rotates. The torque is directly proportional to the sine of the angle between two currents (current in the voltage coil and current in the current coil). The design attempts to make the angle larger (close to 90°) to increase the torque on the disc.

To obtain a better understanding of this device and the eddy currents involved, a finite element model of a watt-hour meter was created. Figure 9.1 shows the geometry of the watt-hour meter. The coils were modeled as non-eddy current conductors and the field was computed using the Biot–Savart law. The current coils are wound around the two limbs of the C-core, and the voltage coil is wound around the central limb of the E-Core. The phase angle of the voltage coil excitation was offset from the excitation in the current coil. The model includes the C-Core (current coil), the E-Core (voltage coil), the conducting disc, and the tie bars (shunts)

Eddy Currents: Theory, Modeling, and Applications, First Edition.
Sheppard J. Salon, M. V. K. Chari, Lale T. Ergene, David Burow, and Mark DeBortoli.

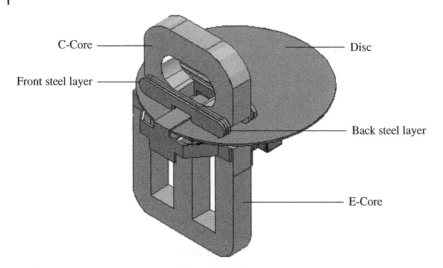

C-Core — Disc

Front steel layer —

— Back steel layer

— E-Core

Figure 9.1 The 3-D model of the watt-hour meter.

across the C-Core. These shunts are located on both sides of the C-Core. Each side of the shunt consists of two layers of steel, one layer of brass, and one layer of aluminum. The C-Core and E-core are made of laminated steel [60].

A three-dimensional second-order finite element mesh with 242 000 elements was created. The disc has one layer of triangular prism elements. Figure 9.2 shows the finite element mesh of the watt-hour meter.

The analysis was done using a magneto-dynamic (steady-state sinusoidal) formulation. Because this is a low-frequency application, one may neglect displacement currents. In the watt-hour meter model, both scalar and vector potential formulation were used in the different regions. The electromagnetic field in the conductors can be derived in two different ways, using vector potentials: $A - V$ and $T - \Omega$ formulations [47] discussed in Section 4.1. One method uses a magnetic vector potential, A, defined by

$$\mathbf{B} = \nabla \times \mathbf{A} \tag{9.1}$$

$$\mathbf{E} = -\frac{\partial \mathbf{A}}{\partial t} - \nabla V \tag{9.2}$$

The other method uses an electric vector potential, $\mathbf{T}$, and magnetic scalar potential, Ω, defined by

$$\mathbf{J} = \nabla \times \mathbf{T} \tag{9.3}$$

$$\mathbf{H} = \mathbf{T} - \nabla \Omega \tag{9.4}$$

Figure 9.2 The 3-D finite element mesh model of the watt-hour meter.

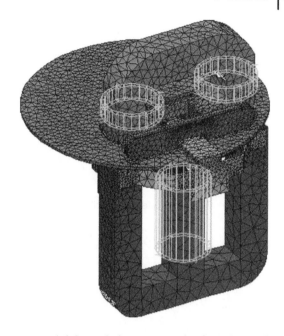

On the disc, the $\mathbf{T} - \Omega$ vector potential formulation was used. There is a 90° phase angle between voltage coil excitation and current coil excitation. By using Faraday's law

$$\nabla \times \rho \nabla \times \mathbf{T} + \mu \frac{\partial \mathbf{T}}{\partial t} - \mu \nabla \frac{\partial \Omega}{\partial t} = 0 \tag{9.5}$$

For eddy current-free regions like air, the magnetic field can be obtained from a magnetic scalar potential as

$$\mathbf{H} = -\nabla \Omega \tag{9.6}$$

The differential equation satisfied by this potential is

$$-\nabla \mu \nabla \Omega = 0 \tag{9.7}$$

Numerical stability requires formulations involving unique potentials. The uniqueness of the potential can be guaranteed by specifying its divergence and its normal or tangential component on the boundary. To use different sets of potentials in the same model, one must ensure that the interface conditions are satisfied. This difficulty can be avoided if the potentials $\mathbf{A}$, and V in the conductors, are coupled with $\mathbf{A}$ outside the conductor, and/or the potentials $\mathbf{T}$ and Ω in the conductors are coupled with Ω in eddy current-free regions. The two regions (conductor and nonconductor) can be coupled by imposing $\mathbf{B} \cdot \hat{\mathbf{u}}_{\mathbf{n}}$ and $\mathbf{H} \times \hat{\mathbf{u}}_{\mathbf{n}}$ continuity. In this problem, the continuity of current ($\mathbf{J} \cdot \hat{\mathbf{u}}_{\mathbf{n}} = 0$) is satisfied by setting the tangential component of the electric vector potential

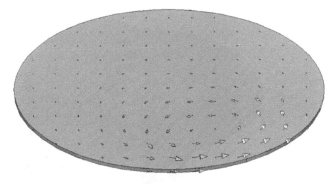

Figure 9.3 The current density distribution (as arrows) of the model for $\omega t = 0$.

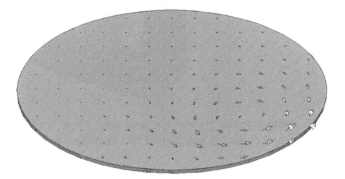

Figure 9.4 The current density distribution (as arrows) of the model for $\omega t = 90°$.

$\mathbf{T} \times \hat{\mathbf{u}}_n$ to zero. This condition and the continuity of Ω ensure that $\mathbf{H} \times \hat{\mathbf{u}}_n$ is continuous. $\mathbf{H}$ and $\mathbf{T}$ differ by the gradient of a scalar and have the same units.

For the problem we are considering, the continuity conditions can be satisfied by setting a gauge condition, for example the Coulomb gauge, $\nabla \cdot \mathbf{T} = 0$.

With this choice, Equation (9.5) becomes

$$\nabla \times \rho \nabla \times \mathbf{T} - \nabla \rho \nabla \cdot \mathbf{T} + \mu \frac{\partial \mathbf{T}}{\partial t} - \mu \nabla \frac{\partial \Omega}{\partial t} = 0 \tag{9.8}$$

The circulating current in the disc will oppose the change of flux linkage. These circulating currents can be seen in Figures 9.3 and 9.4 for different instants of time.

For the torque computation, the virtual work method was used [23]. The torque is calculated by differentiating the system magnetic energy with respect to a virtual rotation angle on an axis parallel to the absolute coordinate system axes.

Table 9.1 summarizes the results obtained for different phase angles between the current and voltage coils. These results were obtained to verify the torque pattern due to the phase angle between voltage coil excitation and current coil excitation. All values are in per unit based on the 90° torque value.

Table 9.1 The torque values for different phase angles.

Angle	90	80	70	45	30	15	5	0
Torque (pu)	1.0	0.9927	0.956	0.738	0.5369	0.297	0.1257	0.0366

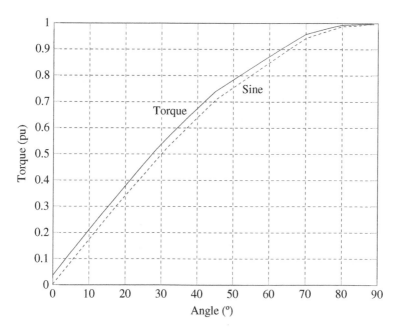

Figure 9.5 The comparison between sine and torque values.

Figure 9.5 shows the comparison between the torque curves for different phase angles and the sine curve. The torque produced on the disc can be written in terms of load current and frequency as

$$T_m = K\frac{I^2}{Z}f\cos\alpha\sin\beta \tag{9.9}$$

where Z is the impedance of the disc, α is the angle between the induced voltages on the disc, and eddy currents, and K is a proportionality constant. As seen, the torque has a sine wave pattern due to the angle of β which is the phase angle between voltage coil excitation and current coil excitation.

The shunts (tie bars) across the C-core consist of sheets of three different materials: steel, brass, and aluminum, as shown in Figure 9.6. We assume that the steel has no conductivity. Conductivity values of $2.6\times10^7\,\mathrm{Sm}^{-1}$ for brass and $4.2\times10^7\,\mathrm{Sm}^{-1}$ for aluminum were used. The brass and aluminum regions were modeled using $\mathbf{T}-\Omega$ formulations to allow for eddy currents. These regions

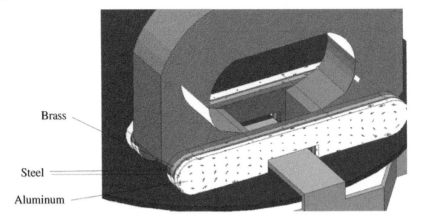

Brass

Steel

Aluminum

Figure 9.6 Eddy currents in the shunt parts.

Figure 9.7 Eddy currents in the screw and aluminum parts.

have the permeability of free space, μ_0. The eddy currents produced in these regions were small in magnitude. However, the effect of the permeable sheets caused the torque value to change significantly. If the torque without the tie bars is normalized to 1.0 pu, the computed torque value is different by 11.2% in the model that includes the tie bars.

The effect of the load adjust screw on the power factor of the watt-hour meter was also studied. Figure 9.7 shows the details of the screw attached to the E-core and the eddy current distribution in it and the aluminum cylinder around it.

Figure 9.8 After adding the light load adjustment piece.

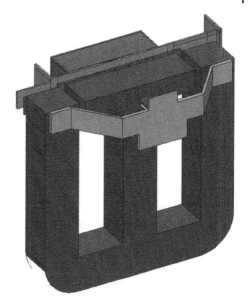

Because of the conductivity of the aluminum, the flux is phase shifted with respect to the E coil flux. The flux through the screw was computed and found to have a phase shift of about 14° compared to the case with no eddy currents in the screw.

With no current coil excitation, any lack of symmetry in the potential coil flux can produce a torque that might be either forward or backward. Since electrical steels are not perfect conductors of magnetic flux, the flux produced by the current coils is not exactly proportional to the current, so that when the meter is carrying a small fraction of its rated load, it tends to run slowly. In addition, there is friction caused by the bearings and the register, which also tends to make the disc rotate at a lower speed than desired with small load currents. To compensate for these tendencies, a controlled driving torque is added to the disc. This is called the "light load adjustment." As the plate is moved circumferentially with respect to the disc, the net driving torque changes and the disc rotation speed changes accordingly. The pieces representing this plate group can be seen in Figure 9.8. This light load adjustment will change the torque by about 10%.

Since the efficiency of this device is low, the force distribution was studied to understand the torque production. Thus the nodal forces were found on the conducting disc.

Knowing the value of the flux density, **B**, and current density, **J**, in the eddy current region is sufficient to calculate the local Lorentz force vector.

$$d\mathbf{F} = \mathbf{J} \times \mathbf{B} \tag{9.10}$$

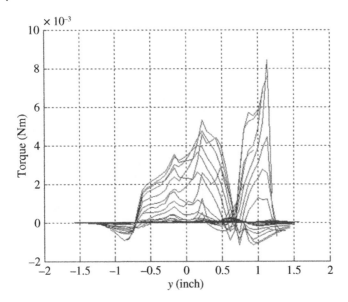

Figure 9.9 The torque distribution over the disc.

Figure 9.9 shows the resulting torque response at the instant of $\omega t = 0$. We see that in addition to positive torque, negative torque is produced at some points on the disc due to the direction of current and flux density.

An electromagnetic braking torque is created by the interaction between a permanent magnet and the conducting disc in this device. This produces an electromagnetic brake in which the braking torque depends on the velocity of the conductor. In the example problem we have a disc that has electrical conductivity ($\sigma = 5 \times 10^7 \, \mathrm{S\,m^{-1}}$) and a permanent magnet region having magnetic properties ($\mu_r = 1.15$ and $B_r = 0.8\,\mathrm{T}$). An air region (nonconducting) surrounds the disc. The thin conducting disc is $0.05\,cm$ thick and $7.6\,cm$ in diameter. The disc and its surrounding region rotate during the solution process, in this case by imposing a speed equal to $80°/\mathrm{s}$. The model with the mesh is shown in Figure 9.10.

In the model, both scalar and vector potential formulations are used in different parts. The $\mathbf{T} - \Omega$ vector potential formulation is used in the conducting disc. Because the nonconducting regions are free of eddy currents, the magnetic field can be described by a magnetic scalar potential.

In this model, the total magnetic scalar potential formulation is used for the magnet and surrounding air.

The geometry consists of independent meshes for the different regions. Each mesh corresponds to a moving or a fixed part. In our case, the fixed part includes a permanent magnet and surrounding air region and the moving mesh is made of

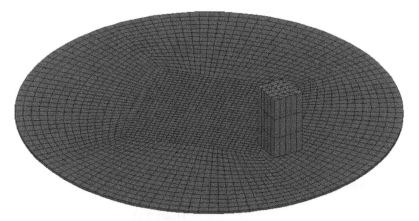

Figure 9.10 The mesh of the model.

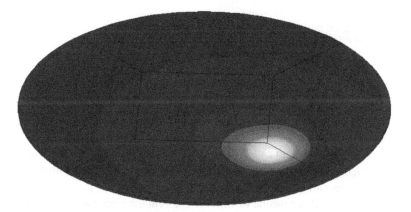

Figure 9.11 Flux density distribution on the disc.

a disc and air gap. The moving mesh is bounded by the fixed mesh. The air region is discretized with a free mesh. The meshes of the disc and magnet are created by an extrusion of the boundary facets.

Global values such as torque or flux vary continuously with the movement of mechanical parts. The equations of the model are solved simultaneously using the finite element system at each time step and then stored. Figure 9.11 shows the flux density distribution on the disc. The effect of the permanent magnet is apparent.

Figure 9.12 shows the eddy current distribution on the conducting region for the same computation time as shown in Figure 9.11. The effect of the magnet can be seen easily on the conducting region. At steady state, these eddy currents remain constant in space as the rotor turns. For the disc region, power dissipation is calculated by integrating the loss density over the disc volume.

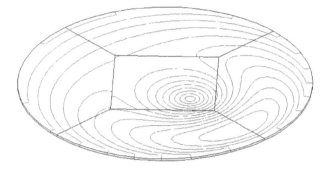

Figure 9.12 The eddy current distribution on the disc.

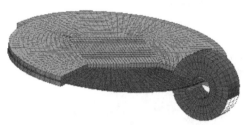

Figure 9.13 The mesh of the model including rotating air gap.

Figure 9.13 shows how the mesh is created in 3D for the model. The braking torque produced by the permanent magnet is expected to be almost equal to the positive torque.

There is no voltage or current coil excitation in this model. The torque is created only by the magnet and conducting disc interaction. The variation of torque vs. time after two cycles can be observed in Figure 9.14.

The variation of calculated torque vs. time for different time step values is apparent. Because of the constant field of the magnet, the braking torque is expected to be a constant at steady state. The transient solution for the braking torque can be seen in Figures 9.14 and 9.15 for different time steps, respectively. Higher velocity requires shorter time steps. Figure 9.15 presents a curve of calculated torque vs. time using a 0.01 ms time step.

There are various difficulties that have critical effects on the numerical solution of the electromagnetic problem. These are reported in the literature and are related to the general problem of moving conductors in a magnetic field. Various solutions have been proposed, usually due to the effects of the $\mathbf{u} \times \mathbf{B}$ terms in the matrix. Since differential equations have time derivative terms, the use of time-stepping algorithms is unavoidable. If the mesh size in the direction of motion exceeds a certain limit, spurious oscillations may occur in time-domain-based numerical solutions. The method used here does not use the $\mathbf{u} \times \mathbf{B}$ term but rather remeshes each time step.

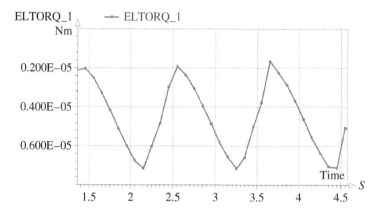

Figure 9.14 Braking torque vs. time $\Delta t = 0.1\,\text{s}$.

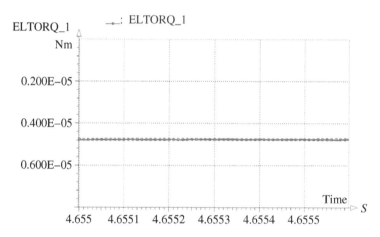

Figure 9.15 Braking torque vs. time $\Delta t = 0.01\,\text{ms}$.

In transient problems with eddy currents, the time step value for every sample point becomes very important for the calculation. Because the finite element algorithm uses a nodal solution, it is obvious that mesh size and time step have an important role in obtaining a stable and accurate solution. The solution is unstable for large time steps when motion is involved.

As the time step is reduced, the torque approaches a constant value as we expect. This procedure can be applied to the three-dimensional models of many different kinds of applications such as watt-hour meters, brakes, dampers of vibration, etc. The discussion above gives some of the modeling considerations relating to mesh and time step.

10

Magnetic Stirring

10.1 Introduction

A practical application of eddy currents for industrial use is the problem of heating and stirring of a liquid metal, steel in this case. In our simplified example, the geometry is axisymmetric. The eddy currents will vary in the r direction. The geometry is illustrated in Figure 10.1.

The billet or load is in the cylindrical region 1. This is where the steel is being heated and stirred by the Lorentz force produced by the interaction of the magnetic field and the eddy currents. The billet is contained in a stainless steel tank, region 2, which will also carry eddy currents. Region 3, which surrounds the stainless steel tank, contains a set of θ directed coils. These coils produce an axial field in the load and will be represented as a θ directed current sheet. The currents in the example are steady-state sinusoidally time-varying at 50 Hz. The outermost region, region 4, is also an annulus and is made of magnetic laminations, which completes the magnetic circuit. We will assume that this region is infinitely permeable and nonconducting. The magnetic properties in all regions are assumed to be linear, homogeneous, and isotropic. We also assume that there is no variation of any quantity in the θ direction. The magnetic field and the eddy currents in the load are functions of r. We will be computing the local loss density and force density as a function of r.

The symbols used in this analysis as well as the numerical values used in the numerical example are as follows:

ℓ length

μ permeability of the billet material

μ_2 permeability of stainless steel sheath

μ_0 free space permeability $(4\pi \times 10^{-7})$

μ_r relative permeability of the billet (assumed to be 100)

μ_{r2} relative permeability of stainless steel sheath (assumed to be 1)

σ conductivity of billet material

Eddy Currents: Theory, Modeling, and Applications, First Edition.
Sheppard J. Salon, M. V. K. Chari, Lale T. Ergene, David Burow, and Mark DeBortoli.
© 2024 The Institute of Electrical and Electronics Engineers, Inc. Published 2024 by John Wiley & Sons, Inc.

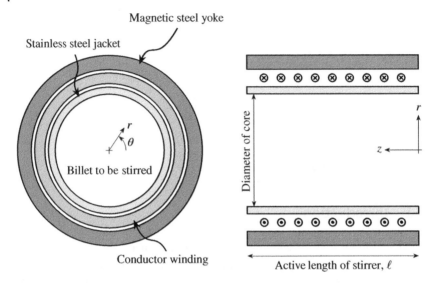

Figure 10.1 Geometry of liquid metal stirrer.

σ_2 conductivity of stainless steel
ρ resistivity of billet material (assumed to be 2.0×10^{-7})
ρ_2 resistivity of stainless steel (assumed to be 5.0×10^{-7})
f frequency (assumed to be 50)
ω angular frequency
J_0 maximum source sheet current density (assumed to be 1)
H_z magnetic field
B_z flux-density in the z direction
J_e eddy current density
k_1 complex constant for the billet (region 1), $\sqrt{\omega \mu \sigma}$
k_2 complex constant for the steel sheath (region 2), $\sqrt{\omega \mu_2 \sigma_2}$
r_0 outer radius of billet (assumed to be 0.1)
r_1 outer radius of the steel sheath (assumed to be 1.2)

10.2 Analysis

The analysis begins with Ampere's law

$$\nabla \times H = J_e = \sigma E \tag{10.1}$$

and

$$\nabla \times \nabla \times H = \sigma \nabla \times E \tag{10.2}$$

From Faraday's law, we have

$$\nabla \times E = -j\omega\mu H \tag{10.3}$$

and

$$\nabla \times \nabla \times H = -j\omega\mu\sigma H \tag{10.4}$$

In cylindrical coordinates, we get

$$\nabla \times H = -\frac{\partial H_z}{\partial r}\hat{u}_\theta \tag{10.5}$$

Neglecting the θ variation of H_z, we can write

$$\nabla \times \nabla \times H = -\frac{1}{r}\frac{\partial}{\partial r}\left(r\frac{\partial H_z}{\partial r}\right) = -j\omega\mu\sigma H_z \tag{10.6}$$

where

$$H_z = H_{z0}\sin(\omega t) \tag{10.7}$$

Expanding Equation (10.6), we have

$$\left(\frac{\partial^2 H_z}{\partial r^2} + \frac{1}{r}\frac{\partial H_z}{\partial r} - j\omega\mu\sigma H_z\right) = 0 \tag{10.8}$$

Equation (10.8) is a Bessel equation of order zero and its solution is given by [43].

$$H_z = C\left[\text{ber}(kr) + j\,\text{bei}(kr)\right] \tag{10.9}$$

For the different regions, we have the following conditions. In region 1, the billet region:

$$H_{z1} = C_1\left[\text{ber}(k_1 r) + j\,\text{bei}(k_1 r)\right] \tag{10.10}$$

In region 2, the stainless steel sheath:

$$H_{z2} = C_2\left[\text{ber}(k_2 r) + j\,\text{bei}(k_2 r)\right] \tag{10.11}$$

Region 3 is the source current region, and region 4 is an infinitely permeable eddy current free region.

We now look at the interfaces. At $r = r_o$, $H_{z1} = H_{z2}$, so that

$$C_1 = \frac{\left[\text{ber}(k_2 r_o) + j\,\text{bei}(k_2 r_o)\right]}{\left[\text{ber}(k_1 r_o) + j\,\text{bei}(k_1 r_o)\right]}C_2 \tag{10.12}$$

At $r = r_1$, $H_{z2} = J_o$, therefore

$$C_2\left[\text{ber}(k_2 r_1) + j\,\text{bei}(k_2 r_1)\right] = J_o \tag{10.13}$$

or

$$C_2 = \frac{J_o}{\left[\text{ber}(k_2 r_1) + j\,\text{bei}(k_2 r_1)\right]} \tag{10.14}$$

Substituting Equation (10.14) in Equation (10.12), we have

$$C_1 = \frac{J_0 \left[\text{ber}(k_2 r_o) + j\,\text{bei}(k_2 r_o)\right]}{\left[\text{ber}(k_2 r_1) + j\,\text{bei}(k_2 r_1)\right]\left[\text{ber}(k_1 r_o) + j\,\text{bei}(k_1 r_o)\right]} \tag{10.15}$$

Therefore,

$$H_{z1} = \frac{J_0 \left[\text{ber}(k_2 r_o) + j\,\text{bei}(k_2 r_o)\right]\left[\text{ber}(k_1 r) + j\,\text{bei}(k_1 r)\right]}{\left[\text{ber}(k_2 r_1) + j\,\text{bei}(k_2 r_1)\right]\left[\text{ber}(k_1 r_o) + j\,\text{bei}(k_1 r_o)\right]} \tag{10.16}$$

$$J_{e1} = -\frac{\partial H_{z1}}{\partial r} \tag{10.17}$$

$$= -\frac{k_1 J_0 \left[\text{ber}(k_2 r_o) + j\,\text{bei}(k_2 r_o)\right]\left[\text{ber}'(k_1 r) + j\,\text{bei}'(k_1 r)\right]}{\left[\text{ber}(k_2 r_1) + j\,\text{bei}(k_2 r_1)\right]\left[\text{ber}(k_1 r_o) + j\,\text{bei}(k_1 r_o)\right]}$$

The power loss is given by

$$P = \int \frac{J_{e1} J_{e1}^* \rho}{2}\, dv \tag{10.18}$$

To simplify the equation, let

$$D = \frac{\text{ber}(k_2 r_o) + j\,\text{bei}(k_2 r_o)}{\text{ber}(k_2 r_1) + j\,\text{bei}(k_2 r_1)} \tag{10.19}$$

Therefore

$$H_{z1} = \frac{J_0 D \left[\text{ber}(k_1 r) + j\,\text{bei}(k_1 r)\right]}{\left[\text{ber}(k_1 r_o) + j\,\text{bei}(k_1 r_o)\right]} \tag{10.20}$$

$$J_{e1} = -\frac{k_1 J_0 D \left[\text{ber}'(k_1 r) + j\,\text{bei}'(k_1 r)\right]}{\left[\text{ber}(k_1 r_o) + j\,\text{bei}(k_1 r_o)\right]} \tag{10.21}$$

For the present numerical example, we see in Figure 10.2, the magnitude of the magnetic field vs. r.

In Figure 10.3, we see the magnitude of the current density as a function of r.

Substituting for the eddy current density, J_{e1}, from Equation (10.21) into Equation (10.18) for power loss, we have

$$P = \int_0^{2\pi} \int_0^{\ell} \int_0^{r_o} k_1^2 J_0^2 DD^* \frac{\rho}{2} \frac{\left[\text{ber}'^2(k_1 r) + \text{bei}'^2(k_1 r)\right]}{\left[\text{ber}'^2(k_1 r_o) + \text{bei}'^2(k_1 r_o)\right]}\, r\, dr\, d\theta\, dz \tag{10.22}$$

Therefore

$$P = \int_0^{r_o} k_1^2 J_0^2 DD^* \rho \pi \ell \frac{\left[\text{ber}'^2(k_1 r) + \text{bei}'^2(k_1 r)\right]}{\left[\text{ber}'^2(k_1 r_o) + \text{bei}'^2(k_1 r_o)\right]}\, r\, dr \tag{10.23}$$

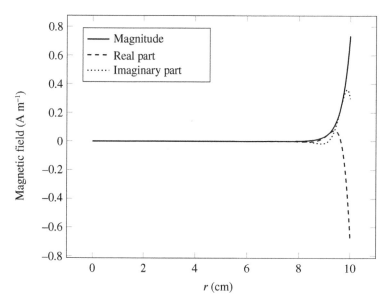

Figure 10.2 Magnetic field components vs. r.

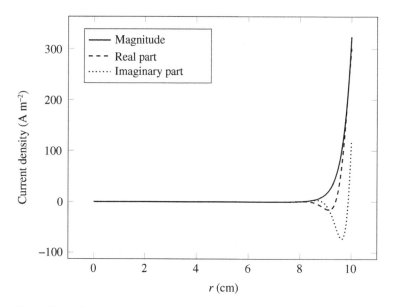

Figure 10.3 Current density components vs. r.

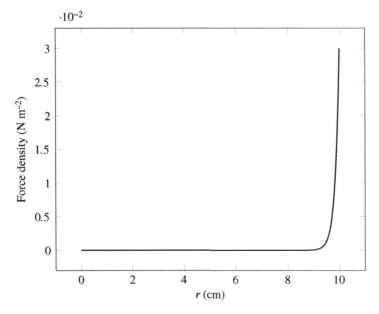

Figure 10.4 Magnitude of the force density vs. r.

which gives for the power,

$$P = k_1^2 J_0^2 DD^* \frac{\rho \pi \ell}{2} \frac{r_0}{k_1} \frac{[\mathrm{ber}'(k_1 r_0)\mathrm{ber}(k_1 r_0) + \mathrm{bei}'(k_1 r_0)\mathrm{bei}(k_1 r_0)]}{\left[\mathrm{ber}'^2(k_1 r_0) + \mathrm{bei}'^2(k_1 r_0)\right]} \quad (10.24)$$

The Lorentz force is given by the expression

$$F = R_e \int \left[J_\theta e^{j(\omega t)} \times B_z e^{-j(\omega t)} \right] dv = \int J_e B_z r\, dr\, d\theta\, dz \quad (10.25)$$

In Figure 10.4, we see the compressive force as a function of r. Substituting for J_e from Equation (10.21) and $B_z = \mu H_z$ with H_z from Equation (10.11) in Equation (10.20), we have

$$
\begin{aligned}
F = &\int_0^{2\pi} \int_0^{\ell} \int_0^{r_0} \\
&\times \frac{-k_1 J_0^2 DD^* \mu \left[\mathrm{ber}'(k_1 r) + j\,\mathrm{bei}'(k_1 r)\right]\left[\mathrm{ber}(k_1 r) - j\,\mathrm{bei}(k_1 r)\right]}{\left[\mathrm{ber}^2(k_1 r_0) + \mathrm{bei}^2(k_1 r_0)\right]} r\, dr\, d\theta\, dz
\end{aligned}
\quad (10.26)
$$

or

$$F = -2\pi \mu k_1 J_0^2 DD^* \ell \int_0^{r_0} \frac{\mathrm{ber}'(k_1 r)\mathrm{ber}(k_1 r) + j\,\mathrm{bei}'(k_1 r)\mathrm{bei}(k_1 r)}{\mathrm{ber}^2(k_1 r_0) + \mathrm{bei}^2(k_1 r_0)} r\, dr \quad (10.27)$$

From [43], it is shown that this equation reduces to the form

$$F = -2\pi\mu J_0^2 DD^* \ell \sum_{k=0}^{\infty} \frac{(k_1 r_0)^{(4k+1)}}{(4k+1)4^{2k}(k!)^2(2k-1)! \left[\text{ber}^2(k_1 r_0) + \text{bei}^2(k_1 r_0)\right]}$$

(10.28)

This model can now be used to study the forces and losses in the liquid metal.

11

Electric Machines

11.1 Eddy Currents in Slot-Embedded Conductors

In this section, we shall consider conductors carrying sinusoidally time-varying currents which are embedded in an iron slot. This class of problem is important for electric machine design and performance calculations, specifically for the determination of eddy current losses in the armature coils of AC machines and for the design of squirrel cage induction motor rotor slots and conductors, where the redistribution of current in the conductor as a function of slip frequency plays an important role in machine performance and can be exploited to customize the torque-speed profile of the machine. The problem was first addressed by Field [61] in 1905 and continues to be treated today in advanced AC machine design books such as Say [62], Lipo [63], Pyrhönen et al. [64], and especially Ostović [65].

11.1.1 Single Conductor in a Rectangular Slot

The first configuration to be analyzed is a single conductor in a simple rectangular slot as shown in Figure 11.1. Any eddy current in the iron is assumed to be negligible for the purpose of calculating eddy current in the conductor (as would be the case for a typical laminated armature core) and the permeability of the iron is assumed to be infinite. Axial current in the conductor then produces flux that travels in the iron on three sides and crosses through the slot perpendicular to the sides of the conductor, producing axial eddy current in the conductor.

Armature coils of AC machines, which carry current at power line frequency, are typically subdivided vertically into separate strands, and the strands themselves may also be transposed such that they do not occupy the same vertical position in the slot along the whole axial extent of the machine. These design features prevent the eddy current losses from becoming intolerable. In the case of the

Eddy Currents: Theory, Modeling, and Applications, First Edition.
Sheppard J. Salon, M. V. K. Chari, Lale T. Ergene, David Burow, and Mark DeBortoli.
© 2024 The Institute of Electrical and Electronics Engineers, Inc. Published 2024 by John Wiley & Sons, Inc.

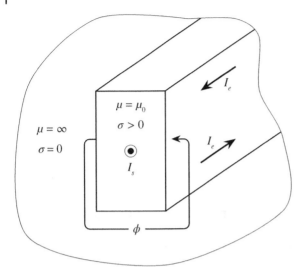

Figure 11.1 Source I_s and eddy I_e currents and cross-slot flux ϕ of a conductor in a slot.

squirrel cage induction motor, the relatively low slip frequency current carried by the rotor conductors at normal running speed does not require fine stranding or transposition. However, the frequency-dependent redistribution of net axial current produced by eddy currents makes the rotor resistance and leakage inductance, and hence the motor torque and current, speed-dependent. Designers exploit this phenomenon by creating special slot and bar shapes to customize motor performance for the needs of the application.

Field [61] showed that the case of the rectangular slot is amenable to straightforward analysis provided it can be assumed that the flux passes straight across the slot. This renders the problem one-dimensional: the magnetic field and flux density in the slot are purely x directed, the current is only z directed, and all three quantities vary only with y. While the assumption of infinite permeability of the iron ensures that the magnetic field is indeed purely x directed at the slot sides, there is no guarantee that this is the case across the entire width of the slot. In strict terms, symmetry about the vertical centerline of the slot is the only other simplifying assumption that can be invoked at this point.

Further analysis by Roth [66], using a two-dimensional vector potential solution, summarized by Hague [67, pp. 313–317] and with results described in Say [62, pp. 53–54] demonstrates that any deviation of the flux lines from a path straight across the slot is negligible in most practical situations. One exception is the case of high-voltage armature coils with thick insulation that creates a large space between the conductors and the slot boundary, where the solution shows a slight "hump" in the flux lines at the center of the slot. The effect is not

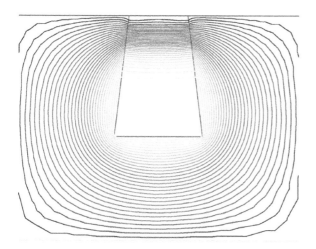

Figure 11.2 Finite element solution of flux lines in a trapezoidal slot for a frequency of 60HZ and conductivity IACS.

extreme, however. Another exception is when slot does not have parallel sides such as a trapezoidal slot. Results from a finite element solution for a slot of this shape are shown in Figure 11.2 where it is apparent that the flux lines are still approximately straight across the slot. We are therefore justified in proceeding with an analysis based on the one-dimensional assumption.

Figure 11.3 shows an elementary configuration of a single conductor in the slot of an electric machine. In general, the conductor does not quite fill the entire slot, due to the need for insulation. We, therefore, consider an equivalent simplified problem, where the conductor does occupy the entire width of the slot, but the current density is adjusted by the ratio of b_c/b_s so that the total current is the same as in the real problem. Then we can write the following field equations for current that varies sinusoidally in time with angular frequency ω, $\mathbf{J} = \hat{\mathbf{u}}_z J_z \cos \omega t$.

$$\nabla \times \mathbf{H} = \mathbf{J} \qquad \text{reduces to} \qquad \frac{\partial H_x}{\partial y} = -\frac{b_c}{b_s} J \tag{11.1}$$

$$\nabla \times \mathbf{E} = -\frac{\partial \mathbf{B}}{\partial t} \qquad \text{reduces to} \qquad \frac{\partial E_z}{\partial y} = -j\omega B_x \tag{11.2}$$

Noting that $\mathbf{J} = \sigma \mathbf{E}$ and $\mathbf{B} = \mu_0 \mathbf{H}$ in the conductor, Equation (11.2) can be written as

$$\frac{\partial J}{\partial y} = -j\omega\sigma\mu_0 H_x \tag{11.3}$$

If we now differentiate Equation (11.1)

$$\frac{\partial^2 H_x}{\partial y^2} = -\frac{b_c}{b_s}\frac{\partial J}{\partial y} \tag{11.4}$$

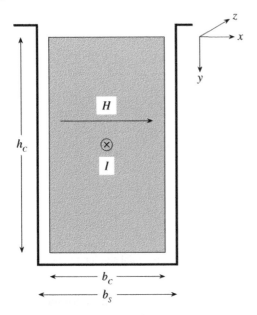

Figure 11.3 Conductor of width b_c in slot of width b_s.

and use the result of Equation (11.3), we obtain

$$\frac{\partial^2 H_x}{\partial y^2} = j\omega\sigma\mu_0 H_x \frac{b_c}{b_s} \tag{11.5}$$

For simplicity going forward, the ratio of the conductor width to the slot width is built into a new modified conductivity, skin depth, and γ as follows:

$$\sigma' = \frac{b_c}{b_s}\sigma \tag{11.6}$$

$$\delta' = \sqrt{\frac{2}{\omega\mu_0\sigma'}} \tag{11.7}$$

$$\gamma = \sqrt{j\omega\mu_0\sigma'} = \frac{1+j}{\delta'} \tag{11.8}$$

Then Equation (11.5) can be written as

$$\frac{\partial^2 H_x}{\partial y^2} = \gamma^2 H_x \tag{11.9}$$

The general solution to this equation can be written in the form

$$H_x(y) = Ce^{\gamma(h_c-y)} - De^{-\gamma(h_c-y)} \tag{11.10}$$

where C and D are constants of integration which are determined by the boundary conditions. At $y = 0$, the top of the conductor, and at $y = h_c$, the bottom of the conductor, we have, respectively

$$H_x(0) = \frac{I}{b_s} \quad \text{and} \quad H_x(h_c) = 0 \tag{11.11}$$

where I is the peak value of the total current in the conductor. Therefore, from Equations (11.10) and (11.11), we have

$$C e^{\gamma h_c} - D e^{-\gamma h_c} = \frac{I}{b_s} \tag{11.12}$$

$$C - D = 0 \tag{11.13}$$

Combining these equations and solving for C and D we obtain

$$\underbrace{C(e^{\gamma h_c} - e^{-\gamma h_c})}_{2\sinh \gamma h_c} = \frac{I}{b_s} \tag{11.14}$$

$$C = \frac{I}{2b_s \sinh \gamma h_c} = D \tag{11.15}$$

Substituting this result into Equation (11.10), we have

$$H_x(y) = \frac{I}{2b_s \sinh \gamma h_c} \underbrace{[e^{\gamma(h_c-y)} - e^{-\gamma(h_c-y)}]}_{2\sinh \gamma(h_c - y)} = \frac{I \sinh\left[\gamma(h_c - y)\right]}{b_s \sinh \gamma h_c} \tag{11.16}$$

We now find the eddy current density in the conductor. From Equations (11.1) and (11.16)

$$J(y) = -\frac{b_s}{b_c} \frac{\partial H_x}{\partial y} = \frac{I\gamma \cosh\left[\gamma(h_c - y)\right]}{b_c \sinh \gamma h_c} \tag{11.17}$$

This is a complex quantity. To obtain expressions for its magnitude, it is convenient to replace γ with its original definition from Equation (11.8)

$$\gamma = \frac{1+j}{\delta'} = \beta(1+j) \tag{11.18}$$

where

$$\beta = \frac{1}{\delta'} \tag{11.19}$$

Then Equation (11.17) can be written as

$$J(y) = \frac{I\beta}{b_c}(1+j) \frac{\cosh\left[\beta(1+j)(h_c - y)\right]}{b_c \sinh\left[\beta(1+j)h_c\right]} \tag{11.20}$$

which separates the real and imaginary parts of the arguments of the hyperbolic function, thereby facilitating use of the identities

$$\sinh(\alpha + j\beta) = \sinh \alpha \cos \beta + j \cosh \alpha \sin \beta \tag{11.21}$$

$$\cosh(\alpha + j\beta) = \cosh \alpha \cos \beta + j \sinh \alpha \sin \beta \tag{11.22}$$

Equation (11.20) can then be written as

$$J(y) = \frac{I\beta}{b_c}(1+j)\frac{\cosh\left[\beta\left(h_c - y\right)\right]\cos\left[\beta\left(h_c - y\right)\right] + j\sinh\left[\beta\left(h_c - y\right)\right]\sin\left[\beta\left(h_c - y\right)\right]}{\sinh\beta h_c \cos\beta h_c + j\cosh\beta h_c \sin\beta h_c} \tag{11.23}$$

The magnitude of the current density is then

$$|J(y)| = \frac{|I|\,\beta\sqrt{2}}{b_c}$$

$$\sqrt{\frac{\cosh^2\left[\beta\left(h_c - y\right)\right]\cos^2\left[\beta\left(h_c - y\right)\right] + \sinh^2\left[\beta\left(h_c - y\right)\right]\sin^2\left[\beta\left(h_c - y\right)\right]}{\sinh^2\beta h_c \cos^2\beta h_c + \cosh^2\beta h_c \sin^2\beta h_c}} \tag{11.24}$$

This expression can be reduced further by applying the identities

$$\cos^2 x = \frac{1 + \cos 2x}{2} \quad \text{and} \quad \sin^2 x = \frac{1 - \cos 2x}{2} \tag{11.25}$$

$$\cosh^2 x = \frac{\cosh 2x + 1}{2} \quad \text{and} \quad \sinh^2 x = \frac{\cosh 2x - 1}{2} \tag{11.26}$$

which yields

$$|J(y)| = \frac{|I|\,\beta\sqrt{2}}{b_c}\sqrt{\frac{\cosh\left[2\beta\left(h_c - y\right)\right] + \cos\left[2\beta\left(h_c - y\right)\right]}{\cosh\left(2\beta h_c\right) - \cos\left(2\beta h_c\right)}} \tag{11.27}$$

If the current I were distributed uniformly over the conductor area, as would be the case for DC current, the current density would be

$$J_{DC} = \frac{|I|}{b_c h_c} \tag{11.28}$$

The current density in the AC case may then be expressed in normalized form relative to the DC case by combining Equations (11.27) and (11.28)

$$\frac{|J(y)|}{J_{DC}} = h_c \beta\sqrt{2}\sqrt{\frac{\cosh\left[2\beta\left(h_c - y\right)\right] + \cos\left[2\beta\left(h_c - y\right)\right]}{\cosh\left(2\beta h_c\right) - \cos\left(2\beta h_c\right)}} \tag{11.29}$$

This function is illustrated in Figure 11.4 which shows the relative magnitude of current density vs. depth in a slot for two different frequencies, with a conductivity of IACS (International Annealed Copper Standard) ($5.8 \times 10^7 \mathrm{Sm}^{-1}$). This result

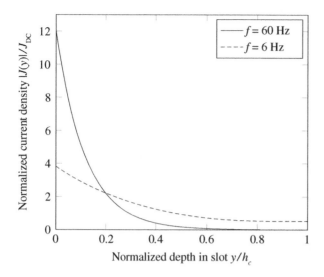

Figure 11.4 Current density vs. slot depth for $\sigma =$ IACS, $b_s = 0.028$m, $b_c = 0.025$m, $h_c = 0.076$m.

is consistent with Figure 11.1, which shows that application of Faraday's Law to cross-slot leakage flux tends to increase the current at the top of the slot and decrease it at the bottom.

The power loss in the volume of the conductor can be determined using Equation (11.27)

$$
\begin{aligned}
P &= \frac{1}{2} \int_0^{h_c} \frac{|J(y)|^2}{\sigma} b_c \ell \, dy \\
&= \frac{|I|^2 \beta^2 \ell}{\sigma b_c} \int_0^{h_c} \frac{\cosh\left[2\beta\left(h_c - y\right)\right] + \cos\left[2\beta\left(h_c - y\right)\right]}{\cosh\left(2\beta h_c\right) - \cos\left(2\beta h_c\right)} dy \\
&= \frac{|I|^2 \beta \ell}{2\sigma b_c} \frac{\sinh\left(2\beta h_c\right) + \sin\left(2\beta h_c\right)}{\cosh\left(2\beta h_c\right) - \cos\left(2\beta h_c\right)}
\end{aligned}
\tag{11.30}
$$

If the current I were distributed uniformly over the conductor area, as would be the case for DC current, the loss in the conductor would be

$$
P_{DC} = \frac{I_{DC}^2 \ell}{\sigma h_c b_c} = \frac{|I^2| \ell}{2\sigma h_c b_c}
\tag{11.31}
$$

where the factor of 2 is due to the definition of I as a peak current in the AC case, and the equivalent DC current from a loss standpoint is the rms value $I/\sqrt{2}$. The ratio of the loss in the AC case to that of the DC case is then:

$$
\frac{P_{AC}}{P_{DC}} = \beta h_c \frac{\sinh\left(2\beta h_c\right) + \sin\left(2\beta h_c\right)}{\cosh\left(2\beta h_c\right) - \cos\left(2\beta h_c\right)}
\tag{11.32}
$$

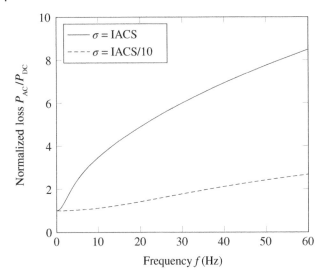

Figure 11.5 Normalized loss vs. frequency for $b_s = 0.028$m, $b_c = 0.025$m, $h_c = 0.076$m.

This expression is plotted in Figure 11.5 for two different conductivities. The greater conductivity results in greater concentration of the current at the top of the slot and higher overall loss with increasing frequency.

This result shows that coil losses due to eddy currents can become intolerable at power line frequencies. In AC machine armatures, such losses are typically mitigated by subdividing the conductor vertically into separate strands which are insulated from each other within the slot (but ultimately all in parallel) as shown Figure 11.6, or by use of conductors with fine, insulated strands called Litz wire.

The form of Equation (11.30) shows that the loss varies as the product βh_c which is the ratio of the conductor height to the effective skin depth. The loss relationship is controlled by the quantity

$$\frac{\sinh\left(2\beta h_c\right) + \sin\left(2\beta h_c\right)}{\cosh\left(2\beta h_c\right) - \cos\left(2\beta h_c\right)} \tag{11.33}$$

which is plotted in Figure 11.7. It is seen that this function has a minimum at $\beta h_c = \pi/2$ which corresponds to a conductor height of $\pi/2$ times the effective skin depth. This is known as the *critical conductor height*,

$$h_{c,crit} = \frac{\pi}{2}\delta' \tag{11.34}$$

11.1.2 Rectangular Hollow Conductor in a Slot

There are applications for a hollow rectangular conductor in a slot. These may exist in the slots of electric machines that are inner cooled with gas or liquid.

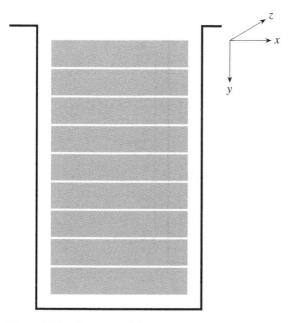

Figure 11.6 Conductor divided into strands.

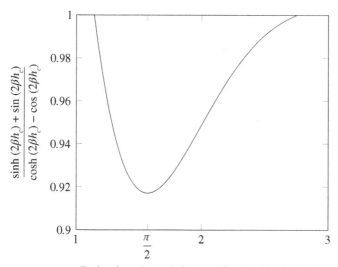

Ratio of conductor height to effective skin depth βh_c

Figure 11.7 Loss function vs. ratio of conductor height to effective skin depth.

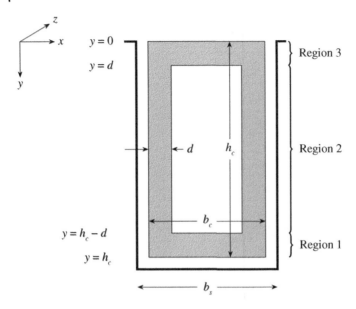

Figure 11.8 Geometry of rectangular hollow conductor.

As shown in Ostović [65, pp. 281–284], the problem may be solved by approximating it as a one-dimensional case. The assumption is that all of the flux is in the x direction as indicated in Figure 11.8.

The solution can be obtained using the relationships developed in Section 11.1.1 for a solid conductor in an infinitely permeable slot. We will start with the general solution of Equation (11.10) which is repeated below with γ replaced by $\beta(1+j)$ per Equation (11.18)

$$H_x(y) = Ce^{\beta(1+j)(h_c-y)} - De^{-\beta(1+j)(h_c-y)} \tag{11.35}$$

If we now consider the hollow rectangular conductor in Figure 11.8, we can view this as a problem with three regions: the bottom conductor (Region 1), the middle two vertical conductors (Region 2), and the top conductor (Region 3). The factor β may be different for the different regions. Recall from Equations (11.19), (11.7), and (11.6) that

$$\beta = \frac{1}{\delta'} = \sqrt{\frac{\omega\sigma'\mu}{2}} = \sqrt{\frac{\omega\sigma b_c\mu}{2b_s}} \tag{11.36}$$

This expression applies directly for Regions 1 and 3, where the conductor width is simply b_c:

$$\beta_1 = \beta_3 = \sqrt{\frac{\omega\sigma b_c\mu}{2b_s}} \tag{11.37}$$

In Region 2 of the hollow conductor, the conductor width is not b_c but rather $2d$; therefore,

$$\beta_2 = \sqrt{\frac{\omega\sigma 2d\mu}{b_s}} \tag{11.38}$$

With our one-dimensional assumption, the magnetic field and current density in the three regions are each described by one equation. These equations have constants C and D which may be different for the three regions. We will denote them as C_1, C_2, C_3, D_1, D_2, and D_3. These six unknowns require six equations, which are found from the boundary and interface conditions. For the magnetic field H, we require that the field be zero at the bottom boundary of Region 1 where $y = h_c$, which yields, from Equation (11.35)

$$C_1 - D_1 = 0 \tag{11.39}$$

We also know that the field at the top of the conductor in Region 3 where $y = 0$, must equal the total current I divided by the slot width b_s. Again from Equation (11.35):

$$C_3 e^{\beta_3(1+j)h_c} - D_3 e^{-\beta_3(1+j)h_c} = \frac{I}{b_s} \tag{11.40}$$

The magnetic field must be continuous at the interface of Regions 1 and 2 ($y = h_c - d$) and the interface of Regions 2 and 3 ($y = d$). Using these conditions in Equation (11.35) yields

$$C_1 e^{\beta_1(1+j)d} - D_1 e^{-\beta_1(1+j)d} = C_2 e^{\beta_2(1+j)d} - D_2 e^{-\beta_2(1+j)d} \tag{11.41}$$

$$C_2 e^{\beta_2(1+j)(h_c-d)} - D_2 e^{-\beta_2(1+j)(h_c-d)} = C_3 e^{\beta_3(1+j)(h_c-d)} - D_3 e^{-\beta_3(1+j)(h_c-d)} \tag{11.42}$$

Finally, we will consider the current density at the interfaces. The general governing equation is obtained from Equation (11.1):

$$J_z(y) = -\frac{b_s}{b_c}\frac{\partial H_x}{\partial y} = \frac{\beta(1+j)b_s}{b_c}\left[Ce^{\beta(1+j)(h_c-y)} + De^{-\beta(1+j)(h_c-y)}\right] \tag{11.43}$$

The current density must be continuous at the interfaces of Regions 1 and 2 and Regions 2 and 3.[1] Using Equation (11.43) at these interfaces:

$$\frac{\beta_1}{b_c}\left[C_1 e^{\beta_1(1+j)d} + D_1 e^{-\beta_1(1+j)d}\right] = \frac{\beta_2}{2d}\left[C_2 e^{\beta_2(1+j)d} + D_2 e^{-\beta_2(1+j)d}\right] \tag{11.44}$$

$$\frac{\beta_2}{2d}\left[C_2 e^{\beta_2(1+j)(h_c-d)} + D_2 e^{-\beta_2(1+j)(h_c-d)}\right] = \frac{\beta_3}{b_c}\left[C_3 e^{\beta_3(1+j)(h_c-d)} + D_3 e^{-\beta_3(1+j)(h_c-d)}\right] \tag{11.45}$$

---

1 In fact, it is the electric field that must be continuous at these boundaries. In the present case, the conductivity is the same in the three regions, so the current density is continuous. In a case where the regions have different conductivities, the current density would be in the ratios of the conductivities.

The six Equations (11.39)–(11.42) and (11.44)–(11.45) can then be placed in matrix equation form to solve for the C and D constants in the three regions simultaneously:

$$[A]\{X\} = \{F\} \tag{11.46}$$

The matrix $[A]$ can be written as follows, using χ for $1 + j$ and h_{cd} for $h_c - d$:

$$
\begin{bmatrix}
1 & -1 & 0 & 0 & 0 & 0 \\
0 & 0 & 0 & 0 & e^{\beta_3 \chi h_c} & -e^{-\beta_3 \chi h_c} \\
e^{\beta_1 \chi d} & -e^{-\beta_1 \chi d} & -e^{\beta_2 \chi d} & e^{-\beta_2 \chi d} & 0 & 0 \\
0 & 0 & e^{\beta_2 \chi h_{cd}} & -e^{-\beta_2 \chi h_{cd}} & -e^{\beta_3 \chi h_{cd}} & e^{-\beta_3 \chi h_{cd}} \\
\dfrac{\beta_1}{b_c} e^{\beta_1 \chi d} & \dfrac{\beta_1}{b_c} e^{-\beta_1 \chi d} & -\dfrac{\beta_2}{2d} e^{\beta_2 \chi d} & \dfrac{\beta_2}{2d} e^{-\beta_2 \chi d} & 0 & 0 \\
0 & 0 & \dfrac{\beta_2}{2d} e^{\beta_2 \chi h_{cd}} & \dfrac{\beta_2}{2d} e^{-\beta_2 \chi h_{cd}} & -\dfrac{\beta_3}{b_c} e^{\beta_3 \chi h_{cd}} & -\dfrac{\beta_3}{b_c} e^{-\beta_3 \chi h_{cd}}
\end{bmatrix}
\tag{11.47}
$$

and $\{X\}$ and $\{F\}$ are

$$\{X\} = \left\{ C_1, D_1, C_2, D_2, C_3, D_3 \right\}^T \tag{11.48}$$

$$\{F\} = \left\{ 0, \frac{I}{b_s}, 0,0, 0,0 \right\}^T \tag{11.49}$$

As a first check of the formulation, it is confirmed that it yields the same result as Equation (11.29) for the limiting case where $2d = b_c$, i.e. a solid rectangular conductor. As shown in Figure 11.9, both formulations give the same current density

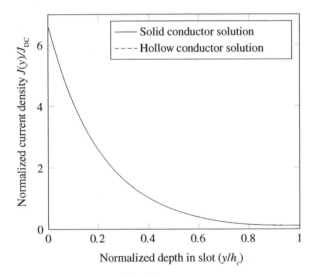

Figure 11.9 Current density vs. depth in slot, for limiting case of solid conductor, computed with solid and hollow conductor formulations, $\sigma = \text{IACS}$, $h_c = 0.04$m, $b_c = b_s = 0.03$m, $d = 0.015$m.

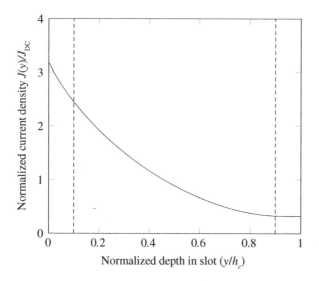

Figure 11.10 Current density vs. depth in slot for hollow conductor, σ = IACS, h_c = 0.04m, $b_c = b_s$ = 0.03m, d = 0.004m. Dashed lines show beginning and end of hollow region.

solution for this case. In Figure 11.10, the current density is shown for a hollow conductor with wall thickness of one-tenth of the conductor height.

11.1.3 Slice Method for Rectangular Slot

There are a number of skin effect problems that can be reduced to a coupled circuit analysis. One practical example is to find the resistance and leakage reactance of a squirrel cage induction motor bar. Induction motor rotor bars may have complicated shapes which make analytical solutions impossible. The effective resistance and inductance are a function of frequency and this is quite important since the starting resistance, where line frequency currents are in the bars, largely determines the starting torque while the reactance at this frequency limits the inrush current. When the induction machine is at load, the frequency on the rotor is quite small (slip frequency) and the resistance and reactance are smaller. The motor efficiency will depend on this value of the resistance.

In order to understand the process, we will first consider the simpler problem of finding the self and mutual inductance in a rectangular slot with conductors carrying uniform current density. This is illustrated in Figure 11.11.

The width of the slot is b and the height of the bottom conductor is h_2 and the top conductor is h_1. The height of the slot above the conductor is h_0. We will make the simplifying assumptions that the sides of the teeth are infinitely permeable and

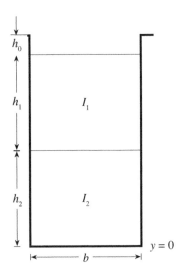

Figure 11.11 Rectangular slot with uniform current.

that the slot is narrow compared to the depth. In this case, the flux will be directed across the slot in the x direction. Applying Ampere's law

$$\oint H \cdot d\ell = I_{\text{enclosed}} \tag{11.50}$$

First, we will assume that there is uniformly-distributed current I_2 in the bottom conductor and no current in the top conductor. We see that by applying Equation (11.50) to different paths, the magnetic field H is zero at the bottom of the slot (no current enclosed) and increases linearly to the top of the conductor (all current enclosed) where it reaches a maximum. The resulting cross-slot flux density $B(y) = \mu_0 H(y)$ is illustrated in Figure 11.12.

$$H(y) = \begin{cases} \dfrac{I_2}{b}\left(\dfrac{y}{h_2}\right) & 0 \le y \le h_2 \\ \dfrac{I_2}{b} & y > h_2 \end{cases} \tag{11.51}$$

We now apply uniform current I_1 to the top conductor. Using Ampere's law, we conclude that there is no field below the top conductor. The field rises linearly in the top conductor and is constant above the top conductor. This is shown in Figure 11.13.

To find the components of inductance attributable to the cross-slot flux, we find the flux linkage of the individual conductors. For the case of the bottom conductor carrying current I_2, we use the flux density distribution of Figure 11.12. All of the flux above the top of the conductor links the entire coil and as such contributes

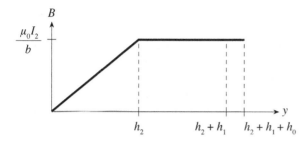

Figure 11.12 Cross-slot flux density for current in bottom conductor.

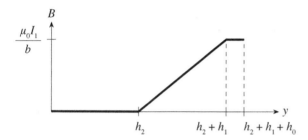

Figure 11.13 Cross-slot flux density for current in top conductor.

directly to the flux linkage. The flux crossing the slot in the region of the conductor itself links a fraction of the conductor. This fraction is zero at the bottom of the conductor and increases linearly to 1.0 at the top of the conductor. The flux density in this region is also increasing linearly from zero to a maximum of $B = \mu_0 I_2/b$. To find the flux linkage of the bottom conductor then we integrate the flux density times the fraction of the bottom conductor linked over the entire slot. Therefore in the squared term in the first integral of Equation (11.52), one factor of y/h_2 represents the variation of flux density over the conductor and the other factor of y/h_2 represents the fraction of the conductor linked. The flux linkage of the lower conductor due to current in the lower conductor only is then

$$\lambda_{22} = \frac{\mu_0 I_2}{b} \left[\int_0^{h_2} \left(\frac{y}{h_2} \right)^2 dy + \int_{h_2}^{h_2+h_1} dy + \int_{h_2+h_1}^{h_2+h_1+h_0} dy \right] \tag{11.52}$$

This gives

$$\lambda_{22} = \frac{\mu_0 I_2}{b} \left(\frac{h_2}{3} + h_1 + h_0 \right) \tag{11.53}$$

A similar analysis is used to obtain the flux linkage of the upper conductor due to current in the upper conductor only:

$$\lambda_{11} = \frac{\mu_0 I_1}{b} \left(\frac{h_1}{3} + h_0 \right) \tag{11.54}$$

Note that the region below a conductor has no effect on the self-flux linkage as there is no self-flux density in this region.

To find the mutual flux linkage, we can either excite the top conductor and find the flux linking the bottom conductor or vice versa. In this case, there is only one factor of y/h in the integral. If we excite the bottom conductor, the flux density in the top conductor is constant but the fraction of conductor linked has this linear variation. If we excite the top conductor, all of the flux completely links the bottom conductor but the flux density is linearly increasing over the top conductor area. The result in either case is

$$\lambda_{12} = \lambda_{21} = \frac{\mu_0 I}{b}\left(\frac{h_1}{2} + h_0\right) \tag{11.55}$$

To find the inductance per unit length, we divide the flux linkage equations by the current:

$$L_{22} = \frac{\mu_0}{b}\left(\frac{h_2}{3} + h_1 + h_0\right) \tag{11.56}$$

$$L_{11} = \frac{\mu_0}{b}\left(\frac{h_1}{3} + h_0\right) \tag{11.57}$$

$$L_{12} = L_{21} = \frac{\mu_0}{b}\left(\frac{h_1}{2} + h_0\right) \tag{11.58}$$

We can now generalize this result to analyze the eddy current behavior of a solid rotor conductor in a squirrel cage induction motor. We will divide the slot into a number of vertical sections (for example 5, as shown in Figure 11.14) and use Equations (11.56)–(11.58). In the bottom-most conductor, assume we have a

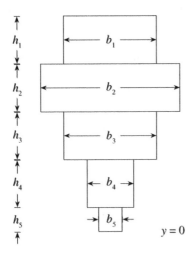

Figure 11.14 Squirrel cage rotor conductor divided into five sections.

uniform current of value I_5. The flux density in this region due only to the current in section 5 is

$$B_{55}(y) = \frac{\mu_0 I_5 y}{b_5 h_5} \tag{11.59}$$

The flux linkage of section 5 conductor, only considering the flux crossing that section, is then

$$\frac{\mu_0 I_5}{b_5} \int_0^{h_5} \left(\frac{y}{h_5} \right)^2 dy = \frac{\mu_0 I_5 h_5}{3 b_5} \tag{11.60}$$

We can now add the contributions of the flux above section 5 and since we are interested in the inductance, we must divide by the current. We get, for the total self-inductance per unit depth

$$L_{55} = \mu_0 \left(\frac{h_5}{3b_5} + \frac{h_4}{b_4} + \frac{h_3}{b_3} + \frac{h_2}{b_2} + \frac{h_1}{b_1} \right) \tag{11.61}$$

Similarly

$$L_{44} = \mu_0 \left(\frac{h_4}{3b_4} + \frac{h_3}{b_3} + \frac{h_2}{b_2} + \frac{h_1}{b_1} \right) \tag{11.62}$$

$$L_{33} = \mu_0 \left(\frac{h_3}{3b_3} + \frac{h_2}{b_2} + \frac{h_1}{b_1} \right) \tag{11.63}$$

$$L_{22} = \mu_0 \left(\frac{h_2}{3b_2} + \frac{h_1}{b_1} \right) \tag{11.64}$$

$$L_{11} = \mu_0 \left(\frac{h_1}{3b_1} \right) \tag{11.65}$$

Following our example, from Equation (11.58) the mutual inductance between sections 4 and 5 is

$$L_{45} = L_{54} = \mu_0 \left(\frac{h_4}{2b_4} + \frac{h_3}{b_3} + \frac{h_2}{b_2} + \frac{h_1}{b_1} \right) \tag{11.66}$$

and then

$$L_{35} = L_{53} = \mu_0 \left(\frac{h_3}{2b_3} + \frac{h_2}{b_2} + \frac{h_1}{b_1} \right) \tag{11.67}$$

$$L_{25} = L_{52} = \mu_0 \left(\frac{h_2}{2b_2} + \frac{h_1}{b_1} \right) \tag{11.68}$$

$$L_{15} = L_{51} = \mu_0 \left(\frac{h_1}{2b_1} \right) \tag{11.69}$$

We also make the important observation that

$$L_{15} = L_{14} = L_{13} = L_{12} \tag{11.70}$$

In other words, it does not make any difference to the flux linkage of section 1, whether there is 1 A in sections 2, 3, 4, or 5. All produce exactly the same results as far as section 1 is concerned. Similar reasoning leads to

$$L_{23} = L_{24} = L_{25} \tag{11.71}$$

and so on.

With all of the circuit elements computed, we can write the coupled circuit equations for the five conducting elements, where $X = \omega L$:

$$\begin{pmatrix} Z_1 & jX_{12} & jX_{13} & jX_{14} & jX_{15} \\ jX_{12} & Z_2 & jX_{23} & jX_{24} & jX_{25} \\ jX_{13} & jX_{23} & Z_3 & jX_{34} & jX_{35} \\ jX_{14} & jX_{24} & jX_{34} & Z_4 & jX_{45} \\ jX_{15} & jX_{25} & jX_{35} & jX_{45} & Z_5 \end{pmatrix} \begin{pmatrix} I_1 \\ I_2 \\ I_3 \\ I_4 \\ I_5 \end{pmatrix} = \begin{pmatrix} V \\ V \\ V \\ V \\ V \end{pmatrix} \tag{11.72}$$

where the diagonal term, $Z_i = R_i + jX_{ii}$.

The resistance of these sections per unit axial length will be

$$R_i = \frac{1}{\sigma_i b_i h_i} \tag{11.73}$$

where σ_i is the conductivity of the section, b_i is the width of the section and h_i is the height of the section.

If there is, for example, a space with no conductor as in a recessed slot, the equations will still work with infinite (or a very high value) of resistance in the open section. The vertical branch in the circuit below can be removed but the height of the section will be included in the inductances.

Equation (11.72) can be conveniently represented by the ladder network of Figure 11.15. The resistances are in all cases the dc resistance of the section. The reactances are made of linear combinations of the self and mutual reactances as

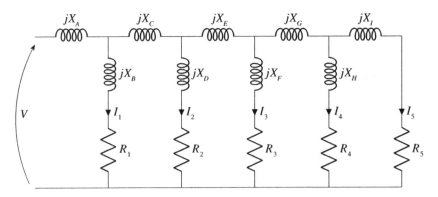

Figure 11.15 Equivalent circuit for five section slot.

follows. If we consider the loop equation for current I_1, we see that the voltage drop is

$$V = R_1 I_1 + j I_1 X_B + j(I_1 + I_2 + I_3 + I_4 + I_5) X_A \tag{11.74}$$

By rearranging this equation as

$$V = R_1 I_1 + j I_1 (X_B + X_A) + j(I_2 + I_3 + I_4 + I_5) X_A \tag{11.75}$$

it is apparent that X_A represents the mutuals from sections 2–5 to section 1:

$$X_A = X_{12} = X_{13} = X_{14} = X_{15} = \omega \mu_0 \frac{h_1}{2b_1} \tag{11.76}$$

and $X_B + X_A$ represents the self term for section 1:

$$X_B + X_A = X_{11} \tag{11.77}$$

or

$$X_B = X_{11} - X_A = X_{11} - X_{12} = -\omega \mu_0 \frac{h_1}{6b_1} \tag{11.78}$$

This satisfies the first row of Equation (11.72).

The remainder of the vertical reactances are

$$X_D = -\omega \mu_0 \frac{h_2}{6b_2} \tag{11.79}$$

$$X_F = -\omega \mu_0 \frac{h_3}{6b_3} \tag{11.80}$$

$$X_H = -\omega \mu_0 \frac{h_4}{6b_4} \tag{11.81}$$

and for the remainder of the horizontal reactances

$$X_C = \omega \mu_0 \left(\frac{h_1}{2b_1} + \frac{h_2}{2b_2} \right) \tag{11.82}$$

$$X_E = \omega \mu_0 \left(\frac{h_2}{2b_2} + \frac{h_3}{2b_3} \right) \tag{11.83}$$

$$X_I = \omega \mu_0 \left(\frac{h_4}{2b_4} + \frac{h_5}{2b_5} \right) \tag{11.84}$$

As a check, consider current only in loop 2. The current passes through reactances X_A, X_C, and X_D. The sum of these three should then be the self reactance of section 2. Adding these three components gives $\omega \mu_0 (h_2/3b_2 + h_1/b_1)$ which is the expected result.

To understand the application of this method, we present a numerical example. In Figure 11.16, we see a motor slot with a partially closed top and a tapered bottom. The slot is assumed to be completely filled with solid copper of conductivity IACS ($5.8 \times 10^7 \text{Sm}^{-1}$), and the frequency is 60 Hz. The conductor is divided into

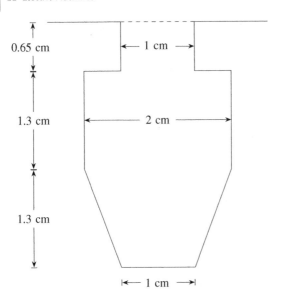

Figure 11.16 Geometry of induction motor slot.

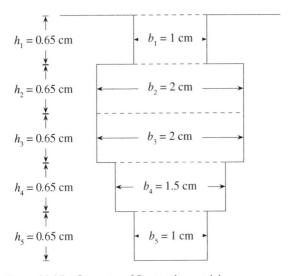

Figure 11.17 Geometry of five section model.

five coupled sections approximated with parallel sides. For the tapered sections, we take the average width. This is shown in Figure 11.17. Using the analysis above we obtain the equivalent circuit of Figure 11.18.

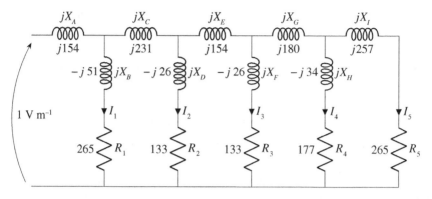

Figure 11.18 Equivalent circuit for the five section slot model (impedances in $\mu\Omega m^{-1}$).

From Equation (11.72) and the equations that precede it, the impedance matrix at 60HZ is found to be

$$Z = \begin{bmatrix} 265+j103 & j154 & j154 & j154 & j154 \\ j154 & 133+j359 & j385 & j385 & j385 \\ j154 & j385 & 133+j513 & j539 & j539 \\ j154 & j385 & j539 & 177+j684 & j719 \\ j154 & j385 & j539 & j719 & 265+j924 \end{bmatrix} \mu\Omega m^{-1}$$

(11.85)

If we now apply $1\,V\,m^{-1}$ to the circuit model we obtain for the five unknown currents

$$I = \begin{pmatrix} 2210-j675 \\ 562-j1569 \\ -389-j729 \\ -337-j65 \\ -146+j113 \end{pmatrix}$$

(11.86)

Adding these terms to find the total current, we find

$$I_{slot} = (1900-j2910)A$$

(11.87)

Since we have applied $1\,V\,m^{-1}$, the inverse of the current is the slot impedance per unit length

$$Z_{slot} = (157+j241)\mu\,\Omega m^{-1}$$

(11.88)

The DC resistance per unit axial length in this case is

$$R_{DC} = \frac{1}{\sum_{i=1}^{5} \sigma_i b_i h_i} = 35\mu\Omega m^{-1}$$

(11.89)

while the AC resistance is seen from Equation (11.88) to be $157\mu\Omega m^{-1}$, giving a ratio of 4.44. If we reduce the frequency to 1 Hz then the resistive part slot of the impedance becomes $35\mu\Omega m^{-1}$ so we get the DC resistance as expected. The reactive part also agrees with the theoretical value of reactance for a uniform current distribution.

11.1.4 Finite Difference Equivalent Circuit Analogy

We have seen in Section 5.1 that the finite difference equations for the two-dimensional magnetic vector potential eddy current problem resulted in an equivalent circuit of resistances and capacitances. In this analog circuit, the nodal voltage is the magnetic vector potential and the eddy current is the current to ground in the capacitances. The resistances in the analog are the magnetic circuit permeances at right angles to the magnetic flux. The current in the analog problem represents the magneto-motive force (total ampere-turns) producing the flux. The circuit elements are illustrated in Figure 11.19.

The resistors are found as the permeance of the flux paths. If we divide the slot into five sections as we did above, the permeances are

$$R_i = \mu_0 \frac{h_i}{b_i} \tag{11.90}$$

per meter of depth.

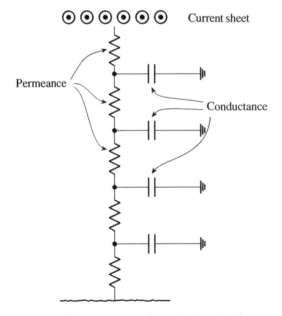

Figure 11.19 Vector potential equivalent circuit.

The capacitances are found as

$$C_i = \sigma_i h_i b_i \tag{11.91}$$

The circuit is solved by a standard nodal admittance analysis. The admittance matrix is a sparse matrix in which the diagonal elements for each node are the sum of all admittances connected to the node. The off-diagonal elements are the negative of the admittances between nodes. We can excite the circuit in a number of ways. For example, we can inject a current into node 1 and see how the current distributes in the slot. At high frequency, the upper capacitor is a low impedance path to ground and much of the current will leave through this path. This illustrates the skin effect in the slot. As the frequency drops toward zero, the currents will divide among the capacitors proportionally to the capacitance. This would result in a uniform current density as we would expect at DC.

The system of equations has the form

$$(I) = (Y)(A) \tag{11.92}$$

where the (A) vector is the magnetic vector potential at the nodes and (I) is the nodal current injection. We solve the set of linear simultaneous equation for the vector potential and then the current density in the slot is found by

$$J = \sigma E = \sigma \frac{\partial A}{\partial t} = j\omega\sigma A \tag{11.93}$$

The total current in each section is the current density times the area or

$$I_i = j\omega\sigma A_i S_i \tag{11.94}$$

where S_i is the cross-sectional area of section i. Note that σS_i is the value of capacitance so the current through the capacitor is the eddy current in the section of bar. The $j\omega$ term is of course for steady-state AC analysis. The model can be used in the time domain for transient problems as well. The current injection in this case is at the top of the slot where we inject 1 A into node 1 and find the currents in the slot sections. We then find the losses and effective resistance and reactance.

The results are very similar to those found in Section 11.1.3 with five slices. We note that for high accuracy, we should use more than five nodes or slice sections, as the slot depth is several times the skin depth, and we should have multiple divisions per skin depth for a good representation.

The magnitude of the currents at each node (the capacitor current), is plotted in Figure 11.20. When these five currents are added (as phasors), the sum is 1.0 A.

11.1.5 Slot with Arbitrary Shape

If the one-dimensional approximation is valid, we can obtain a solution for slots that have fairly arbitrary cross sections. This can be done in an approximate way

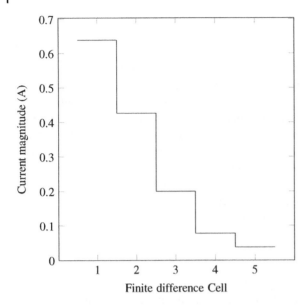

Figure 11.20 Magnitude of the current in each of five finite difference cells.

by making an equivalent circuit in which the slot is divided into a number of slices that can have arbitrary widths and depths as shown in Figure 11.21.

The magnetic field at the kth layer is produced by all currents below that layer. From Ampere's law, we find

$$H_k b_k = \sum_{p=1}^{k} i_p \tag{11.95}$$

The flux density at the top of the kth layer is then

$$B_k = \frac{\mu_0}{b_k} \sum_{p=1}^{k} i_p \tag{11.96}$$

We can write this in terms of the permeance of each layer where h_k is the height of the layer and ℓ is the length of the conductor.

$$\mathcal{P}_k = \frac{\mu_0 h_k \ell}{b_k} \tag{11.97}$$

Consider now the electromotive force (emf) induced in the loop formed by sections k and $k+1$, as illustrated in Figure 11.22. Considering the loop going from the center of layer k to the center of layer $k+1$, we now find the flux linking that loop. We will consider the flux linking the loop to be from three sources. The

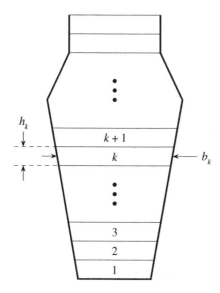

Figure 11.21 Non-uniform slot divided into slices.

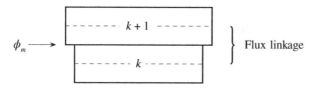

Figure 11.22 Flux linking loop formed by slice k and $k + 1$.

currents below layer k will produce flux linkage of

$$\phi_k = \left(\frac{\mathcal{P}_k + \mathcal{P}_{k+1}}{2} \right) \sum_{p=1}^{k-1} i_p \tag{11.98}$$

The flux linkage produced by the kth layer of current is found by referring to Figure 11.23. The flux density increases linearly over the slice region, going from zero at the bottom to $\mu_0 i_k / b_k$ at the top. Since we consider the loop from the midpoint of the layer, we need only consider the flux above that point. The average flux density is 3/4 of the peak value at the top. We can multiply this flux density by 1/2 the permeance of layer k to find the flux linking the loop up to the bottom of layer $k + 1$. The current in layer k also produces flux linkage in the bottom half of layer $k + 1$ which is included in the loop. The flux density in the region i_k is

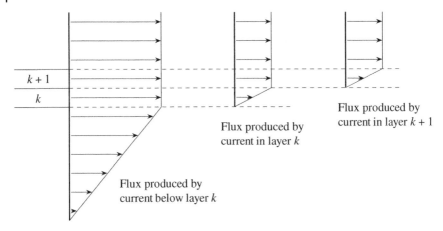

Figure 11.23 Flux produced by three sources.

constant and crosses half the area of layer $k + 1$. The contribution of the current in layer k is then

$$\phi_k = \left(\frac{3}{8}\mathcal{P}_k + \frac{1}{2}\mathcal{P}_{k+1}\right) i_k \tag{11.99}$$

The final contribution to the flux linkage is that produced by the current in the $k + 1$ layer. This produces flux in the lower half of the $k + 1$ slice. The average flux density produced by i_{k+1} in this region is $1/4$ of the maximum flux density produced by that current. The final term then is

$$\phi_k = \left(\frac{1}{8}\mathcal{P}_{k+1}\right) i_{k+1} \tag{11.100}$$

The total flux producing an emf in the loop is then the sum of the three contributions.

$$\frac{d}{dt}\left(\phi_{k,k+1}\right) + R_k i_k - R_{k+1} i_{k+1} = 0 \tag{11.101}$$

The resistance of each slice is the DC resistance and is found as

$$R_k = \frac{\ell}{\sigma h_k b_k} \tag{11.102}$$

where σ is the conductivity of the material.

While Equation (11.101) is valid for general excitation types, we will consider here steady-state ac excitation, in which case we replace d/dt by $j\omega$. Now the governing equation becomes

$$j\omega\left[\frac{\mathcal{P}_k + \mathcal{P}_{k+1}}{2}\sum_{p=1}^{k-1} i_p + \left(\frac{3}{8}\mathcal{P}_k + \frac{1}{2}\mathcal{P}_{k+1}\right) i_k + \frac{1}{8}\mathcal{P}_{k+1} i_{k+1}\right] + R_k i_k - R_{k+1} i_{k+1} = 0 \tag{11.103}$$

A permeance times ω can be thought of as a reactance and the equation becomes

$$j\frac{X_k + X_{k+1}}{2}\sum_{p=1}^{k-1}i_p + j\left(\frac{3}{8}X_k + \frac{1}{2}X_{k+1}\right)i_k + \frac{1}{8}jX_{k+1}i_{k+1} + R_k i_k - R_{k+1}i_{k+1} = 0$$

(11.104)

If we have n slices, there are $n-1$ loops and therefore $n-1$ equations. To make the problem well-posed, we add a constraint equation so that the sum of all $n-1$ loop currents is equal to a specified value. We accomplish this by setting all of the terms in the nth row of the system matrix equal to 1 and then setting the nth term of the input vector to the specified total slot current. This gives a final $n \times n$ system and makes the matrix non-singular.

The final system of equations has the form

$$(Z)(I_{\text{loop}}) = (I_{\text{in}})$$

(11.105)

where the matrix Z has the form

$$Z = \begin{pmatrix} R_1 + \frac{3}{8}jX_1 + \frac{1}{2}jX_2 & -R_2 + jX_2/8 & \cdots & 0 & 0 \\ j(X_2 + X_3)/2 & R_2 + \frac{3}{8}jX_2 + \frac{1}{2}jX_3 & -R_3 + \frac{1}{8}jX_3 & \cdots & 0 \\ \cdots & \cdots & \cdots & \cdots & \cdots \\ j(X_2 + X_3)/2 & j(X_3 + X_4)/2 & \cdots & R_{n-1} + \frac{3}{8}jX_{n-1} + \frac{1}{2}jX_n & -R_n + \frac{1}{8}jX_n \\ 1 & 1 & 1 & \cdots & 1 \end{pmatrix}$$

(11.106)

The I_{in} vector is all zeros, except for the nth term which is the complex total current in the slot. (In practice, the last row of the Z matrix may be scaled to the same order of magnitude as the other entries in the matrix, to better condition the matrix for solving, with the last entry of I_{in} compensated accordingly.)

Two numerical examples will illustrate the process. We first consider the solid rectangular conductor considered in Section 11.1.1. The conductor is 0.03 m wide and 0.04 m deep with conductivity $\sigma = \text{IACS} = 5.8 \times 10^7 \text{Sm}^{-1}$. The frequency is 60HZ. In Figure 11.24, the normalized current density vs. depth is plotted for the slice method and the closed-form solution given in Equation (11.29). As can be seen, the solutions are almost identical.

Now consider a trapezoidal conductor in a slot as shown in Figure 11.25. In this example, the slot is 1.0 m long, the slot width at the top is 0.02 m, and the width at the bottom is 0.03 m. The slot is 0.04 m deep. The conductivity of the slot conductor is $\sigma = \text{IACS} = 5.8 \times 10^7 \text{Sm}^{-1}$. The frequency is 60HZ and the total input current is $1.0 + j0.0$ A.

Figure 11.26 shows the current density in each layer with a total of 25 layers. The plot shows the real and imaginary components. As a check, the sum of all 25 currents is $1.00000 - j55.511 \times 10^{-18}$ A.

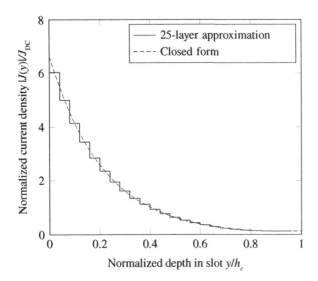

Figure 11.24 Current density vs. depth for rectangular slot, 25-layer approximation.

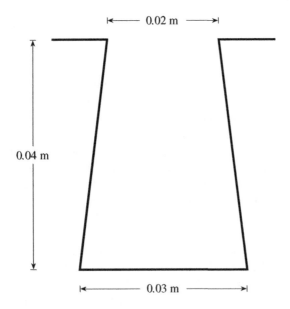

Figure 11.25 Slot with trapezoidal cross section.

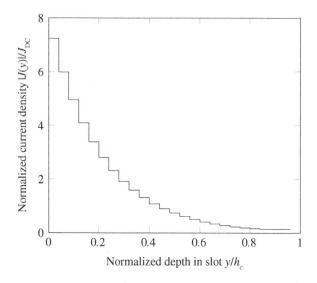

Figure 11.26 Current density in 25 layers of trapezoidal slot shown in Figure 11.25.

11.2 Solid Rotor Electric Machines

In this section, we shall discuss eddy current effects in rotating conducting structures subjected to a time-varying magnetic field from a stationary region such as an electric machine stator. The excitation current will be assumed to be sinusoidally distributed spatially. The analytical technique developed will be applied to a rotating electric machine for evaluating the magnetic and electric fields and the power loss.

In our example, the rotating electric machine has a conducting rotating structure, the rotor, enclosed by a stationary structure, the stator. Discrete conductors carrying electric currents can be placed both in the rotor and in the stator. The presence of these discrete conductors in the rotor and stator presents difficulties in electromagnetic analysis of these machines, especially for an asynchronous machine.

The separation of variables method can only be applied to field variables in space and time if the boundaries of the field problem are invariant in time. In the magnetostatic analysis of synchronous machines, time invariance of the field is obtained by rotating the rotor in synchronism with the rotating field produced by the stator. The field equations are then solved at one instant in time, which is valid at all instants of time. Some slight variations in the field may occur due to tooth ripple as the rotor advances through each slot pitch.

In an asynchronous machine, such as the induction motor, the problem can only be rigorously solved by time-stepping methods using numerical analysis techniques [23, 46]. For steady-state asynchronous problems, however, the time-stepping technique is time-consuming and is not the preferred choice. An alternate method is to devise a way of separating the space and time variables. This requires adopting a coordinate system, or frame of reference, which is fixed to the rotor (a rotating coordinate system), so that the rotor geometry can be made stationary. The basic problem, however, is to discover how the field equations are modified in a rotating coordinate system. In this section, we describe the method of deriving the field equations modified for a rotating coordinate system, and we present a technique for separating time and space variables. This example presents the closed-form solution of two-dimensional eddy current problems in cylindrical structures like a solid rotor induction motor. A closed-form solution of magnetic fields in induction motors provides a convenient method of field analysis, which can help in sizing studies, in design and evaluation of performance and in the determination of equivalent circuit parameters. Multilayered cylindrical geometry is used for the example. A two-dimensional parabolic partial differential equation is solved for each layer. From the analytical solution, eddy currents and eddy current losses are calculated in the solid conducting region and compared to the results of a finite element model. A summary of the main assumptions and limitations of this analysis is that:

- The machine is long and end-effects are neglected;
- The field problem is considered two-dimensional and linear;
- The permeability and conductivity are time invariant and independent of the field;
- Saturation and hysteresis effects are neglected;
- The stator is laminated and eddy currents are neglected;
- The stator source current is represented by a current sheet on the inner bore of the stator;
- The rotor is solid and the permeability and conductivity are isotropic and single valued;
- The field solution is obtained in cylindrical coordinates;
- The vector potential **A** has only a z-directed component equal to A_z, and;
- The flux-density has two components in the radial and peripheral directions: B_r and B_θ.

11.2.1 Induction Motor Model

The cross-sectional model of the solid rotor induction motor is shown in Figure 11.27. Region 1 is the rotor region with a relative permeability of μ_{r1}. This region is an eddy current region. Torque is created by the rotor magnetic field which arises from eddy currents in the rotor. Region 2, lying between r_1 and

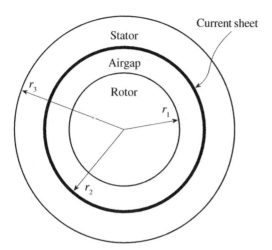

Figure 11.27 Induction motor model.

r_2, is free space with relative permeability of 1.0. This region represents the air gap between the stator and rotor. Region 3, lying between r_2 and r_3, is the stator region with a relative permeability of μ_{r3}. Beyond the stator, where $r > r_3$, is free space. The stator magnetic field passes through the rotor, crossing the air gap. The source current density, J_s, is applied as a current sheet between the air gap and the stator core [68, 69].

Region 1 is characterized by the two-dimensional diffusion equation in polar coordinates.

$$\frac{1}{\mu_{r1}\mu_0}\left[\frac{1}{r}\frac{\partial}{\partial r}\left(r\frac{\partial A}{\partial r}\right) + \frac{1}{r^2}\frac{\partial^2 A}{\partial \theta^2}\right] = j\omega\sigma A \tag{11.107}$$

Using the separation of variables technique described in Appendix B, Equation (11.107) can be formulated as two equations.

$$r^2\frac{\partial^2 R(r)}{\partial r^2} + r\frac{\partial R(r)}{\partial r} - (j\omega\mu_r\mu_0\sigma r^2 + n^2)R(r) = 0 \tag{11.108}$$

$$\frac{\partial^2 \Theta(\theta)}{\partial \theta^2} + n^2\Theta(\theta) = 0 \tag{11.109}$$

where the vector potential solution is given by

$$A = R(r)\Theta(\theta) \tag{11.110}$$

Using Equations (11.108) and (11.109) to find $R(r)$ and $\Theta(\theta)$ and substituting into Equation (11.110), we obtain the solution for the diffusion equation (11.107) as

$$A = \sum_{n=1}^{\infty}\left\{C_n\left[\mathrm{ber}_n(kr) + j\mathrm{bei}_n(kr)\right]\right.$$
$$\left. + D_n\left[\mathrm{ker}_n(kr) + j\mathrm{kei}_n(kr)\right]\right\}\left[E_n\cos n\theta + F_n\sin n\theta\right] \tag{11.111}$$

Because the field at $r = 0$ is finite and has only cosine components, we have $D_n = 0$ and $F_n = 0$. Therefore,

$$A = \sum_{n=1}^{\infty} \left\{ C_n \left[\text{ber}_n(kr) + j\text{bei}_n(kr) \right] \cos n\theta \right\} \tag{11.112}$$

where $k = \sqrt{\omega \mu_r \mu_0 \sigma}$.

Examining each region of the motor, the vector potential is evaluated as follows. In Region 4, which is the free space region outside the motor, where $r > r_3$,

$$A_4 = \sum_{n=1}^{\infty} D_{4n} r^{-n} \cos n\theta \tag{11.113}$$

In Region 3, which is the stator core region with no eddy currents, where $r_2 < r < r_3$,

$$A_3 = \sum_{n=1}^{\infty} C_{3n} r^{n} + D_{3n} r^{-n} \cos n\theta \tag{11.114}$$

In Region 2, which is the air-gap region where, $r_1 < r < r_2$,

$$A_2 = \sum_{n=1}^{\infty} C_{2n} r^{n} + D_{2n} r^{-n} \cos n\theta \tag{11.115}$$

Lastly, in Region 1, which is the eddy current region, where $0 < r < r_1$,

$$A_1 = \sum_{n=1}^{\infty} C_{1n} \left[\text{ber}_n(kr) + j\text{bei}_n(kr) \right] \cos n\theta \tag{11.116}$$

At each of the region boundaries, the conditions on the normal component of the **B** field and on the tangential components of the **H** field must be satisfied. Specifically, at $r = r_3$, $B_{r3} = B_{r4}$, and $H_{\theta 3} = H_{\theta 4}$. At $r = r_2$, $B_{r3} = B_{r2}$, and $H_{\theta 3} - H_{\theta 2} = J_s$. And, at $r = r_1$, $B_{r2} = B_{r1}$, and $H_{\theta 2} = H_{\theta 1}$.

From

$$\mathbf{B} = \nabla \times \mathbf{A} = \frac{1}{r} \frac{\partial \mathbf{A}}{\partial \theta} \hat{\mathbf{u}}_{\mathbf{r}} - \frac{\partial \mathbf{A}}{\partial \theta} \hat{\mathbf{u}}_{\theta} \tag{11.117}$$

expressions for the normal and tangential components of the flux density are found as

$$B_r = \frac{1}{r} \frac{\partial \mathbf{A}}{\partial \theta} \tag{11.118}$$

$$B_\theta = \frac{\partial \mathbf{A}}{\partial r} \tag{11.119}$$

Evaluating Equations (11.118) and (11.119) to find the field components in each of the regions of the motor, we find in Region 4,

$$B_{r4} = \sum_{n=1}^{\infty} -n D_{4n} r^{-n-1} \sin n\theta \tag{11.120}$$

$$H_{\theta 4} = \sum_{n=1}^{\infty} \frac{n}{\mu_0} D_{4n} r^{-n-1} \cos n\theta \tag{11.121}$$

In Region 3,

$$B_{r3} = \sum_{n=1}^{\infty} - n(C_{3n} r^{n-1} + D_{3n} r^{-n-1}) \sin n\theta \tag{11.122}$$

$$H_{\theta 3} = \sum_{n=1}^{\infty} - \frac{n}{\mu_{r3} \mu_0} (C_{3n} r^{n-1} - D_{3n} r^{-n-1}) \cos n\theta \tag{11.123}$$

In Region 2,

$$B_{r2} = \sum_{n=1}^{\infty} - n(C_{2n} r^{n-1} + D_{2n} r^{-n-1}) \sin n\theta \tag{11.124}$$

$$H_{\theta 3} = \sum_{n=1}^{\infty} - \frac{n}{\mu_0} (C_{2n} r^{n-1} - D_{2n} r^{-n-1}) \cos n\theta \tag{11.125}$$

In Region 1,

$$B_{r1} = \sum_{n=1}^{\infty} - \frac{nC_{1n}}{r} \left[\text{ber}_n(kr) + j \, \text{bei}_n(kr) \right] \sin n\theta \tag{11.126}$$

$$H_{\theta 1} = \sum_{n=1}^{\infty} - \frac{nk}{\mu_{r1} \mu_0} C_{1n} \left[\text{ber}'_n(kr) + j \, \text{bei}'_n(kr) \right] \cos n\theta \tag{11.127}$$

Using the boundary conditions at $r = r_3$, one finds

$$\sum_{n=1}^{\infty} - nD_{4n} r_3^{-n-1} \sin n\theta = \sum_{1}^{n} - n \left(C_{3n} r_3^{n-1} + D_{3n} r_3^{-n-1} \right) \sin n\theta \tag{11.128}$$

$$\sum_{n=1}^{\infty} \frac{n}{\mu_0} D_{4n} r_3^{-n-1} \cos n\theta = \sum_{n=1}^{\infty} - \frac{n}{\mu_{r3} \mu_0} \left(C_{3n} r_3^{n-1} - D_{3n} r_3^{-n-1} \right) \cos n\theta \tag{11.129}$$

Solving Equations (11.128) and (11.129) we obtain

$$C_{3n} = (1 - \mu_{r3}) \frac{D_{4n}}{2} r_3^{-2n} \tag{11.130}$$

$$D_{3n} = (1 + \mu_{r3}) \frac{D_{4n}}{2} \tag{11.131}$$

Substituting C_{3n} and D_{3n} into the expressions for B_{r3} and $H_{\theta 3}$, we get

$$B_{r3} = \sum_{n=1}^{\infty} - \frac{nD_{4n} r^{-n-1}}{2} \left[(1 - \mu_{r3}) \left(\frac{r}{r_3} \right)^{2n} + (1 + \mu_r) \right] \sin n\theta \tag{11.132}$$

$$H_{\theta 3} = \sum_{n=1}^{\infty} - \frac{nD_{4n} r^{-n-1}}{2\mu_{r3} \mu_0} \left[(1 - \mu_{r3}) \left(\frac{r}{r_3} \right)^{2n} - (1 + \mu_r) \right] \cos n\theta \tag{11.133}$$

A similar analysis follows from using the boundary conditions at $r = r_1$.

$$\sum_{n=1}^{\infty} - n(C_{2n} r_1^{n-1} + D_{2n} r_1^{-n-1}) \sin n\theta =$$

$$\sum_{n=1}^{\infty} -\frac{nC_{n1}}{r_1}\left[\text{ber}_n(kr_1)+j\,\text{bei}_n(kr_1)\right]\sin n\theta \tag{11.134}$$

$$\sum_{n=1}^{\infty} -\frac{n}{\mu_0}(C_{2n}r_1^{n-1}-D_{2n}r_1^{-n-1})\cos n\theta =$$

$$\sum_{n=1}^{\infty} -\frac{nk}{\mu_{r1}\mu_0}C_{1n}\left[\text{ber}_n'(kr_1)+j\,\text{bei}_n'(kr_1)\right]\cos n\theta \tag{11.135}$$

Solving for C_{2n} and D_{2n} from the above set of simultaneous equations,

$$C_{2n} = \frac{C_{1n}}{2}r_1^{-n+1}\left[\frac{P}{r_1}+Q\frac{k}{\mu_{r1}}\right] \tag{11.136}$$

$$D_{2n} = \frac{C_{1n}}{2}r_1^{n+1}\left[\frac{P}{r_1}-Q\frac{k}{\mu_{r1}}\right] \tag{11.137}$$

where

$$P = \text{ber}_n(kr_1)+j\,\text{bei}_n(kr_1) \tag{11.138}$$

$$Q = \text{ber}_n'(kr_1)+j\,\text{bei}_n'(kr_1) \tag{11.139}$$

Substituting C_{2n} and D_{2n} into the expressions for B_{r2} and $H_{\theta2}$,

$$B_{r2} = \sum_{n=1}^{\infty} -\frac{nC_{1n}}{2}\left[\left(\frac{P}{r_1}+Q\frac{k}{\mu_{r1}}\right)\left(\frac{r}{r_1}\right)^{n-1}+\left(\frac{P}{r_1}-Q\frac{k}{\mu_{r1}}\right)\left(\frac{r}{r_1}\right)^{-n-1}\right]\sin n\theta \tag{11.140}$$

$$H_{\theta2} = \sum_{n=1}^{\infty} -\frac{nC_{1n}}{2\mu_0}\left[\left(\frac{P}{r_1}+Q\frac{k}{\mu_{r1}}\right)\left(\frac{r}{r_1}\right)^{n-1}-\left(\frac{P}{r_1}-Q\frac{k}{\mu_{r1}}\right)\left(\frac{r}{r_1}\right)^{-n-1}\right]\cos n\theta \tag{11.141}$$

The following two expressions will be used to help evaluate the boundary conditions at $r = r_2$.

$$E = \left[(1-\mu_{r3})\left(\frac{r_2}{r_3}\right)^{2n}+(1+\mu_{r3})\right] \tag{11.142}$$

$$F = \left[(1-\mu_{r3})\left(\frac{r_2}{r_3}\right)^{2n}-(1+\mu_{r3})\right] \tag{11.143}$$

$$\sum_{n=1}^{\infty}\frac{-nD_{4n}}{2}r_2^{-n-1}E\sin n\theta =$$

$$\sum_{n=1}^{\infty} -\frac{nC_{1n}}{2}\left[\left(\frac{P}{r_1}+Q\frac{k}{\mu_{r1}}\right)\left(\frac{r_2}{r_1}\right)^{n-1}+\left(\frac{P}{r_1}-Q\frac{k}{\mu_{r_1}}\right)\left(\frac{r_2}{r_1}\right)^{-n-1}\right]\sin n\theta \tag{11.144}$$

$$\sum_{n=1}^{\infty}\frac{-nD_{4n}}{2\mu_{r1}\mu_0}r_2^{-n-1}F\cos n\theta$$

$$-\sum_{n=1}^{\infty} -\frac{nC_{1n}}{2\mu_0}\left[\left(\frac{P}{r_1}+Q\frac{k}{\mu_{r1}}\right)\left(\frac{r_2}{r_1}\right)^{n-1}-\left(\frac{P}{r_1}-Q\frac{k}{\mu_{r_1}}\right)\left(\frac{r_2}{r_1}\right)^{-n-1}\right]\cos n\theta =$$

$$\sum_{n=1}^{\infty} J_s \cos n\theta \tag{11.145}$$

Therefore,

$$D_{4n}2r_2^{-n-1}E - C_{1n}\left[\left(\frac{P}{r_1}+Q\frac{k}{\mu_{r1}}\right)\left(\frac{r_2}{r_1}\right)^{n-1}+\left(\frac{P}{r_1}-Q\frac{k}{\mu_{r_1}}\right)\left(\frac{r_2}{r_1}\right)^{-n-1}\right] = 0 \tag{11.146}$$

$$\frac{D_{4n}}{\mu_{r3}}r_2^{-n-1}F$$

$$-\sum_{n=1}^{\infty} -C_{1n}\left[\left(\frac{P}{r_1}+Q\frac{k}{\mu_{r1}}\right)\left(\frac{r_2}{r_1}\right)^{n-1}-\left(\frac{P}{r_1}-Q\frac{k}{\mu_{r_1}}\right)\left(\frac{r_2}{r_1}\right)^{-n-1}\right] = -2\mu_0 J_s \tag{11.147}$$

From the first equation, we have

$$C_{1n} = \frac{D_{4n}r_2^{-n-1}E}{\left[\left(\frac{P}{r_1}+Q\frac{k}{\mu_{r1}}\right)\left(\frac{r_2}{r_1}\right)^{n-1}+\left(\frac{P}{r_1}-Q\frac{k}{\mu_{r_1}}\right)\left(\frac{r_2}{r_1}\right)^{-n-1}\right]} \tag{11.148}$$

Substituting Equation (11.148) into Equation (11.147) yields

$$\frac{D_{4n}}{\mu_{r3}}r_2^{-n-1}F$$

$$-\frac{D_{4n}r_2^{-n-1}E\left[\left(\frac{P}{r_1}+Q\frac{k}{\mu_{r1}}\right)\left(\frac{r_2}{r_1}\right)^{n-1}-\left(\frac{P}{r_1}-Q\frac{k}{\mu_{r_1}}\right)\left(\frac{r_2}{r_1}\right)^{-n-1}\right]}{\left(\frac{P}{r_1}+Q\frac{k}{\mu_{r1}}\right)\left(\frac{r_2}{r_1}\right)^{n-1}+\left(\frac{P}{r_1}-Q\frac{k}{\mu_{r_1}}\right)\left(\frac{r_2}{r_1}\right)^{-n-1}} = -2\mu_0 J_s \tag{11.149}$$

$$D_{4n} = \frac{-2\mu_0 J_s r_2^{n+1}}{\frac{F}{\mu_{r3}} - E\frac{(P/r_1+Qk/\mu_{r1})(r_2/r_1)^{n-1}-(P/r_1)-Qk/\mu_{r1})(r_2/r_1)^{-n-1}}{(P/r_1+Qk/\mu_{r1})(r_2/r_1)^{n-1}+(P/r_1)-Qk/\mu_{r1})(r_2/r_1)^{-n-1}}} \tag{11.150}$$

To simplify the expressions for the integration constants, let

$$U = \left(\frac{P}{r_1}+\frac{Qk}{\mu_{r1}}\right)\left(\frac{r_2}{r_1}\right)^{n-1}-\left(\frac{P}{r_1}-\frac{Qk}{\mu_{r_1}}\right)\left(\frac{r_2}{r_1}\right)^{-n-1} \tag{11.151}$$

$$V = \left(\frac{P}{r_1}+\frac{Qk}{\mu_{r1}}\right)\left(\frac{r_2}{r_1}\right)^{n-1}+\left(\frac{P}{r_1}-\frac{Qk}{\mu_{r_1}}\right)\left(\frac{r_2}{r_1}\right)^{-n-1} \tag{11.152}$$

Using U and V in the expression for D_{4n} in Equation (11.150) we get

$$D_{4n} = \frac{-2\mu_0 J_s r_2^{n+1}}{F/\mu_{r3} - EU/V} \tag{11.153}$$

Likewise, using U and V for the remaining integration constants yields the following expressions.

$$C_{2n} = \frac{C_{1n} r_1^{-n+1}}{2}\left(\frac{P}{r_1} + \frac{Qk}{\mu_{r1}}\right) = \frac{-\mu_0 J_s E r_1^{-n+1}\left(P/r_1 + Qk/\mu_{r1}\right)}{V\left(F/\mu_{r3} - EU/V\right)} \tag{11.154}$$

$$D_{2n} = \frac{C_{1n} r_1^{n+1}}{2}\left(\frac{P}{r_1} + \frac{Qk}{\mu_{r1}}\right) = \frac{-\mu_0 J_s E r_1^{n+1}\left(P/r_1 - Qk/\mu_{r1}\right)}{V\left(F/\mu_{r3} - EU/V\right)} \tag{11.155}$$

$$C_{3n} = \frac{(1-\mu_{r3})D_{4n} r_3^{-2n}}{2} = \frac{-(1-\mu_{r3})\mu_0 J_s r_3^{-2n} r_2^{n+1}}{F/\mu_{r3} - EU/V} \tag{11.156}$$

$$D_{3n} = \frac{(1+\mu_{r3})D_{4n}}{2} = \frac{-(1+\mu_{r3})\mu_0 J_s r_2^{n+1}}{F/\mu_{r3} - EU/V} \tag{11.157}$$

Now that all the integration constants have been determined, the vector potentials, the radial components of the flux-densities and the tangential components of the magnetic fields in each region are obtained as follows.

$$A_1 = \sum_{n=1}^{\infty} \frac{-2\mu_0 J_s E\left[\text{ber}_n(kr) + j\,\text{bei}_n(kr)\right]}{V\left(F/\mu_{r3} - EU/V\right)}\cos n\theta \tag{11.158}$$

$$B_{r1} = \sum_{n=1}^{\infty} \frac{2n\mu_0 J_s E\left[\text{ber}_n(kr) + j\,\text{bei}_n(kr)\right]}{Vr\left(F/\mu_{r3} - EU/V\right)}\sin n\theta \tag{11.159}$$

$$H_{\theta1} = \sum_{n=1}^{\infty} \frac{2n J_s E k\left[\text{ber}_n'(kr) + j\,\text{bei}_n'(kr)\right]}{V\mu_{r1}\left(F/\mu_{r3} - EU/V\right)}\cos n\theta \tag{11.160}$$

$$A_2 = \sum_{n=1}^{\infty} \frac{-\mu_0 J_s E r_1\left[\left(\frac{P}{r_1} + \frac{Qk}{\mu_{r1}}\right)\left(\frac{r}{r_1}\right)^{n} - \left(\frac{P}{r_1} - \frac{Qk}{\mu_{r1}}\right)\left(\frac{r}{r_1}\right)^{-n}\right]}{V\left(F/\mu_{r3} - EU/V\right)}\cos n\theta \tag{11.161}$$

$$B_{r2} = \sum_{n=1}^{\infty} \frac{n\mu_0 J_s E\left[\left(\frac{P}{r_1} + \frac{Qk}{\mu_{r1}}\right)\left(\frac{r}{r_1}\right)^{n-1} + \left(\frac{P}{r_1} - \frac{Qk}{\mu_{r1}}\right)\left(\frac{r}{r_1}\right)^{-n-1}\right]}{V\left(F/\mu_{r3} - EU/V\right)}\sin n\theta \tag{11.162}$$

$$H_{\theta2} = \sum_{n=1}^{\infty} \frac{n J_s E\left[\left(\frac{P}{r_1} + \frac{Qk}{\mu_{r1}}\right)\left(\frac{r}{r_1}\right)^{n-1} + \left(\frac{P}{r_1} - \frac{Qk}{\mu_{r1}}\right)\left(\frac{r}{r_1}\right)^{-n-1}\right]}{V\left(F/\mu_{r3} - EU/V\right)}\cos n\theta \tag{11.163}$$

$$A_3 = \sum_{n=1}^{\infty} \frac{-\mu_0 J_s r_2^{n+1} \left[(1 - \mu_{r3})r_3^{-2n}r^n + (1 + \mu_{r3})r^{-n}\right]}{F/\mu_{r3} - EU/V} \cos n\theta \qquad (11.164)$$

$$B_{r3} = \sum_{n=1}^{\infty} \frac{n\mu_0 J_s r_2^{n+1} \left[(1 - \mu_{r3})r_3^{-2n}r^{n-1} + (1 + \mu_{r3})r^{-n-1}\right]}{F/\mu_{r3} - EU/V} \sin n\theta \qquad (11.165)$$

$$H_{\theta 3} = \sum_{n=1}^{\infty} \frac{n J_s r_2^{n+1} \left[(1 - \mu_{r3})r_3^{-2n}r^{n-1} - (1 + \mu_{r3})r^{-n-1}\right]}{V\mu_{r3}\left(F/\mu_{r3} - EU/V\right)} \cos n\theta \qquad (11.166)$$

$$A_4 = \sum_{n=1}^{\infty} \frac{-2\mu_0 J_s r_2}{F/\mu_{r3} - EU/V} \left(\frac{r}{r_2}\right)^{-n} \cos n\theta \qquad (11.167)$$

$$B_{r4} = \sum_{n=1}^{\infty} \frac{2n\mu_0 J_s}{F/\mu_{r3} - EU/V} \left(\frac{r}{r_2}\right)^{-n-1} \sin n\theta \qquad (11.168)$$

$$H_{\theta 4} = \sum_{n=1}^{\infty} \frac{-2n J_s}{F/\mu_{r3} - EU/V} \left(\frac{r}{r_2}\right)^{-n-1} \cos n\theta \qquad (11.169)$$

Now that we have an expression for the vector potential, we can find the eddy currents and the power loss. The eddy current density is found as

$$J_e = -j\omega\sigma A_1 = \sum_{n=1}^{\infty} \frac{2j\omega\mu_0\sigma J_s E \left[\text{ber}_n(kr) + j\,\text{bei}_n(kr)\right]}{V\left[\frac{F}{\mu_{r3}} - \frac{EU}{V}\right]} \cos(n\theta) \qquad (11.170)$$

Figure 11.28 illustrates the variation of eddy current in the rotor as a function of the radius.

$$P = \int \frac{J_e J_e^*}{2\sigma}\, dv \qquad (11.171)$$

where $\omega = s\omega_f$, f is the fundamental frequency; s is the per unit slip, $J_s = \frac{\sigma V_b}{l}$; ℓ is the stack length of the stator, and V_b is the bar voltage.

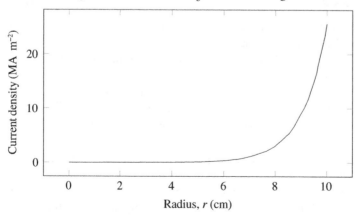

Figure 11.28 Eddy current density variation in the solid rotor with rotor radius.

After deriving the current density Equation (11.170), the power loss in eddy current region can be obtained by using Equation (11.171) in polar coordinates:

$$P = 2p \int_0^{2\pi} \int_0^l \int_0^{r_1} \frac{J_e J_e^*}{2\sigma} r\,dr\,d\theta\,dz \tag{11.172}$$

Substituting J_e from the Equation (11.170) and its conjugate into Equation (11.172) for power loss, one obtains, after some algebraic manipulation

$$P = 4ps^2\omega^2\mu_0^2\sigma J_s^2 E^2 \frac{\int_0^{2\pi} \int_0^l \int_0^{r_1} \left[\text{ber}_n^2(kr) + j\text{bei}_n^2(kr)\right] \cos^2 n\theta\ r\,dr\,d\theta\,dz}{VV^* \left[\frac{F}{\mu_{r3}} - \frac{EU}{V}\right]\left[\frac{F}{\mu_{r3}} - \frac{EU}{V}\right]^*} \tag{11.173}$$

Solving for the integral in Equation (11.173) using reference [43], the loss expression becomes

$$P = 4ps^2\omega^2\mu_0^2\sigma J_s^2 E^2 \pi l \frac{\frac{r_1}{k}\left[\text{ber}_n(kr_1)\text{bei}_n'(kr_1) + \text{ber}_n'(kr_1)\text{bei}_n(kr_1)\right]}{VV^* \left[\frac{F}{\mu_{r3}} - \frac{EU}{V}\right]\left[\frac{F}{\mu_{r3}} - \frac{EU}{V}\right]^*} \tag{11.174}$$

11.2.2 Finite Element Model

A finite element analysis of the solid rotor motor was performed to verify the closed-form solution. The same magnetic material is used in both the rotor and the stator ($\mu_{r1} = \mu_{r3} = 750$). The flux density crossing the air gap is calculated for different operating conditions of the motor. Since the rotor is solid, the conductivity is defined for this region ($\sigma = 3 \times 10^6\,\text{S m}^{-1}$). The current source of the motor is defined as the fundamental of the stator current [68].

$$J_s = J_s \cos n\theta \tag{11.175}$$

The line current density is applied to the model and is shown on the motor model with the finite element mesh in Figure 11.29.

The supply frequency is 50HZ. The phase currents of the motor are shown in Figure 11.30 in per unit.

The solid rotor induction machine was analyzed at steady state for different speeds. The equi-flux lines for different slip values are given in Figure 11.31. The skin effect and phase delay caused by the eddy currents can be seen at different slips.

The flux density variation in the air gap for different slip values is presented in Figure 11.32. The four-pole distribution is apparent.

In the analytical model of the solid rotor machine, we computed the fundamental of the flux density. Figure 11.33 shows the normal component of the flux density and its fundamental. The flux density was calculated by the finite element method.

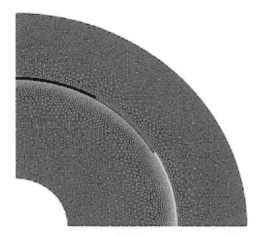

Figure 11.29 Mesh model of the motor with the injected current density.

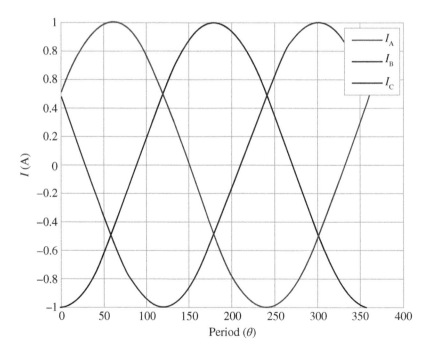

Figure 11.30 FEM results of the phase currents.

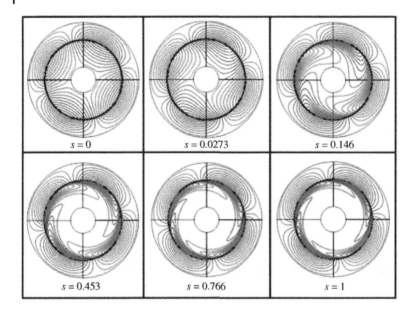

Figure 11.31 Eddy currents for different slips.

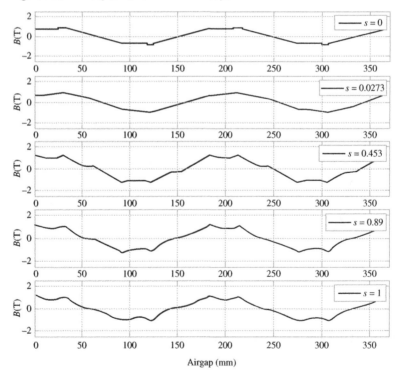

Figure 11.32 Flux density in the air gap for different slip values.

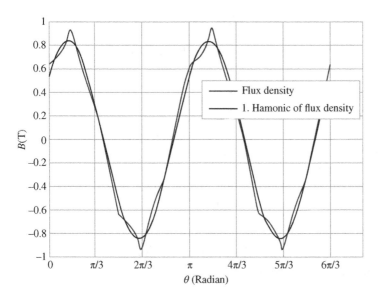

Figure 11.33 Air-gap flux density and its first harmonic with FEM.

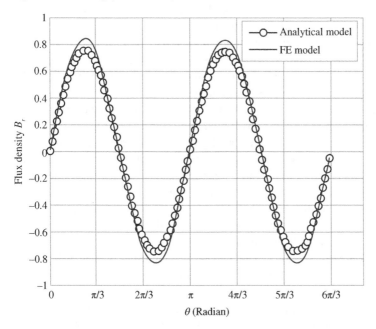

Figure 11.34 Comparison of the analytical and numerical results of the flux density around the air gap.

Table 11.1 Current density and power loss comparison.

Method	J_e (per unit)	Power loss (per unit)
Finite element method	1.00	1.00
Analytical method	1.08	1.12

The finite element results for the solid rotor induction motor are compared with the closed-form solution in Figure 11.34. The finite element result is compared with the analytical result for air-gap flux density which agree quite well.

The finite element and analytical results of eddy current density and eddy current loss for the solid rotor are given in Table 11.1.

11.3 Squirrel Cage Induction Motor Analysis by the Finite Element Method

The squirrel cage induction motor is perhaps the most widely used electric motor due to its low cost and high reliability. It is also a motor in which eddy currents play a major role. Not only are eddy currents important in determining the stray losses, such as losses in the stator and rotor laminations (see Section 2.1), and the stator copper losses, which include skin effect and proximity effect (see Section 2.3) but the eddy currents are responsible for producing the main torque of the motor. The squirrel cage has no power supply attached to it.[2]

The finite element method, described in Section 6.1, is a very popular technique to analyze induction machines in particular and rotating machines in general. It is versatile and can deal with complicated geometry and nonlinear nonhomogeneous materials. It is also common practice to couple the electric circuits in the motor directly to the finite element field model by means of the magnetic vector potential [23]. The analysis can be done in the time domain with a moving rotor which involves re-meshing at each time step [23]. This is rather expensive and typically many cycles must be computed until a steady state is reached.

The analysis presented here is done in the steady-state ac regime, sometimes referred to as *magneto-dynamic*. The analysis is done using phasor variables for the magnetic vector potential and the circuit current and voltage.

---

2 There is another type of induction motor which has a winding on the rotor, the wound-rotor induction motor, and this winding is connected through slip rings either to a variable impedance or to a power supply for the case of a doubly-fed induction motor. These motors can also be treated by the finite element method but are not considered here.

Using a time harmonic excitation with an angular frequency ω, the diffusion equation in terms of magnetic vector potential is given in Equation (11.176).

$$\nabla \times \left(\frac{1}{\mu} \nabla \times \mathbf{A} \right) + j\sigma\omega\mathbf{A} + \sigma\nabla V = 0 \tag{11.176}$$

Ohm's law is written as:

$$\mathbf{J} = -j\sigma\omega\mathbf{A} - \sigma\nabla V \tag{11.177}$$

where A is the magnetic vector potential, μ is the permeability of the material, σ is the conductivity of the material, and $\mathbf{J}$ is the current density.

In the case of the induction motor, where the rotor is operating at slip s, the rotor currents are subjected to the pulsation frequency $s\omega$, which must be considered in the governing Equations (11.176) and (11.177).

The finite element solution gives the complex vector potential, $\mathbf{A}$, at the nodes. The eddy current density is [23, 46]

$$\mathbf{J} = -j\sigma\omega\mathbf{A} \tag{11.178}$$

The instantaneous eddy current loss can be written in terms of current density

$$P = \iiint_{\Omega} \frac{1}{\sigma} \mathbf{J}^2 dxdydz \tag{11.179}$$

To understand the methodology used here, we can refer to the classical induction machine equivalent circuit and the method used to refer the rotor quantities to the stator. In a standard three-phase induction motor, the stator conductors are wound in slots in a laminated magnetic core. The stator currents produce a rotating magnetic field in the air gap whose fundamental rotates at synchronous speed. This field is approximately sinusoidally distributed in space. It is this field that induces voltage and current on the rotor. When the rotor is at standstill, the frequency of current on the rotor and stator are the same and the machine resembles a transformer with a short-circuited secondary (the rotor). However, at rated speed the rotor is rotating near synchronous speed. The per-unit variation from synchronous speed is called the *slip* and is defined as

$$s = \frac{\omega_s - \omega_m}{\omega_s} \tag{11.180}$$

where ω_s and ω_m are the synchronous speed and mechanical speed in radians/second, respectively.

An observer on the rotor, therefore, sees a time variation of magnetic field at a frequency which is the difference between synchronous speed and the mechanical speed of the rotor. This is equivalent to multiplying the stator frequency by the slip. For example, if the stator is excited with 60 Hz current and the rotor has a

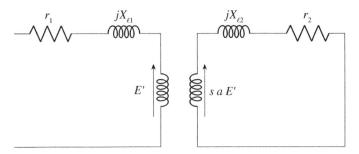

Figure 11.35 Equivalent circuit showing an ideal transformer.

slip of $s = 0.03$, then the frequency on the rotor is $60 \times 0.03 = 1.8\,\text{Hz}$. This can be represented by the equivalent circuit shown in Figure 11.35.

The ideal transformer, shown in the middle of the figure, represents the electromagnetic coupling between the rotor and stator. The two share the common air-gap flux. To find the EMF on the rotor, we first apply the turns ratio between stator and rotor. This is indicated by the symbol a. This represents the effective turns of the stator divided by the effective turns on the rotor as required by Faraday's law. This turns ratio is not important to us in the finite element context, as the correct number of turns is included in the finite element analysis and the solution of the field problem will automatically account for the turns. In the induction motor case, we also have the issue that the frequency on the rotor is $s\omega$ and therefore, since the flux is common, the induced voltage must be multiplied by the slip. The secondary current is then

$$I_2 = \frac{E_2}{r_2 + js\omega_s\ell_2} \tag{11.181}$$

If we now write the secondary EMF in terms of the primary EMF, we have

$$I_2 = \frac{sE_1}{r_2 + js\omega_s\ell_2} \tag{11.182}$$

where the EMF is multiplied by the slip. In order to refer the secondary quantities to the primary, effectively getting rid of the ideal transformer and connecting the primary and secondary together, we can divide both numerator and denominator by the slip. This gives

$$I_2 = \frac{E_1}{r_2/s + j\omega_s\ell_2} \tag{11.183}$$

This process results in the standard representation of the induction motor shown in Figure 11.36.

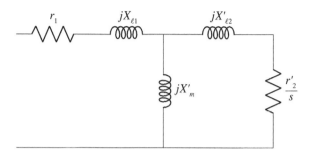

Figure 11.36 Equivalent circuit of induction motor with rotor quantities referred to the stator.

Now the rotor quantities are referred to the stator both by the turns ratio (automatically done in the finite element model) and by the frequency. In the circuit, we divide the rotor resistance by the slip, and in the finite element model, we divide all rotor resistivities by the slip. The validity of this step can be justified by considering the skin depth on the rotor. In the physical device, we have on the rotor

$$\delta = \sqrt{\frac{2}{s\omega\mu\sigma}} \tag{11.184}$$

So we see that if we multiply the physical rotor conductivity by the slip (or divide the resistivity by the slip) we obtain the same skin depth as in the physical rotor, so that the resistance, inductance, and losses will be the same. It is this process that allows us to model the induction motor operating at speed in steady state in the frequency domain.

Some approximations are necessary due to the use of phasor analysis. In the physical induction motor, other frequencies exist apart from the fundamental frequency on the stator and slip frequency on the rotor. Due to the layout of the stator winding, there will be traveling harmonics of flux in the air gap. The most important are the 5th and 7th and the 11th and 13th. These harmonics rotate in the air gap, with the 5th and 11th moving backward or opposite in direction to the rotor, and the 7th and 13th rotating in the direction of the rotor. These rotating harmonics induce current on the rotor at a frequency dependent on the rotor speed and the speed and direction of the harmonic waves themselves. These effects can not be included in our phasor analysis, since only the fundamental frequency is considered. There are also effects which are caused by the rotor motion. An observer on the rotor will pass by the stator slots and teeth at a frequency determined by the number of stator slots per pole and the rotor mechanical speed. This ripple in the flux density will induce currents at the tooth passing frequency

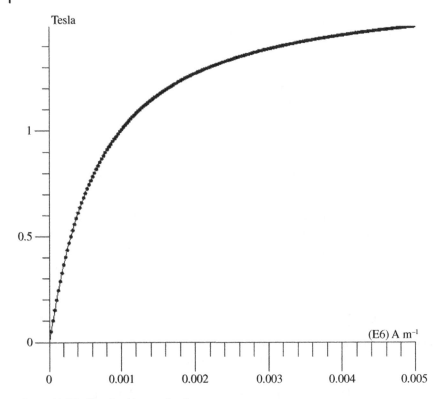

Figure 11.37 The $B - H$ curve for the rotor.

and its harmonics. These effects are not included in our magneto-dynamic analysis. Also, since the rotor is not spinning, the three stator phases see a fixed rotor slot-tooth pattern. This means that each stator phase will have a somewhat different impedance and slightly different current. This is normally a small effect, but if it is important, then several solutions with the rotor in different positions over a tooth pitch can be made and then averaged.

Another factor that should be considered is the impedance of the end windings. The finite element analysis is two-dimensional and as such we consider only axial current and magnetic field in the plane of the cross section. The end turns on the rotor and stator are added as "lumped" circuit elements. For the stator, the

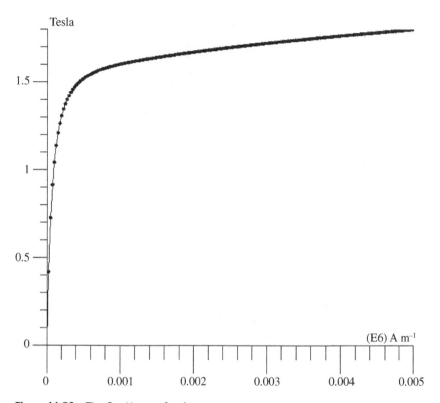

Figure 11.38 The $B - H$ curve for the stator.

values are found from normal design formulae [70, 71]. For the rotor, the resistance and inductance of the end ring segments are inserted between the squirrel cage bars [72].

The machine used for the studies is a three-phase, 4-pole, Y-connected squirrel cage rotor induction motor. The motor has 36 stator slots and 28 rotor slots. The saturation curves for the rotor and stator laminations are shown in Figures 11.37 and 11.38. Figure 11.39 shows the cross section of the motor. Figure 11.40 shows one pole of the motor which will be modeled. Figures 11.41 and 11.42 show the finite element mesh for one pole and an enlarged view including a section of the air gap. Figure 11.43 shows the flux plot for rated load.

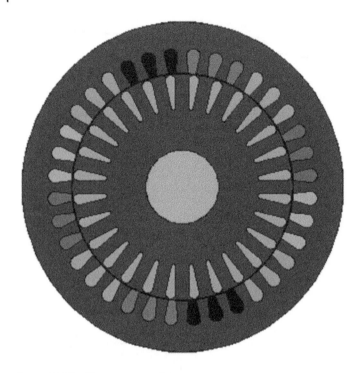

Figure 11.39 The geometry of the motor.

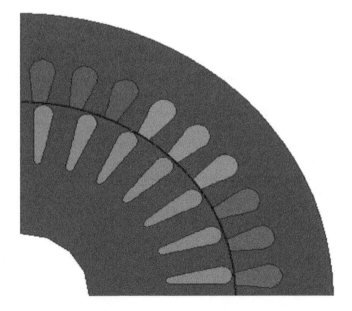

Figure 11.40 Enlargement of the motor geometry.

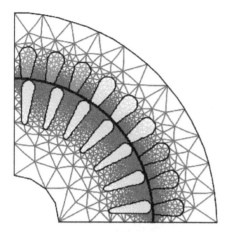

Figure 11.41   The finite element mesh of the motor.

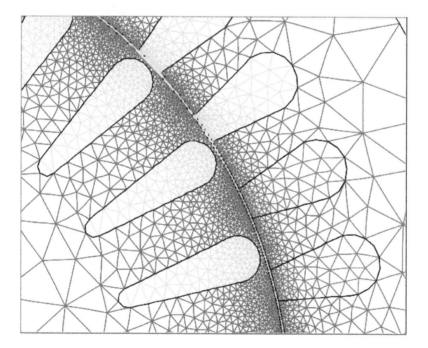

Figure 11.42   Enlargement of the air-gap region.

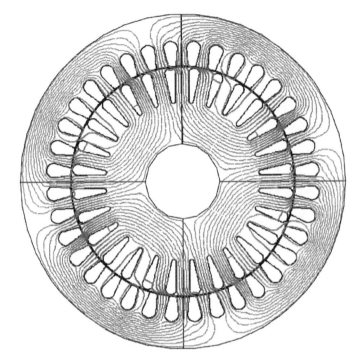

Figure 11.43 Equi-flux lines for rated load laminated rotor.

12

Transformer Losses

12.1 Foil Wound Transformer

As a practical example of an application in which one conductor dimension is much larger than the skin depth and another is much smaller than a skin depth, we consider the case of a foil wound or sheet wound transformer [16, 17], see Figure 12.1. In this problem, the sheets, usually made of copper or aluminum, are quite thin in the radial direction and long in the axial direction. The radial field in the transformer window will produce circulating currents in the sheet. In a transformer, we expect most of the flux in the winding region to be in the axial direction. Since the foil winding is very thin in the direction normal to the flux, the eddy currents produced by the axial flux are less of a concern. We can get a good qualitative understanding of the circulating current by using the analysis which we have done in rectangular coordinates for conducting plates.

The circulating current problem can be treated as one-dimensional because we are ignoring any variation in current across the conductor. Using symmetry around the mid-plane, we can write the eddy current density as

$$J(y) = \frac{H_0}{\delta} e^{-\left(\frac{h}{2} - y\right)/\delta} e^{j\left(\frac{\pi}{4} - \frac{\frac{h}{2} - y}{\delta}\right)} \tag{12.1}$$

For copper at 60 Hz, we have $\delta = 0.00853$ m. Considering a sheet of height 0.2 m, we see, in Figure 12.2, the magnitude of the circulating current density vs. height for half of the sheet. By evaluating the square of the current density and integrating over the length, we find that almost 90 % of the loss occurs in the first skin depth.

Eddy Currents: Theory, Modeling, and Applications, First Edition.
Sheppard J. Salon, M. V. K. Chari, Lale T. Ergene, David Burow, and Mark DeBortoli.
© 2024 The Institute of Electrical and Electronics Engineers, Inc. Published 2024 by John Wiley & Sons, Inc.

We have only considered the circulating current. If we add the load current back in, we see, in Figure 12.3, the magnitude of the total current. The reader will note that, except for the region of around 2–3 skin depths from the end, the current density is fairly constant and equal to the applied current density, 1.0 A m^{-2} in this case.

We also note that at around three skin depths from the end of the winding (27 mm), the current density magnitude has a minimum. This can be explained by considering the phase shift of the eddy currents. We have seen that the phase shifts one radian in one skin depth. This means that in π skin depths, the phase shift is 180° and the current reverses direction. The eddy current is then 180° out to phase with the load current. In looking at the current density magnitude in Figure 12.3, this is just where the minimum occurs. We also see that in the region of 2–3 skin depths from the end of the winding, the losses can be more than a factor of 2 greater than the average loss in the conductor which should be considered in the design.

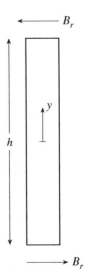

Figure 12.1 Foil wound transformer schematic.

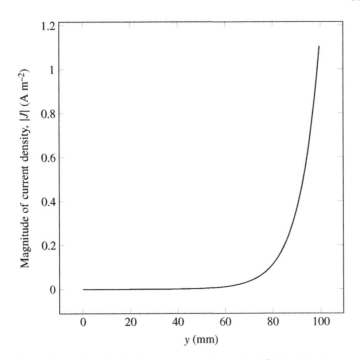

Figure 12.2 Magnitude of eddy current density for 60 Hz and copper.

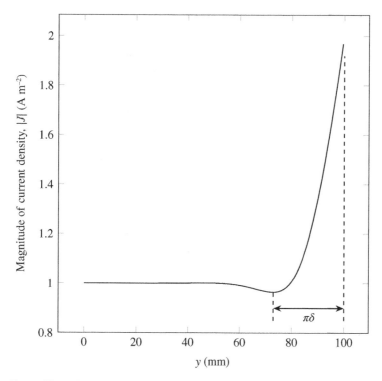

Figure 12.3 Scaled total current magnitude.

12.2 Phase Shifting Transformers

We have discussed the losses in machine windings and the theory of stranded conductors [16, 17]. This will now be applied to a multi-phase rectifier transformer typical in motor drive systems. In a two-winding transformer, the flux pattern in the region of the copper winding is mostly vertical (axial), except near the winding ends where the flux turns radially. This is illustrated in Figure 12.4.

When using stranded windings, the strands are oriented as shown in the figure, with the narrow dimension in the radial direction and the wide dimension in the axial direction. In this way, for most of the winding, the losses due to eddy currents will be rather small. Near the ends, we will see an increase in eddy losses since the radial flux will impinge on the wide dimension of the strand. Once a field solution is found, the calculation of eddy losses is done using the resistance limited formula (see Section 2.1) for the loss per unit depth

$$W = \frac{\omega^2 B_0^2 d^3}{24\rho} \tag{12.2}$$

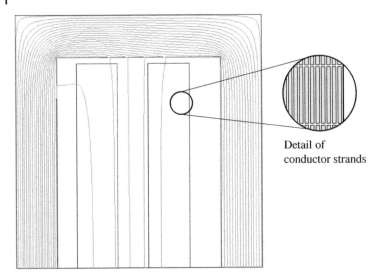

Detail of
conductor strands

Figure 12.4 Flux map of two winding transformer.

Figure 12.5 3-Phase, 36 pulse phase-shifting rectifier transformer.

The axial direction of flux in the two-winding transformer is due to the fact that the primary currents and secondary currents are 180° out of phase. In the 3-phase, 36 pulse phase-shifting rectifier transformer shown in Figure 12.5, there are 18 low voltage windings in the extended delta secondary. Each winding has current which is phase displaced by 10°. These are arranged in a complicated way in the transformer which results in the secondary winding having many phase changes as we go up the winding. This produces a significant radial flux. If the same type of strand is used in this configuration, the losses due to the radial flux can get quite high. There is no closed-form solution that can be used to find the field in this case. The finite element model can give the designer the radial and axial flux components and then we can use Equation (12.2) and find the losses in the different sections in the winding.

Many manufacturers will use continuously transposed conductors (CTC) which are small strands that are transposed as the conductor travels through the transformer. Figures 12.6 and 12.7 show the flux distribution in the transformer.

The example illustrates that even though we can use relatively simple expression to find the losses, these expressions require the field distribution, and the field distribution is extremely complicated and requires numerical methods to describe.

Figure 12.6 Real part of the flux distribution in transformer.

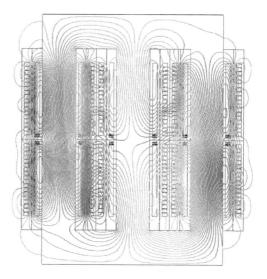

Figure 12.7 Imaginary part of the flux distribution in transformer.

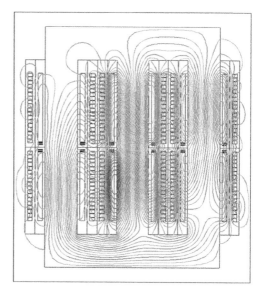

Appendix A

Bessel Functions

The differential equation known as *Bessel's* equation is given as [42]

$$x^2\frac{d^2y}{dx^2} + x\frac{dy}{dx} + (x^2 + v^2)y = 0 \tag{A.1}$$

The solution of this equation is given as

$$y = AJ_v(x) + BY_v(x) \tag{A.2}$$

where A and B are constants and J and Y are Bessel functions of the first and second kind respectively. The order of the Bessel function is $v = 0, 1, 2, \ldots$.

By substituting $x = jx$ into Bessel's equation, we obtain the *Modified Bessel equation*

$$x^2\frac{d^2y}{dx^2} + x\frac{dy}{dx} - (x^2 + v^2)y = 0 \tag{A.3}$$

which has solutions

$$y = CI_v(x) + DK_v(x) \tag{A.4}$$

where I and K are modified Bessel functions of the first and second kind, respectively.

A generalized form of the modified Bessel equation is written as

$$x^2\frac{d^2y}{dx^2} + x\frac{dy}{dx} - (\beta^2x^2 + v^2)y = 0 \tag{A.5}$$

whose solution is now

$$y = CI_v(\beta x) + DK_v(\beta x) \tag{A.6}$$

where β is a constant.

In Section 3.1, we obtained an expression for the magnetic flux density using the time-harmonic form of the diffusion equation as

$$r^2\frac{d^2B_\theta}{dr^2} + r\frac{dB_\theta}{dr} - (1 + j\omega\mu\sigma r^2)B_\theta = 0 \tag{A.7}$$

Eddy Currents: Theory, Modeling, and Applications, First Edition.
Sheppard J. Salon, M. V. K. Chari, Lale T. Ergene, David Burow, and Mark DeBortoli.
© 2024 The Institute of Electrical and Electronics Engineers, Inc. Published 2024 by John Wiley & Sons, Inc.

The reader can check that this is a modified Bessel equation, and our solutions are modified Bessel functions of order 1. Since the coefficient β contains $j = \sqrt{-1}$, we can find the real and imaginary parts of the Bessel functions.

We have

$$I_\nu(x) = j^{-\nu} J_\nu(jx) = J_\nu(j^{3/2}x) \tag{A.8}$$

Then the solution becomes

$$y = CJ_\nu(j^{3/2}x) + DK_\nu(j^{1/2}x) \tag{A.9}$$

This gives us the Kelvin functions

$$\begin{aligned}
\text{ber}_\nu &= \Re e\, J_\nu(j^{3/2}x) \\
\text{bei}_\nu &= \Im m\, J_\nu(j^{3/2}x) \\
J_\nu(j^{3/2}x) &= \text{ber}_\nu(x) + j\,\text{bei}_\nu(x)
\end{aligned} \tag{A.10}$$

Similarly

$$\begin{aligned}
\text{ker}_\nu &= \Re e\, j^{-\nu} K_\nu(j^{1/2}x) \\
\text{kei}_\nu &= \Im m\, j^{-\nu} K_\nu(j^{1/2}x) \\
j^{-\nu} K_\nu(j^{1/2}x) &= \text{ker}_\nu(x) + j\,\text{kei}_\nu(x)
\end{aligned} \tag{A.11}$$

Appendix B

Separation of Variables

Separation of variables is a technique that can be used with linear homogeneous equations [14]. In our examples in Chapters 2 and 3, we have used this method to find the solutions to the diffusion equation with Dirichlet boundary conditions. The diffusion equation involves variables that are a function of time and of space. In the separation of variables technique, we assume that the solution can be written as a product of two functions. One is a function of time but not of space, and one a function of space but not of time.

B.1 One-Dimensional Separation of Variables in Rectangular Coordinates

For a one-dimensional example in Cartesian coordinates, we have

$$\frac{\partial F}{\partial t} = \gamma^2 \frac{\partial^2 F}{\partial x^2} \tag{B.1}$$

Assume now that

$$F(x, t) = X(x)T(t) \tag{B.2}$$

There are many functions that satisfy Equation (B.1). Since the equation is linear, the sum of all the solutions is also a solution.

Substituting Equation (B.2) into (B.1)

$$X(x)T'(t) = \gamma^2 X''(x)T(t) \tag{B.3}$$

which gives

$$\frac{X''(x)}{X(x)} = \frac{1}{\gamma^2}\frac{T'(t)}{T(t)} \tag{B.4}$$

Eddy Currents: Theory, Modeling, and Applications, First Edition.
Sheppard J. Salon, M. V. K. Chari, Lale T. Ergene, David Burow, and Mark DeBortoli.
© 2024 The Institute of Electrical and Electronics Engineers, Inc. Published 2024 by John Wiley & Sons, Inc.

The left-hand side of the equation is only a function of x and not of t. The right-hand side of the equation is only a function of t and not a function of x. In order for these to be equal, each side must be a constant. So we have

$$\frac{X''(x)}{X(x)} = \frac{1}{\gamma^2} \frac{T'(t)}{T(t)} = k \tag{B.5}$$

We now have ordinary differential equations instead of partial differential equations. The solution to Equation (B.5) (for $k \neq 0$) is

$$X(x) = c_1 e^{\sqrt{k}x} + c_2 e^{-\sqrt{k}x} \tag{B.6}$$

or we can use

$$X(x) = A \sin(\gamma x) + B \cos \gamma x \tag{B.7}$$

and

$$T(t) = c_3 e^{\gamma^2 t} \tag{B.8}$$

Since the constants c_1, c_2, and c_3 are arbitrary, we have a large number of solutions and since the system is linear, the sum of all these solutions is also a solution.

$$F(x, t) = \sum_i c_i X_i(x) T_i(t) \tag{B.9}$$

To eliminate many of these solutions, we now apply the boundary conditions and initial conditions. The process may be slightly different from this point on and will have to be adapted in order to match these conditions depending on the specific application.

As an example, let us consider a bar of length ℓ in which the top and bottom are held at a value $F(0, t) = F(\ell, t) = 0$. Let us also assume that the bar at $t = 0^-$ is initially at a uniform value F_0. F then becomes

$$F(x, t) = \sum_{m=1}^{\infty} \left(A_m \sin \gamma_m x + B_m \cos \gamma_m x \right) e^{-\gamma^2 t} \tag{B.10}$$

The only way that the boundary conditions can be satisfied is that

$$B_m = 0, \quad \gamma_m = \frac{m\pi}{\ell} \tag{B.11}$$

Then initially, for $0 < x < \ell$

$$F_0 = \sum_{m=1}^{\infty} A_m \sin \left(\frac{m\pi x}{\ell} \right) \tag{B.12}$$

We now use the orthogonality properties of the sine functions and multiply equation (B.12) by $\sin(p\pi x/\ell)$ and integrate from 0 to ℓ. The result is that all terms $m \neq p$ vanish, as do all terms with even m. This gives

$$A_m = \frac{4F_0}{m\pi} \quad m = 1, 3, 5, \ldots \tag{B.13}$$

The final solution is then

$$F(x, t) = \frac{4F_0}{\pi} \sum_{n=0}^{\infty} \frac{1}{2n+1} e^{-\gamma(2n+1)^2 \pi^2 t / \ell^2} \sin\left(\frac{(2n+1)\pi x}{\ell}\right) \tag{B.14}$$

We can use $m = 2n + 1$, so the summation goes from 0 to ∞.

B.2 Two-Dimensional Separation of Variables in Cylindrical Coordinates

As another example, consider the two-dimensional time-harmonic diffusion equation in cylindrical coordinates,

$$\frac{\partial^2 A}{\partial r^2} + \frac{1}{r}\frac{\partial A}{\partial r} + \frac{1}{r^2}\frac{\partial^2 A}{\partial \theta^2} = j\omega\mu_r\mu_0\sigma A_s \tag{B.15}$$

We shall express the magnetic vector potential, A, as the product of two functions: one of r and one of θ respectively, such that

$$A = R(r)\Theta(\theta) \tag{B.16}$$

Substituting Equation (B.16) into Equation (B.15) yields

$$\Theta(\theta)\frac{\partial^2 R(r)}{\partial r^2} + \Theta(\theta)\frac{1}{r}\frac{\partial R(r)}{\partial r} + \frac{R(r)}{r^2}\frac{\partial^2\Theta(\theta)}{\partial\theta^2} = j\omega\mu_r\mu_0\sigma R(r)\Theta(\theta) \tag{B.17}$$

Dividing through by $R(r)\Theta(\theta)$, Equation (B.17) becomes

$$\frac{1}{R(r)}\frac{\partial^2 R(r)}{\partial r^2} + \frac{1}{R(r)}\frac{1}{r}\frac{\partial R(r)}{\partial r} + \frac{1}{\Theta(\theta)}\frac{1}{r^2}\frac{\partial^2\Theta(\theta)}{\partial\theta^2} = j\omega\mu_r\mu_0\sigma \tag{B.18}$$

or

$$\frac{r^2}{R(r)}\frac{\partial^2 R(r)}{\partial r^2} + \frac{r}{R(r)}\frac{\partial R(r)}{\partial r} + \frac{1}{\Theta(\theta)}\frac{\partial^2\Theta(\theta)}{\partial\theta^2} = j\omega\mu_r\mu_0\sigma r^2 \tag{B.19}$$

Equation (B.19) can be split into two equations, one as a function of r only and the other as a function of θ only, so that

$$r^2\frac{\partial^2 R(r)}{\partial r^2} + r\frac{\partial R(r)}{\partial r} - (j\omega\mu_r\mu_0\sigma r^2 + n^2)R(r) = 0 \tag{B.20}$$

$$\frac{\partial^2\Theta(\theta)}{\partial\theta^2} + n^2\Theta(\theta) = 0 \tag{B.21}$$

Equation (B.20) is a Bessel equation of order n, whose solution is of the form

$$R(r) = \sum_{n=1}^{\infty} C_n[\text{ber}_n(kr) + j\,\text{bei}_n(kr)] + D_n[\text{ker}_n(kr) + j\,\text{kei}_n(kr)] \tag{B.22}$$

where $k = \omega\mu_r\mu_0\sigma$

The solution of Equation (B.21) is given as

$$\Theta(\theta) = \sum_{n=1}^{\infty} E_n \cos n\theta + F_n \sin n\theta \tag{B.23}$$

Combining Equations (B.22) and (B.23), the solution for the vector potential, A, is obtained as

$$A = R(r)\Theta(\theta)$$

$$= \sum_{n=1}^{\infty} \left(C_n[\text{ber}_n(kr) + j\,\text{bei}_n(kr)] + D_n[\text{ker}_n(kr) + j\,\text{kei}_n(kr)] \right)$$

$$\times [E_n \cos n\theta + F_n \sin n\theta] \tag{B.24}$$

In free space, where the conductivity is zero, Equation (B.15) reduces to Laplace's equation

$$\frac{\partial^2 A}{\partial r^2} + \frac{1}{r}\frac{\partial A}{\partial r} + \frac{1}{r^2}\frac{\partial^2 A}{\partial \theta^2} = 0 \tag{B.25}$$

The solution of this equation is obtained by separation of variables as before and is given by

$$A = \sum_{n=1}^{\infty} [C_n r^n + D_n r^{-n}][E_n \cos n\theta + F_n \sin n\theta] \tag{B.26}$$

Appendix C

The Error Function

The error function is a solution to the diffusion problem in rectangular coordinates [40]. We define the error function as

$$\text{erf}(x) = \frac{2}{\sqrt{\pi}} \int_0^x e^{-\gamma^2} \, d\gamma \tag{C.1}$$

From Equation (C.1), we can see that

$$\text{erf}(0) = 0$$
$$\text{erf}(\infty) = 1$$
$$\text{erf}(-x) = -\text{erf}(x) \tag{C.2}$$

A couple of approximations can be useful. For small values of x,

$$\text{erf}(x) = \frac{2}{\sqrt{\pi}} \sum_{k=0}^{\infty} \frac{-1^k x^{2k+1}}{(2k+1)k!} \tag{C.3}$$

For large values of x, we can evaluate the series

$$\text{erf}(x) = 1 - \frac{e^{-x^2}}{\sqrt{\pi}} \left(\frac{1}{x} - \frac{1}{2x^3} + \frac{1 \cdot 3}{2^2 x^5} - \frac{1 \cdot 3 \cdot 5}{2^3 x^7} \cdots \right) \tag{C.4}$$

We also define the complementary error function as

$$\text{erfc}(x) = 1 - \text{erf}(x) \tag{C.5}$$

Eddy Currents: Theory, Modeling, and Applications, First Edition.
Sheppard J. Salon, M. V. K. Chari, Lale T. Ergene, David Burow, and Mark DeBortoli.
© 2024 The Institute of Electrical and Electronics Engineers, Inc. Published 2024 by John Wiley & Sons, Inc.

Appendix D

Replacing Hollow Conducting Cylinders with Line Currents Using the Method of Images

We would like to replace two hollow current-carrying cylinders of radii a and separated by $2h$, as shown in Figure D.1, with equivalent line currents so that the fields exterior to the cylinders are the same as those produced by the cylinders themselves.

Referring to Figure D.2, we have a cylindrical coordinate system and a long axial current directed in the positive z direction.

From Ampere's law, we find that the magnetic flux density from the long wire with current I is

$$B_\theta = \frac{\mu_0 I}{2\pi r} \hat{\mathbf{u}}_\theta \tag{D.1}$$

Recalling that the magnetic vector potential is defined such that $\nabla \times A = B$, and for this application, the flux density has only a θ component and does not vary in the z or θ directions, we find

$$-\frac{\partial A_z}{\partial r} = \frac{\mu_0 I}{2\pi r} \tag{D.2}$$

This implies that

$$A = -\frac{\mu_0 I}{2\pi} \ln r \, \hat{\mathbf{u}}_z \tag{D.3}$$

In the application we are interested in, the cylindrical conductor carries current in a thin layer at the surface and the flux at the surface is tangential. In terms of the magnetic vector potential, we require that the surface of each cylinder be a flux line. This is equivalent to enforcing the condition that the surface is an equipotential.

Referring to Figure D.3, we define the geometry. Assuming that the equivalent line currents are on the x axis and are separated by a distance $2h'$, we now show that the resulting equipotentials are circular surfaces. If we can find conditions such that the circles are centered at the center of the physical cylinders, the uniqueness theorem allows us to replace the cylinders by the line sources.

Eddy Currents: Theory, Modeling, and Applications, First Edition.
Sheppard J. Salon, M. V. K. Chari, Lale T. Ergene, David Burow, and Mark DeBortoli.
© 2024 The Institute of Electrical and Electronics Engineers, Inc. Published 2024 by John Wiley & Sons, Inc.

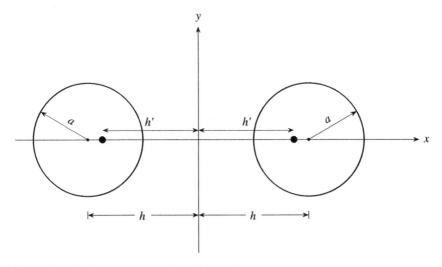

Figure D.1 Hollow cylinders and equivalent line currents.

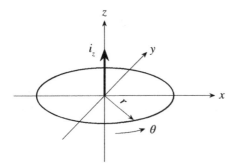

Figure D.2 Cylindrical coordinate system and infinitely long current in z.

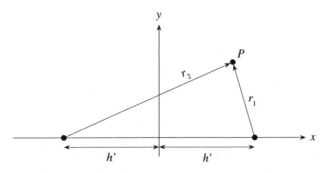

Figure D.3 Coordinates of the equivalent current filaments.

At a particular observation point, P, the potential is

$$A = \frac{-\mu_0 I}{2\pi} \ln r_1 + \frac{\mu_0 I}{2\pi} \ln r_2 = \frac{-\mu_0 I}{2\pi} \ln \frac{r_1}{r_2} \tag{D.4}$$

From Equation (D.4), we can see that for a surface of constant potential, the ratio $\frac{r_1}{r_2}$ must be constant. Let us assume that the constant vector potential is A_k. Then we have

$$\frac{r_2}{r_1} = e^{\frac{2\pi A_k}{\mu_0 I}} \tag{D.5}$$

From Figure D.3, we have

$$r_1^2 = (h' - x)^2 + y^2 \tag{D.6}$$

and

$$r_2^2 = (h' + x)^2 + y^2 \tag{D.7}$$

For a particular value of the potential, we can describe the loci of points corresponding to a constant value as

$$C^2 = \frac{r_2^2}{r_1^2} = \frac{(x + h')^2 + y^2}{(x - h')^2 + y^2} \tag{D.8}$$

This can be written as

$$x^2 - 2xh' \frac{C^2 + 1}{C^2 - 1} + h'^2 + y^2 = 0 \tag{D.9}$$

Comparing Equation (D.9) to the expression for a circle of radius a located on the x axis at a point $x = h$, we have

$$(x - h)^2 + y^2 - a^2 = x^2 - 2xh + (h^2 - a^2) + y^2 = 0 \tag{D.10}$$

To make Equations (D.9) and (D.10) equal, we must have

$$-2h = -2h' \frac{C^2 + 1}{C^2 - 1} \tag{D.11}$$

and

$$h'^2 = h^2 - a^2 \tag{D.12}$$

This is the result we have used in Section 8.1.

Appendix E

Inductance of Parallel Wires

The flux density produced by a long conductor is found from Ampere's law as

$$B_\theta = \frac{\mu_0}{2\pi r} I_z \tag{E.1}$$

The total flux between points at a distance r_1 and r_2 produced by current I is

$$\lambda = \frac{\mu_0 I}{2\pi} \ln \frac{r_2}{r_1} \tag{E.2}$$

To find the flux linkage of each conductor produced by any other conductor, we refer to Figure E.1.

We want to find the flux linkage of conductor k produced by current in conductor i up to point P. We see that flux line 1 and flux line 2 do not link this path, while flux lines 3 and 4 do. By considering the figure, we note that the flux between the conductor k and point P is the same flux crossing the line from conductor i from $d_{k,P}$ to point P. Therefore, we can use Equation (E.2) using the distance from conductor i to point P.

To include all of the flux, we let P go to infinity. This causes a problem since the logarithm also goes to infinity as the argument goes to infinity. In fact, the inductance of an isolated wire is infinite. There must always be a return circuit. We will use the properties of the logarithm to address this issue. We can write Equation (E.2) for conductor 1 as

$$\lambda_1 = \frac{\mu_0}{2\pi} \left(I_1 \ln \left(\frac{d_{1,P}}{r_1'} \right) + I_2 \ln \left(\frac{d_{2,P}}{d_{1,2}} \right) \dots I_n \ln \left(\frac{d_{N,P}}{d_{1,N}} \right) \right) \tag{E.3}$$

We can now break the expression into two separate terms.

$$\lambda_1 = \frac{\mu_0}{2\pi} \left(I_1 \ln \left(d_{1,P} \right) + I_2 \ln \left(d_{2,P} \right) \dots I_n \ln \left(d_{N,P} \right) \right)$$
$$- \left(I_1 \ln \left(r_1' \right) + I_2 \ln \left(d_{1,2} \right) \dots I_n \ln \left(d_{1,N} \right) \right) \tag{E.4}$$

With the assumption that the sum of all currents is zero, we have

$$I_1 + I_2 + I_3 + \dots + I_N = 0 \tag{E.5}$$

Eddy Currents: Theory, Modeling, and Applications, First Edition.
Sheppard J. Salon, M. V. K. Chari, Lale T. Ergene, David Burow, and Mark DeBortoli.
© 2024 The Institute of Electrical and Electronics Engineers, Inc. Published 2024 by John Wiley & Sons, Inc.

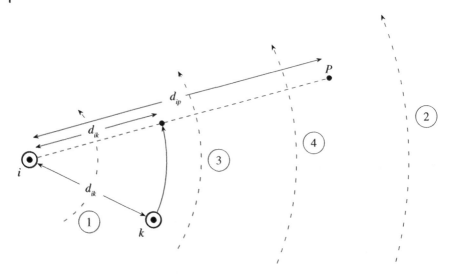

Figure E.1 Flux linkage of parallel wires.

We can now add the following expression which is equal to zero.

$$I_1 \ln d_{1,P} + I_2 \ln d_{1,P} + I_3 \ln d_{1,P} + \cdots + I_N \ln d_{1,P} = 0 \tag{E.6}$$

Recombining, we have for Equation (E.4)

$$\lambda_1 = \frac{\mu_0}{2\pi} \left(I_1 \ln \left(d_{1,P}/d_{1,P} \right) + I_2 \ln \left(d_{2,P}/d_{1,P} \right) + \cdots + I_n \ln \left(d_{N,P}/d_{1,P} \right) \right)$$
$$- \left(I_1 \ln \left(r_1' \right) + I_2 \ln \left(d_{1,2} \right) \ldots I_n \ln \left(d_{1,N} \right) \right) \tag{E.7}$$

As we let the point P go to infinity and the distance becomes much larger than the spacing between conductors, the fraction $\frac{d_{i,P}}{d_{1,P}}$ approaches 1 and the natural logarithm of 1 is zero. The first term then vanishes and we are left with the flux linkage of conductor 1 as

$$\lambda_1 = \frac{\mu_0}{2\pi} \left(I_1 \ln \left(1/r_1' \right) + I_2 \ln \left(1/d_{1,2} \right) + \cdots + I_n \ln \left(1/d_{1,N} \right) \right) \tag{E.8}$$

The term r' is the geometric mean radius (GMR), which is the effective radius of the conductor that corrects for the internal flux linkage. For circular conductors, $r' = 0.7788r$ where r is the conductor radius. For square conductors, the GMR is $0.44705a$, where a is the dimension of a side of the square (see Section 7.1). We repeat the process for all conductors and obtain a system of equations of the form

$$(\lambda) = (L)(I) \tag{E.9}$$

Appendix F

Shape Functions for First-Order Hexahedral Element

In this appendix, we present the shape functions for a first-order, hexahedral element shown in Figure F.1. For the development of these shape functions, the index, i, represents a node number. So, for this 8-node element, i will be $1, 2, \ldots, 8$.

In the local coordinate system of the element (ξ, η, λ), the shape function for each node can be written as

$$\zeta_i = \frac{(1 + \xi_i \xi)(1 + \eta_i \eta)(1 + \lambda_i \lambda)}{8} \tag{F.1}$$

where

$$\xi_i = \begin{cases} +1 & i = 1, 2, 5, 6 \\ -1 & i = 3, 4, 7, 8 \end{cases} \tag{F.2}$$

$$\eta_i = \begin{cases} +1 & i = 2, 3, 6, 7 \\ -1 & i = 1, 4, 5, 8 \end{cases} \tag{F.3}$$

$$\lambda_i = \begin{cases} +1 & i = 5, 6, 7, 8 \\ -1 & i = 1, 2, 3, 4 \end{cases} \tag{F.4}$$

The derivatives of each function with respect to each direction of the local system are

$$\frac{\partial \zeta_i}{\partial \xi} = \frac{\zeta_i (1 + \eta_i \eta)(1 + \lambda_i \lambda)}{\partial \xi} \tag{F.5}$$

$$\frac{\partial \zeta_i}{\partial \eta} = \frac{\zeta_i (1 + \xi_i \xi)(1 + \lambda_i \lambda)}{\partial \xi} \tag{F.6}$$

$$\frac{\partial \zeta_i}{\partial \lambda} = \frac{\zeta_i (1 + \eta_i \eta)(1 + \xi_i \xi)}{\partial \xi} \tag{F.7}$$

The global coordinates (x, y, z) are found from the shape functions.

$$x = \sum_{i=1}^{n} \zeta_i x_i \quad y = \sum_{i=1}^{n} \zeta_i y_i \quad z = \sum_{i=1}^{n} \zeta_i z_i \tag{F.8}$$

Eddy Currents: Theory, Modeling, and Applications, First Edition.
Sheppard J. Salon, M. V. K. Chari, Lale T. Ergene, David Burow, and Mark DeBortoli.
© 2024 The Institute of Electrical and Electronics Engineers, Inc. Published 2024 by John Wiley & Sons, Inc.

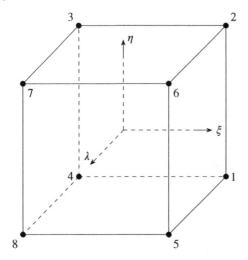

Figure F.1 First-order hexahedral element.

And, the derivatives of the global coordinates with respect to the local coordinates are

$$\frac{\partial x}{\partial \xi} = \sum_{i=1}^{n} \frac{\partial \zeta_i}{\partial \xi} x_i \qquad \frac{\partial y}{\partial \xi} = \sum_{i=1}^{n} \frac{\partial \zeta_i}{\partial \xi} y_i \qquad \frac{\partial z}{\partial \xi} = \sum_{i=1}^{n} \frac{\partial \zeta_i}{\partial \xi} z_i \qquad \text{(F.9)}$$

$$\frac{\partial x}{\partial \eta} = \sum_{i=1}^{n} \frac{\partial \zeta_i}{\partial \eta} x_i \qquad \frac{\partial y}{\partial \eta} = \sum_{i=1}^{n} \frac{\partial \zeta_i}{\partial \eta} y_i \qquad \frac{\partial z}{\partial \eta} = \sum_{i=1}^{n} \frac{\partial \zeta_i}{\partial \eta} z_i \qquad \text{(F.10)}$$

$$\frac{\partial x}{\partial \lambda} = \sum_{i=1}^{n} \frac{\partial \zeta_i}{\partial \lambda} x_i \qquad \frac{\partial y}{\partial \lambda} = \sum_{i=1}^{n} \frac{\partial \zeta_i}{\partial \lambda} y_i \qquad \frac{\partial z}{\partial \lambda} = \sum_{i=1}^{n} \frac{\partial \zeta_i}{\partial \lambda} z_i \qquad \text{(F.11)}$$

The Jacobian is

$$|J| = \begin{vmatrix} \dfrac{\partial x}{\partial \xi} & \dfrac{\partial y}{\partial \xi} & \dfrac{\partial z}{\partial \xi} \\[2mm] \dfrac{\partial x}{\partial \eta} & \dfrac{\partial y}{\partial \eta} & \dfrac{\partial z}{\partial \eta} \\[2mm] \dfrac{\partial x}{\partial \lambda} & \dfrac{\partial y}{\partial \lambda} & \dfrac{\partial z}{\partial \lambda} \end{vmatrix} \qquad \text{(F.12)}$$

References

1 Harit Majmudar. *Electromechanical Energy Converters*. Allyn and Bacon, Boston, Massachusetts, 1965.

2 P. Hammond. P.D. versus E.M.F. *Students' Quarterly Journal*, 30(117):3–6, September 1959.

3 W. Lewin. *Non-Conservative Fields – Do Not Trust Your Intuition*. MIT Physics, March 2002.

4 B.H. McGuyer. Symmetry and voltmeters. *American Journal of Physics*, 80:101, 2021.

5 H.W. Nicholson. What does the voltmeter read? *American Journal of Physics*, 73:1194–1196, 2005. doi: 10.1119/1.1997171.

6 E. Lanzara and R. Zangara. Potential difference measurements in the presence of a varying magnetic field. *Physics Education*, 30(2):85, 1995. doi: 10.1088/0031-9120/30/2/006.

7 A. Hernandez. EMF and potential difference: An illustrative example. *European Journal of Physics*, 5(1):13, 1984. doi: 10.1088/0143-0807/5/1/004.

8 P.C. Peters. The role of induced EMF's in simple circuits. *American Journal of Physics*, 52(3):208–211, 1983.

9 R.H. Romer. What do voltmeters measure?: Faraday's law in a multiply connected region. *American Journal of Physics*, 50:1089–1093, 1982.

10 F. Reif. Generalized Ohm's law, potential difference, and voltage measurements. *American Journal of Physics*, 50:1048–1049, 1982. doi: 10.1119/1.12944.

11 W.K.H. Panofsky and M. Phillips. *Classical Electricity and Magnetism, Second Edition*. Dover Publications, 2005.

12 Heinz E. Knoepfel. *Magnetic Fields: A Comprehensive Theoretical Treatise for Practical Use*. John Wiley & Sons, Inc., New York, 2000. ISBN 0-471-32205-9.

13 D. Hillel. *Introduction to Soil Physics*. Academic Press, San Diego, California, 1982.

14 J. Crank. *The Mathematics of Diffusion, Second Edition*. Clarendon Press, Oxford, 1975.

15 G.W. Carter. *The Electromagnetic Field in its Engineering Aspects, Second Edition*. Longmans, London, 1967.

16 S.V. Kulkarni and S.A. Khaparde. *Transformer Engineering*. CRC Press, 2013.

17 William T. McLyman. *Transformer and Inductor Design Handbook*. Marcel Dekker, Inc., 2006.

Eddy Currents: Theory, Modeling, and Applications, First Edition.
Sheppard J. Salon, M. V. K. Chari, Lale T. Ergene, David Burow, and Mark DeBortoli.
© 2024 The Institute of Electrical and Electronics Engineers, Inc. Published 2024 by John Wiley & Sons, Inc.

18 Jiří Lammeraner and Miloš Štafl. *Eddy Currents*. CRC Press, Cleveland, Ohio, 1966. ISBN 978-0-592-05016-4.

19 P.D. Agarwal. Eddy-current losses in solid and laminated iron. *Transactions of the American Institute of Electrical Engineers, Part I: Communication and Electronics*, 78(2):169–181, 1959.

20 H.M. McConnell. Eddy-current phenomena in ferromagnetic materials. *Transactions of the American Institute of Electrical Engineers, Part I: Communication and Electronics*, 73(3):226–235, 1954.

21 E. Rosenberg. Eddy-currents in solid iron. *Electrotech Maschinenbau*, 41:317–325, 1923.

22 N.A. Demerdash and D.H. Gillott. A new approach for determination of eddy currents and flux penetration in non-linear ferromagnetic material. *IEEE Transactions on Magnetics*, 74:682–685, 1974.

23 S.J. Salon. *Finite Element Analysis of Electrical Machines*. Kluwer Academic Publishers, 1995.

24 S.J. Salon and H.B. Hamilton. Calculation of induced field current and voltage in solid rotor turbine generators. *IEEE Transactions on Power Apparatus and Systems*, 97:1918–1924, 1978.

25 D. O'Kelly. Flux penetration in a ferromagnetic material including hysteresis and eddy-current effects. *Journal of Physics D: Applied Physics*, 5(1):203–213, January 1972. doi: 10.1088/0022-3727/5/1/329.

26 R.L. Stoll. *The Analysis of Eddy Currents*. Clarendon Press, Oxford, 1974.

27 H. Poritsky and R.P. Jerrard. Eddy-current losses in a semi-infinite solid due to a nearby alternating current. *Transactions of the American Institute of Electrical Engineers, Part I: Communication and Electronics*, 73(2):97–106, 1954.

28 M.P. Jain and L.M. Ray. Field pattern and associated losses in aluminum sheet in presence of strip bus bars. *IEEE Transactions on Power Apparatus and Systems*, PAS-89(7):1525–1539, 1970.

29 M.P. Jain and L.M. Ray. Proximity effects in cylindrical bus enclosures. *IEEE Transactions on Power Apparatus and Systems*, PAS-92(4):1174–1179, 1973.

30 P.J. Lawrenson, P. Reece, and M.C. Ralph. Tooth-ripple losses in solid poles. *Proceedings of the Institution of Electrical Engineers*, 113(4):657–662, 1966.

31 R.L. Stoll and P. Hammond. Calculation of the magnetic field of rotating machines. Part 4: Approximate determination of the field and the losses associated with eddy currents in conducting surfaces. *Proceedings of the Institution of Electrical Engineers*, 112(11):2083–2094, 1965.

32 R.L. Stoll and P. Hammond. Calculation of the magnetic field of rotating machines. Part 5: Field in the end region of turbogenerators and the eddy-current loss in the end plates of stator cores. *Proceedings of the Institution of Electrical Engineers*, 113(11):1793–1804, 1966.

33 S.K. Sen and B. Adkins. The application of the frequency-response method to electrical machines. *Proceedings of the IEE - Part C: Monographs*, 103(4):378–391, 1956.

34 H.M. McConnell and E.F. Sverdrup. The induction machine with solid iron rotor [includes discussion]. *Transactions of the American Institute of Electrical Engineers. Part III: Power Apparatus and Systems*, 74(3):343–349, 1955.

35 R.L. Winchester. Stray losses in the armature end iron of large turbine generators [includes discussion]. *Transactions of the American Institute of Electrical Engineers. Part III: Power Apparatus and Systems,* 74(3):381–391, 1955.

36 K.K. Lim and P.J. Lawrenson. Universal loss chart for the calculation of eddy-current losses in thick, steel plates. *Proceedings of the IEE,* 117(4):857–864, 1970.

37 A. Press. Resistance and reactance of massed rectangular conductors. *Physics Review,* VIII(4):417–422, 1912.

38 P. Silvester. *Modern Electromagnetic Fields.* Prentice-Hall, Inc., Englewood Cliffs, New Jersey, 1968.

39 Milton Abramowitz and Irene A. Stegun, editors. *Handbook of Mathematical Functions with Formulas, Graphs, and Mathematical Tables.* National Bureau of Standards Applied Mathematics Series - 55, USA, 1972.

40 N.N. Lebedev. *Special Functions and Their Applications.* Dover Publications, 1972.

41 Y.L. Luke. *Integrals of Bessel Functions.* McGraw Hill, New York, 1962.

42 G. Watson. *A Treatise on the Theory of Bessel Functions.* Cambridge University Press, 1922.

43 N.W. McLachlan. *Bessel Functions for Engineers.* Clarendon Press (Oxford University Press), London, 1961.

44 M.P. Perry. *Low Frequency Electromagnetic Design.* CRC Press, Boca Roton, Florida, London, New York, 1985.

45 H.A. Haus and J.R. Melcher. *Electromagnetic Fields and Energy.* Prentice-Hall, Inc., Englewood Cliffs, New Jersey, 1989.

46 M.V.K. Chari and S.J. Salon. *Numerical Methods in Electromagnetism.* Academic Press, 2000.

47 O. Biro and K. Preis. Finite element analysis of 3D eddy currents. *IEEE Transactions on Magnetics,* 26(2):418–423, March 1990.

48 Charles Jordan. *Calculus of Finite Differences.* Chelsea Publishing Company, New York, 1950.

49 M.N.O. Sadiku. *Numerical Techniques in Electromagnetics.* CRC Press, Boca Roton, Florida, Ann Arbor, Michigan, London, 1992.

50 A. Al-Khafaji and J. Tooley. *Numerical Methods in Engineering Practice.* Holt, Reinhart and Winston, New York, 1986.

51 E.I. King. Equivalent circuits for two dimensional fields: I – The static field. *IEEE Transactions on Power Apparatus and Systems,* 85:927–935, 1966.

52 E.I. King. Equivalent circuits for two dimensional fields: II – The sinusoidally time varying field. *IEEE Transactions on Power Apparatus and Systems,* 85:936–945, 1966.

53 J. Roberts. Analogue treatment of eddy-current problems involving two-dimensional fields. *Proceedings of the IEE - Part C: Monographs,* 107(11):11–18, 1959.

54 C.J. Carpenter. Finite-element network models and their application to eddy-current problems. *Proceedings of the IEE,* 122(4):455–462, April 1975.

55 A. Konrad and P. Silvester. A finite element program package for axisymmetric scalar field problems. *Computer Physics Communications,* 5(6):437–455, 1973. ISSN 0010-4655. doi: 10.1016/0010-4655(73)90081-7. URL https://www.sciencedirect.com/science/article/pii/0010465573900817.

56 Edward B. Rosa and Frederick W. Grover. Formulas and tables for the calculation of mutual and self-inductance. *Bulletin of the Bureau of Standards*, 8(1), 1916. doi: 10.6028/bulletin.185.

57 C.S. Biddlecombe, C.J. Collie, J. Simkin, and C.W. Trowbridge. The integral equation method applied to eddy currents. Technical report, Rutherford Laboratory, 1976.

58 B.H. McDonald and A. Wexler. *Finite Elements in Electrical and Magnetic Field Problems*, Chapter 9: Mutually Constrained Partial Differential and Integral Equation Field Formulation. John Wiley & Sons, Inc., 1980.

59 G.H. Brown, C.N. Hoyler, and R.A. Bierwirth. *Theory and Application of Radio-Frequency Heating*. D. Van Nostrand, New York, 1947.

60 L.T. Ergene. *Advances in Modeling of Electromechanical Devices with Moving Eddy Current Regions*. PhD thesis, Rensselaer Polytechnic Institute, Troy, New York, 2003.

61 A.B. Field. Eddy currents in large slot-wound conductors. *Transactions of the AIEE*, XXIV:761–788, 1905.

62 M.G. Say. *Alternating Current Machines, Fifth Edition*. Longman Scientific & Technical, Essex, UK, 1983. ISBN 978-0-470-27451-4.

63 Thomas A. Lipo. *Introduction to AC Machine Design*. Wiley IEEE Press, Hoboken, New Jersey, 2017. ISBN 978-1-119-35216-7.

64 Juha Pyrhönen, Tapani Jokinen, and Valéria Hrabovcová. *Design of Rotating Electrical Machines, Second Edition*. John Wiley & Sons, Ltd., Chichester, UK, 2014. ISBN 978-1-118-58157-5.

65 Vlado Ostović. *The Art and Science of Rotating Field Machine Design: A Practical Approach*. Springer, 2017.

66 E. Roth. Etude analytique du champ propre d'une encoche. *Revue Générale d'Electricité*, 22:417–424, 1927.

67 B. Hague. *Electromagnetic Problems in Electrical Engineering*. Oxford University Press, London, 1929.

68 M.V.K. Chari and G. Bedrosian. Electromagnetic field analysis of eddy currents effects in rotating electrical apparatus and machinery. *IEEE Transactions on Magnetics*, 18(6):1713–1715, 1982.

69 M. Marković and T. Perriard. An analytical solution for the torque and power of a solid-rotor induction motor. *Proceedings of IEEE International Electric Machines and Drives Conferences*, pages 1053–1057, 2011.

70 P.L. Alger. *Induction Machines: Their Behavior and Uses*. Gordan and Breach, New York, 1970. Chapters 9 & 10.

71 Thomas A. Lipo. *Introduction to AC Machine Design*. University of Wisconsin Press, 2007.

72 D.G. Dorrell. Calculation and effects of end-ring impedance in cage induction motors. *IEEE Transactions on Magnetics*, 41(3):1176–1183, 2005. doi: 10.1109/TMAG.2004.843337.

Index

Eddy Currents: Theory, Modeling, and Applications, First Edition.
Sheppard J. Salon, M. V. K. Chari, Lale T. Ergene, David Burow, and Mark DeBortoli.
© 2024 The Institute of Electrical and Electronics Engineers, Inc. Published 2024 by John Wiley & Sons, Inc.